# CELL ADHESION

## BIOPROCESS TECHNOLOGY

*Series Editor*

**W. Courtney McGregor**

Xoma Corporation
Berkeley, California

1. Membrane Separations in Biotechnology, *edited by W. Courtney McGregor*
2. Commercial Production of Monoclonal Antibodies: A Guide for Scale-Up, *edited by Sally S. Seaver*
3. Handbook on Anaerobic Fermentations, *edited by Larry E. Erickson and Daniel Yee-Chak Fung*
4. Fermentation Process Development of Industrial Organisms, *edited by Justin O. Neway*
5. Yeast: Biotechnology and Biocatalysis, *edited by Hubert Verachtert and René De Mot*
6. Sensors in Bioprocess Control, *edited by John V. Twork and Alexander M. Yacynych*
7. Fundamentals of Protein Biotechnology, *edited by Stanley Stein*
8. Yeast Strain Selection, *edited by Chandra J. Panchal*
9. Separation Processes in Biotechnology, *edited by Juan A. Asenjo*
10. Large-Scale Mammalian Cell Culture Technology, *edited by Anthony S. Lubiniecki*
11. Extractive Bioconversions, *edited by Bo Mattiasson and Olle Holst*
12. Purification and Analysis of Recombinant Proteins, *edited by Ramnath Seetharam and Satish K. Sharma*
13. Drug Biotechnology Regulation: Scientific Basis and Practices, *edited by Yuan-yuan H. Chiu and John L. Gueriguian*
14. Protein Immobilization: Fundamentals and Applications, *edited by Richard F. Taylor*
15. Biosensor Principles and Applications, *edited by Loïc J. Blum and Pierre R. Coulet*
16. Industrial Application of Immobilized Biocatalysts, *edited by Atsuo Tanaka, Tetsuya Tosa, and Takeshi Kobayashi*
17. Insect Cell Culture Engineering, *edited by Mattheus F. A. Goosen, Andrew J. Daugulis, and Peter Faulkner*
18. Protein Purification Process Engineering, *edited by Roger G. Harrison*
19. Recombinant Microbes for Industrial and Agricultural Applications, *edited by Yoshikatsu Murooka and Tadayuki Imanaka*
20. Cell Adhesion: Fundamentals and Biotechnological Applications, *edited by Martin A. Hjortso and Joseph W. Roos*

*ADDITIONAL VOLUMES IN PREPARATION*

Bioreactor System Design, *edited by Juan A. Asenjo and José C. Merchuk*

Gene Expression in Recombinant Microorganisms, *edited by Alan Smith*

# CELL ADHESION

## Fundamentals and Biotechnological Applications

edited by

**Martin A. Hjortso**
*Louisiana State University*
*Baton Rouge, Louisiana*

**Joseph W. Roos**
*Ethyl Corporation*
*Richmond, Virginia*

Marcel Dekker, Inc. New York • Basel • Hong Kong

**Library of Congress Cataloging-in-Publication Data**

Cell adhesion : fundamentals and biotechnological applications / edited by Martin A. Hjortso, Joseph W. Roos.
p. cm. — (Bioprocess technology ; v. 20)
Includes bibliographical references and index.
ISBN 0-8247-8945-8
1. Cell adhesion. 2. Bioreactors. 3. Biotechnology.
I. Hjortso, Martin A. II. Roos, Joseph W.
III. Series.
TP248.25.C42C45 1994
660'.63—dc20 94-22882
CIP

The publisher offers discounts on this book when ordered in bulk quantities. For more information, write to Special Sales/Professional Marketing at the address below.

This book is printed on acid-free paper.



Marcel Dekker, Inc.
270 Madison Avenue, New York, New York 10016

Current printing (last digit):
10 9 8 7 6 5 4 3 2 1

PRINTED IN THE UNITED STATES OF AMERICA

# Series Introduction

Bioprocess technology encompasses all of the basic and applied sciences as well as the engineering required to fully exploit living systems and bring their products to the marketplace. The technology that develops is eventually expressed in various methodologies and types of equipment and instruments built up along a bioprocess stream. Typically in commercial production, the stream begins at the bioreactor, which can be a classical fermentor, a cell culture perfusion system, or an enzyme bioreactor. Then comes separation of the product from the living systems and/or their components followed by an appropriate number of purification steps. The stream ends with bioproduct finishing, formulation, and packaging. A given bioprocess stream may have some tributaries or outlets and may be overlaid with a variety of monitoring devices and control systems. As with any stream, it will both shape and be shaped with time. Documenting the evolutionary shaping of bioprocess technology is the purpose of this series.

Now that several products from recombinant DNA and cell fusion techniques are on the market, the new era of bioprocess technology is well established and validated. Books of this series represent developments in various segments of bioprocessing that have paralleled progress in the life sciences. For obvious proprietary reasons, some developments in industry, although validated, may be published only later, if at all. Therefore, our continuing series will follow the growth of this field as it is available from both academia and industry.

*W. Courtney McGregor*

# Preface

Over the past several years, cell adhesion has progressed from a phenomenon of peripheral interest to biochemical engineers to become a useful tool in bioreactor design and bioseparations. Along with these innovations have arisen increased demands for better understanding of the phenomena that underlie cell adhesion. The aim of this book is to present the student or researcher in biochemical engineering with an introduction to the basic phenomena that govern cell adhesion and to provide descriptions of bioengineering processes in which cell adhesion plays an important part.

Cell adhesion has already been studied extensively because of its medical significance. The ability of different cells to adhere or aggregate plays a crucial role in such processes as microbial infections, spread of cancer through the body, and plaque formation on teeth. However, the abundant literature that has come out of the biomedical research is not always particularly useful to the biochemical engineer, since the medical perspective is much different from the engineering point of view. Engineering applications of cell adhesion are presented in the research literature, but to date no comprehensive reference work has been available on this topic. In view of this, it appeared timely to assemble a book dedicated to the engineering aspects of cell adhesion.

To the biochemical engineer, cell adhesion is a potential problem as well as a valuable tool. Cell adhesion may lead to biofouling of process equipment but is purposely employed in immobilized bioreactors to influence reactor performance. In the case of anchorage-dependent animal cells, adhesion to a solid support is essential for cultivation. Cell-cell adhesion or flocculation may be undesirable during a fermentation but is often an important step in the final separation of cells and broth. In all cases, judicious engineering practice requires an understanding of cell adhesion and the controlling physical and chemical factors.

In putting together this book, we had two objectives: to detail the fundamental aspects of cell adhesion and to present contemporary engineering aspects of cell adhesion.

Cell adhesion mechanisms can be generally classified as two types, occurring via either specific or nonspecific interactions, both of which are idealized descriptions of the adhesion process. On a fundamental level, all types of adhesion are mediated by the same forces, van der Waals and electrostatic forces between molecules on the cell surface and molecules on the adhesion surface. However, the distribution of these forces is perceived differently in the two cases. In nonspecific adhesion, the adhesion forces are viewed as being continuously distributed over the contact area, making it possible to describe the strength of the adhesion in terms of an adhesive energy, the free energy reduction per area of contact. In specific adhesion, on the other hand, the forces are concentrated into discrete bonds between receptors on the cell surface and complementary ligands on the adhesion surface. Bonds are formed only when the shapes of the two molecules match each other in such a way that groups on the two molecules can approach one another closely enough for van der Waals and electrostatic forces to become effective. It is the steric hindrance that prevents formation of bonds between noncomplementary ligands and receptors that give rise to the specificity of this type of adhesion. The two introductory chapters emphasize the characteristics of specific adhesion. Limited discussion is directed toward nonspecific adhesion and generally addresses situations where it can be viewed as a limiting case of specific adhesion. In general, it is felt that aspects of nonspecific adhesion can be dealt with more appropriately in light of particular applications that exploit the phenomena. Discussions of these applications, where appropriate, are included in the remaining chapters.

The second half of the book is application-oriented. In a series of chapters, leading researchers discuss cell adhesion as applied to problems of interest to biochemical engineers. Cell adhesion is discussed in several chapters as a tool for bioreactor design. The use of biofilm formation in bioreactors is discussed for three separate cell types: microbial, animal, and plant cells. Each type of reactor has different characteristics dictated by the requirements of the adhering cells. In another chapter, cell-cell adhesion or flocculation is discussed. Flocculation, a procedure that has been used for centuries as a method to isolate biomass from broth, increases cell sedimentation rates. Recently, flocculation has found many new applications in the manipulation of bioprocesses. While both cell aggregation and biofilm formation can result from any cell adhesion mechanism, individual applications of cell adhesion may require careful preparation of the adhesion surface. This is the topic of the last chapter, in which different supports and their chemical modification are discussed.

It would be satisfying if the processes described in the application chapters could be modeled in terms of fundamental events and kinetics. However, cell adhesion models based on fundamental processes have not yet reached a stage where this is altogether possible. Consequently, engineering models of processes involving cell adhesion often use concepts and a nomenclature specific to the area. So, although a common nomenclature is used in the introductory chapters, we have not attempted to keep this nomenclature through the application chapters.

*Martin A. Hjortso*
*Joseph W. Roos*

# Contents

# Contributors

**Jean Archambault** Chemical Engineering Section, Department of Engineering, University of Quebec at Trois-Rivières, Trois-Rivières, Quebec, Canada

**William G. Characklis**† The Center for Biofilm Engineering, Montana State University, Bozeman, Montana

**Robert H. Davis** Department of Chemical Engineering, University of Colorado, Boulder, Colorado

**Martin A. Hjortso** Department of Chemical Engineering, Louisiana State University, Baton Rouge, Louisiana

**Manfred R. Koller** Aastrom Biosciences, Inc., Ann Arbor, Michigan

**Eleftherios T. Papoutsakis** Department of Chemical Engineering, Northwestern University, Evanston, Illinois

**Brent M. Peyton** Pacific Northwest Laboratory, Richland, Washington

**Joseph W. Roos** Ethyl Corporation, Richmond, Virginia

**William H. Scouten** Biotechnology Center, Utah State University, Logan, Utah

---

†Deceased.

# 1
# Kinetics of Ligand-Receptor Bond Formation

**Joseph W. Roos**
*Ethyl Corporation, Richmond, Virginia*

**Martin A. Hjortso**
*Louisiana State University, Baton Rouge, Louisiana*

## 1 INTRODUCTION

Specific adhesion is mediated by pairs of molecules that fit each other much like two pieces in a jigsaw puzzle. Only molecules with the right shape will be able to form the specific bond, and even small changes in the shape may greatly affect the strength of the bond. The archetypical specific interaction is the bond formed between an antigen and an antibody, but many other kinds of specific bonds are known such as those between cell surface proteins and components of the extracellular matrix, carbohydrate and lectins, transport proteins and their substrates, sensory receptors and their target compounds, and hormonal receptors and their hormones. The molecule that resides in the cell membrane is usually referred to as the receptor and the complementary molecule is called the ligand. In cell-cell interactions, we will, somewhat arbitrarily refer to the molecules on one of the cells as the receptors and the molecules on the other cell will be called ligands. We will also restrict ourselves to systems where the interaction is mediated by only one variety of ligand-receptor pair which all possess the same kinetic properties. Often this is the case, though several important systems are known in which multiple receptor types interact with the ligand during the adhesion process.

A large number of cell adhesion models have been proposed in the literature. However, only a few of these models make explicit use of the assumption that the adhesion is caused by specific bonds. These models may be more applicable to cell adhesion phenomena caused by factors other than specific forces. In this chapter, we restrict ourselves to a discussion of the fundamental processes that govern the dynamics of the specific bond formation. In the next chapter, cell adhesion models that incorporate these processes are discussed.

Consider a cell, placed in a fluid environment with a surface to which it can adhere through the formation of discrete bonds. The adhesion surface can be another cell or an inaminate surface. For adhesion to occur the cell must approach sufficiently close to the surface for the ligands and receptors to interact and form a bond. The number of discrete bonds that form determines whether the cell remains attached when subjected to removal forces and depends on the overall kinetics of bond formation and the time available for bond formation. Several factors, such as the fluid environment of the cell, the rate of bond formation, the mechanical properties of the cell membrane, and the characteristics of the adhesion surface, influence the kinetics and the time available for bond formation. In all situations the fate of individual bonds is a dynamic problem. The process that leads to specific cell adhesion may be best described by considering the fundamental mechanisms.

## 2 LIGAND-RECEPTOR BOND FORMATION

It is unlikely that the receptors and ligands are in an appropriate position to form a large number of bonds immediately after a cell encounters an adhesion surface. For the discrete bonding between the cell and the surface to occur, the receptors and ligands must move into the proper position. This process is represented as a two-step mechanism (1–4).

$$R + L \underset{d_-}{\overset{d_+}{\rightleftharpoons}} R - L \underset{k_-}{\overset{k_+}{\rightleftharpoons}} RL$$

The first step represents the movement of the receptor, $R$, and the ligand, $L$, into the proper position for bond formation. A receptor and ligand pair in this state is termed an encounter complex, $R - L$. Two rate constants, $d_+$ and $d_-$, describe the formation and breakup of the encounter complex, respectively.

The second process is the reversible formation of the ligand-receptor bond. This is described by the intrinsic rate constants for bond formation and bond breakup, $k_+$ and $k_-$, respectively.

In analysis of ligand-receptor interaction and binding, a quasi-steady-state assumption, similar to the well-known Michaelis-Menten assumptions used in enzyme kinetics, is almost always employed. If the concentration of the encounter complex is assumed to reach its steady-state value quickly, compared to the rates at which the concentrations of receptor, ligands, or ligand-receptor bonds change, the apparent rate constants for ligand-receptor bond formation becomes

$$k_f = \frac{d_+ k_+}{d_- + k_+} \tag{1a}$$

and the apparent rate constant for bond breakup becomes,

$$k_r = \frac{d_- k_-}{d_- + k_+} \tag{1b}$$

The quasi-steady-state assumption is not valid for a short time after initial contact of the cell with the adhesion surface. It may also be invalid short periods after an adhering cell is perturbed in such a way that the kinetics of ligand-receptor binding are altered. Imposition of fluid forces on the adhering cell or a sudden addition of a soluble, competing ligand or receptor are two perturbations that could fit this role.

Using the apparent rate constants, the rate of change in ligand-receptor bond density between a cell membrane and an adhesion surface is given by

$$\frac{dC_B}{dt} = k_f C_{RF} C_{LF} - k_r C_B \tag{2}$$

where $C_B$, $C_{RF}$, and $C_{LF}$ are the concentration of bonds, free or unbound receptors, and free or unbound ligands, respectively. These concentrations can be either surface or volume concentrations, depending on the type of control volume used. To avoid confusion in the following, surface concentrations or densities will be indicated by $C$'s while volume concentrations will be indicated by $\tilde{C}$'s.

This simple representation was employed in the theoretical framework for specific cell adhesion introduced by Bell (1). The concept has found application in the analysis of cell adhesion and determination of probable bond number (5,6).

In the following sections, a brief outline is presented on evaluating the rate constants for encounter complex formation and the rate constants for ligand-receptor binding. Expressions for the rate constants in terms of fundamental properties of the system, such as diffusion coefficients, temperature, and membrane viscosity, provide insight into how these parameters affect the kinetics and results in a better understanding of the ligand-receptor bond formation process. These expressions may also help to determine when certain simplifications of the apparent rate constants are permissible, such as when bond formation is reaction or diffusion limited.

## 2.1 Rate Constants for Encounter Complex Formation

The formation of an encounter complex between the receptor and ligand is dependent on the mobilities of the reactants. The generally accepted fluid mosaic model of a cell membrane attributes characteristic mobilities to components of the cell membrane (7). The translational movement of a receptor or ligand associated with a cell or artificial membrane is restricted to the plane of the membrane. A ligand or receptor that is immobilized to a surface by covalent attachment would display no movement.

To illustrate the range of motion available to a receptor or ligand in a membrane, consider the motion of a protein associated with a membrane. Kotyk et al. (8) classified the movements of proteins in membranes as being of four basic types. The first is rotational oscillation of amino acid aliphatic side chains. Because of the limited range of this type of movement and its small correlation time, 0.1 to 10 ns, it is probably not significant in determining the rate of encounter complex formation. Kotyk et al. speculate that such movement may aid in altering protein conformation.

The second category of mobility displayed by proteins in a membrane is the bending or vibration of a portion of the protein. The frequency of such bending is on the order of 1 to $10^7$/s (8). An example of the importance of this mobility in specific adhesion is an investigation of adhesion between immunoglobulin-covered latex spheres and a surface (9,10). In this system, the only movement possible by the receptor and ligand is the bending of the immunoglobulins. This bending is sufficient to allow formation of encounter complex and subsequent adhesion of the latex spheres.

The rotation of the protein about an axis normal to the membrane is movement of the third type. Often a ligand or receptor is asymmetric with respect to its binding region (11,12), and the ability of such a protein to rotate is essential for proper alignment of reactants in the encounter complex. Such rotation has been characterized for a number of proteins in membranes and found to have correlation times of $1–2 \times 10^4$ μs (8).

The final type, arguably the most important, is the translational diffusion in the membrane plane. It is translational diffusion that allows receptors and ligands to enter into encounter complexes without the translational movement of the interacting cell and adhesion surfaces (1,4,13–15). Also, after initial contact between the cell and the adhesion surface, translational diffusion of receptors can lead to receptor accumulation in the membrane area near the surface (16–18).

The rate constants for encounter complex formation must represent the various modes of component mobility. For the case of the latex spheres with immobilized immunoglobulins adhering to an immunoglobulin-covered support, the bending of the immunoglobulins was sufficient to form the encounter complex. For cells or surfaces where the receptor or ligand is mobile, the translational and rotational movement of these components must also be considered. While, at first glance, the prospect of considering several types of mobility suggests a complex problem, considerable simplicity is introduced by only considering the rate-limiting process. Only the mechanism that controls the rate of encounter complex formation is used to estimate a representative rate constant. As the initial step in obtaining an expression for the rate constants of encounter complex formation, the case where translational movement is rate limiting is considered.

To estimate the rate constants for encounter complex formation through receptor or ligand translation, a mean encounter time for a ligand receptor system is calculated (3,19). For receptors in a cell membrane and immobilized ligands, this encounter time is

$$t_+ = \frac{s_l^4}{2D_m(s_l^2 - s_c^2)}\left(\ln\left(\frac{s_l}{s_c}\right) - \frac{1}{4}\left(3 - \left(\frac{s_c}{s_l}\right)^2\right)\left(1 - \left(\frac{s_c}{s_l}\right)^2\right)\right) \tag{3}$$

It is a function of the diffusion coefficient of the receptor in the membrane, $D_m$, and the radius of the encounter complex, $s_c$. The surface density of the unbound ligand is represented by $s_l$, the mean separation distance between unbound ligand. Using this result, the net flux of ligand into the encounter complex becomes,

$$J^+ = \frac{\pi s_l^2 C_{RF}}{t_+} \tag{4}$$

where $C_{RF}$ is the surface concentration of free receptors in the contact area. The rate constant for the formation of the encounter complex is determined from this flux as

$$d_+ = \frac{J^+}{C_{RF}} = \frac{2\pi D_m\left(1 - \left(\frac{s_c}{s_l}\right)^2\right)}{\left(\ln\left(\frac{s_l}{s_c}\right) - \frac{1}{4}\left(3 - \left(\frac{s_c}{s_l}\right)^2\right)\left(1 - \left(\frac{s_c}{s_l}\right)^2\right)\right)} \tag{5}$$

DeLisi (3) defined a rate constant for encounter complex formation as the inverse of the encounter time given in Eq. (3). This expression does not account for the flux across the entire boundary around the site for encounter complex formation and does not have the appropriated dimensions for use in Eqs. (1) and (2).

The rate constant for breakup of the encounter complex, $d_-$, is determined in a similar fashion to that used to obtain an estimate for $d_+$. The mean time for a receptor to move out of the encounter complex, $t_-$, and the flux of receptor across the encounter complex boundary, $s_c$, are used. If the encounter complex can only contain one receptor, the rate constant becomes

$$d_- = \frac{8D_m}{s_c^2} \tag{6}$$

Other expressions for the forward rate constant have been used. Bell (1) considered adhesion of two cells, in which case both ligands and receptors can be mobile, and used expressions of the form

$$d_+ = 2\pi(D_{m1} + D_{m2}) \tag{7a}$$

$$d_- = 2(D_{m1} + D_{m2})/s_c^2 \tag{7b}$$

where $D_{mi}$ is the diffusion coefficient on cell $i$.

Lauffenburger and DeLisi (4) have proposed that the rate constants for encounter complex formation and breakup can be estimated as

$$d_+ = \frac{2\pi(D_{m1} + D_{m2})}{\ln\left(\dfrac{s_l}{s_c}\right) - \dfrac{3}{4}} \tag{8a}$$

$$d_- = \frac{2(D_{m1} + D_{m2})}{s_c^2\left(\ln\left(\dfrac{s_l}{s_c}\right) - \dfrac{3}{4}\right)} \tag{8b}$$

These expressions were formulated for the limiting condition where $s_l \gg s_c$. This condition arises when the density of unbound ligands or receptors is low.

Up to this point, encounter complex formation for symmetric ligands and receptors has been addressed. Here the entire surface described by the encounter complex radius is reactive. On an asymmetric ligand or receptor only a portion of the surface defined by the encounter complex radius is reactive. The approach used to define $d_+$ and $d_-$ for symmetric receptors or ligands is also valid for encounter complex between an asymmetric ligand or receptor if the characteristic time for receptor rotation is small compared to the characteristic time of translational motion. When the characteristic time for rotation is of the same order of magnitude as or much larger than that for translation, the effects of the asymmetric reactive surface would lower the rate constant for encounter complex formation. An estimate of the effect of an asymmetric receptor or ligand on $d_+$ has been proposed for the case where the characteristic time for rotation is much larger than that of translation (3).

The regions of interactions or the constraints on interaction between some ligands and receptors are well documented. However, the development of a general expression for rate constants of encounter complex formation that incorporate orientational constraints or the asymmetric nature of ligands and receptors has not been carried out. The remainder of the discussion about encounter complex formation and mobility of ligand and receptors will deal only with translational movement.

### 2.1.1 Diffusion Coefficients for Membrane Bound Receptor or Ligand

A theoretical basis is available for estimating translational diffusion coefficient in a membrane, $D_m$. The theory provides good estimates of the diffusion coefficient under limiting conditions (20) and also provides a background for understanding factors that alter the mobility of membrane-bound proteins.

Expressions for the diffusion coefficient of a protein in an infinitely dilute system have been derived (21,22). The approach treats the membrane as a continuous, two-dimensional viscous fluid and uses hydrodynamic principles to describe the movement of cylindrical proteins that do not interact with one another. The diffusion coefficients are found to be functions of temperature, $T$, the viscosity of the membrane fluid, $\mu_m$, and the outer fluid, $\mu$, usually water, the radius of the protein cylinder, $a$, and the membrane thickness, $h$. The diffusion coefficient for translation, estimated by the Saffman-Delbruck equation, is (21)

$$D_T = \frac{k_B T}{4\pi\mu_m h}\left(\ln\frac{\mu_m h}{\mu a} - \gamma\right) \tag{9}$$

where $k_B$ is Boltzmann's constant and $\gamma$ is Euler's constant.

The rotational diffusion coefficient can also be estimated using the Saffman-Delbruck equation (21).

$$D_r = \frac{k_B T}{4\pi\mu_m a^2 h} \tag{10}$$

The theory has been tested for proteins at low density in a lipid bilayer (20). The values for $D_m$, measured for a range of bulk fluid viscosities, agreed quite well with those predicted by the Saffman-Delbruck equation. The measured diffusion coefficients were used in the Saffman-Delbruck equations for translational and rotational diffusion to predict the radius of the protein. Again, the prediction agreed reasonably well with a value determined from the proteins structural model. The measured translational diffusion coefficient, however, started to differ from the predicted value as the bilayers' protein content was increased.

The Saffman-Delbruck equation is limited to dilute systems. However, as just discussed and observed by others, the translational diffusion coefficient is a function of protein concentration in the membrane (20,23). This effect appears to be caused by increased membrane viscosity and protein-protein interactions. The membrane protein concentration and interaction between proteins have been included in a number of models for prediction of protein mobility (24–29).

While protein mobility is found to be a function of the protein content, differences in mobility also exist between proteins in an artificial bilayer and a cell membrane (30). The proteins in a cell membrane may interact with a membrane-associated cytoskeleton or the extracellular matrix. This interaction alters diffusion of the protein in the membrane (31,32) or immobilizes a fraction of the protein (17).

Some membrane proteins have relatively high diffusion coefficients but are restricted to certain areas of the membrane (30,33). The localization of proteins implies some restriction on the mobility of these proteins. This restriction may not alter the ability of the proteins to diffuse over short distances on the order of 300–600 $\eta$m but can prevent diffusion over the entire membrane (34). This can be significant in specific cell adhesion because it will prevent unbound receptors from diffusing into or out of the area of the membrane that mediates adhesion.

If not restricted, receptors have been reported to cluster in the area in contact with the adhesion surface over time, increasing the local concentration (16–18). Membrane-bound molecules that do not mediate specific adhesion can also accumulate in the region of contact (16). This accumulation is expected to alter the rate constants for encounter complex formation, through both the protein density effect on the local translational diffusion coefficient and the decreased mean separation distance between unbound receptors. Solutions and simulations of the mean encounter time with variable translational diffusion coefficient have been formulated (35,36).

The translational diffusion coefficient, $D_m$, has been measured for several proteins in cell membranes. These values range from $10^{-8}$ $cm^2/s$ to $10^{-12}$ $cm^2/s$ (8,17,30–32). The diffusion coefficient for proteins in an artificial membrane or a membrane bleb was in the range of $10^{-8}$ $cm^2/s$ to $10^{-9}$ $cm^2/s$ (23,30,32). The larger diffusion constants determined for proteins in artificial membranes or blebs suggests that proteins in cell membranes interact with other components, leading to lowered mobility.

## 2.2 Rate Constants for Ligand-Receptor Binding

The intrinsic rate constants describe the formation of the bond between a ligand and a receptor in an encounter complex. Hydrogen bonding and van der Waals and electrostatic interactions between components of the receptor and ligand lead to the bond formation. The van der Waals and electrostatic interactions become stronger as two binding structures come close together. Hydrogen bonding between two chemical groups requires a separation distance of about 2–3 Å. Hydrophobic interactions are also often used to describe binding. But as has been pointed out (37), these forces can be accounted for through the van der Waals force concept, and there is therefore no need to introduce the additional notion of hydrophobic interactions.

The forces that lead to ligand-receptor bond formation are the same as those that result in nonspecific cell adhesion. The specificity of adhesion mediated by ligand-receptor binding is attributed to steric restraints imposed on the binding sites. Only a compound with complementary structure can approach within the distance necessary to form a bond. While movement of receptor or ligand structure may be necessary to form the bonds, the imposition of steric restrictions is considered independent of encounter complex formation. It is when the ligand and receptor are in the encounter complex that steric restriction become important. It is the effect of all these forces that the intrinsic reaction rate constants represent.

Uncoupling encounter complex formation and steric hindrance simplifies the description of the bond formation process. Consider the ligand-receptor system of starch and the maltoporin of *Escherichia coli*. In one study, different maltoporins that varied by a point mutation were used to mediate cell adhesion to immobilized starch (38). In this case equilibrium binding was measured using the same ligand. Based on estimates of receptor and ligand size, rate constants for encounter complex formation can be calculated. A unique equilibrium rate constant was then determined for each mutant receptor. This constant describes the effect of the amino acid substitutions on the binding site and steric restriction to the binding site. Until a better understanding of the structures of the various ligand-receptor systems is obtained, this approach provides a method to obtain intrinsic reaction rate constants for study of cell adhesion.

## 2.3 Estimating Equilibrium and Rate Constant for Receptors

Several experimental techniques are employed to obtain data on ligand-receptor binding kinetics (39–41). In these systems, the movement of at least one component, the ligand or receptor, occurs in three dimensions. As already discussed, this is not the case during specific cell adhesion where receptor and ligand movement is restricted to two dimensions. However, the intrinsic rate constants can be estimated from kinetic information obtained in these experiments.

Analysis of the ligand-receptor binding data yields qualitative and quantitative information that is important in the study of specific cell adhesion. From the ligand-receptor binding kinetics, the type of interaction between the ligand and receptor, the effect of multivalent ligand and receptors, and the role of multiple ligand or receptor populations can be evaluated (39,40,42,43,45–48). This allows formulation of appropriate models for the ligand-receptor binding process during specific cell adhesion.

The following is a brief description of the treatment of the ligand-receptor binding data for univalent ligands and receptors that display non-cooperative binding. This analysis leads to determination of intrinsic rate constants or equilibrium constants for binding. This framework is applicable to many ligand-receptor interactions of interest in the study of specific cell adhesion. It also provides the basis for incorporation of more complex binding kinetics.

The intrinsic rate constants are determined from the measured apparent rate constants using Eqs. (1a) and (1b) and the rate constants for encounter complex formation (1,12). The form of the expressions for the rate constant of encounter complex formation depends on the experimental method used to estimate the apparent rate constant.

In some methods, ligand and solubilized receptor are mixed together and the binding is directly measured (40). The apparent rate constants are calculated based on the concentration of ligand and receptor and the rate of binding. The rate constants for encounter complex formation are estimated as follows (2,4,12):

$$d_+ = 4\pi s_c D \tag{11a}$$

and

$$d_- = 3\frac{D}{s_c^2} \tag{11b}$$

where $D$ is the sum of the translational diffusion coefficients for the receptor and ligand in solution.

The effects of orientational constraints and rotational diffusion on the interaction between two components in solution have been addressed (12,49–52). However, there is not a general treatment of the problem; each different configuration is unique. One approach incorporates orientational constraints in the rate constants for encounter complex formation as a proportionality constant (3). The value of this constant should be determined separately for individual cases.

When the receptor is not free in solution, but is expressed on the surface of a cell or incorporated into a liposome, the physical picture is somewhat altered. The apparent rate constant is then calculated based on the ligand and cell (liposome) concentration. This rate constant is based on the whole cell, and the apparent rate constant becomes a function of the density of unoccupied receptors on the cell.

To evaluate this dependence, consider receptors, evenly dispersed over a spherical cell surface, separated by patches of membrane. Because a ligand that reaches the cell surface does not necessarily do so in the vicinity of a receptor, the flux of ligand into an encounter complex is less than the flux to the membrane surface. Similarly, the flux away from the sphere is also lower than expected based on the rate of breakup of the encounter complex. The flux of ligand into receptors located on a membrane was calculated by Berg and Purcell (19), and their concepts have been extended to define the rate constants for encounter complex formation (3,53–55). The rate constants for encounter complex formation and breakup are expressed as

$$d_+ = 4\pi r D\left(1 - \frac{\pi r}{N_{RF}s_c + \pi r}\right) \tag{12a}$$

and

$$d_- = 3\frac{D}{s_c^2}\left(\frac{\pi r}{N_{RF}s_c + \pi r}\right) \tag{12b}$$

where $r$ is the radius of the cell, $D$ is the ligand diffusion coefficient, and $N_{RF}$ is the number of free receptors on the cell surface. The quantity $\pi r/(N_{RF}s_c + \pi r)$ is the probability that a ligand at the cell surface diffuses into the bulk solution before forming an encounter complex with a cell surface receptor. The rate constants are nonlinear functions of the number of free receptors on the surface.

For the cell or liposome, the rate at which ligand-receptor bonds form is

$$\frac{d\tilde{C}_B}{dt} = \frac{d_+ k_+}{d_- + k_+}\tilde{C}_{\text{cell}}\tilde{C}_{LF} - \frac{d_- k_-}{d_- + k_+}\tilde{C}_B \tag{13}$$

where $\tilde{C}_B$ is the volume concentration of ligand-receptor bonds on the cell or liposome, $\tilde{C}_{\text{cell}}$ is the volume concentration of cells or liposomes, and $\tilde{C}_{LF}$ the volume concentration of free ligand.

The rate at which the ligand-receptor bond forms becomes a linear function of receptor concentration under two limiting conditions, reaction limitation and when $N_{RF}s_c \ll \pi r$. If the ligand-receptor binding is reaction limited, $d_- \gg k_+$, the apparent rate constants become

$$k_f = k_+ K_{\text{diff}} N_{RF} \tag{14a}$$

and

$$k_r = k_- \tag{14b}$$

where $K_{\text{diff}} = \frac{4}{3} s_c^3$.

The term $K_{\text{diff}}$ is the "equilibrium constant," $d_+/d_-$, for encounter complex formation. Note that the constant is independent of the diffusion coefficients and a linear function of the number of free receptors.

When $N_{RF} s_c \ll \pi r$, the apparent rate constants for encounter complex formation are defined as

$$k_f = \frac{4\pi s_c D k_+}{\frac{3D}{s_c^2} + k_+} N_{RF} \tag{15a}$$

and

$$k_r = \frac{\frac{3D}{s_c^2} k_-}{\frac{3D}{s_c^2} + k_+} \tag{15b}$$

In either case, using the apparent rate constants defined in Eqs. (14a) and (14b) or Eqs. (15a) and (15b) in the balance on ligand receptor bonds [Eq. (13)] yields an expression for the rate of ligand-receptor bond formation that is proportional to the product $(N_{RF}\tilde{C}_{\text{cell}})$ or the volume concentration of free receptors.

If the system is allowed to reach equilibrium, the expression for the equilibrium constant is the same for solubilized receptor or membrane-bound receptor interacting with the ligand. For either case, the fraction of receptors not involved in a ligand-receptor bond is

$$\frac{N_{RF}}{N_R} = \frac{1}{1 + K\tilde{C}_L} \tag{16}$$

where

$$K = \left(\frac{4}{3} s_c^3\right) \frac{k_+}{k_-} \tag{17}$$

The rate constants discussed above address the situation where the membrane acts only to restrict free diffusion of the ligand. However, there is evidence that for some ligand receptor systems, ligands in solution associate with the cell membrane and then diffuse rapidly across the membrane to the receptor (56). In this situation, expressions for the rate constants describing the encounter complex formation must be modified (35,36).

The apparent rate constants and equilibrium binding constants have been measured for several ligand-receptor systems of interest in specific cell adhesion. Equilibrium constants for several ligands in the extracellular matrix and their receptors have been tabulated (11). Estimates on the receptor density are also reported.

The apparent rate constants and equilibrium constants for over 40 antibody-antigen systems have been tabulated by Pecht (12). The equilibrium constants for these multivalent interactions ranged over seven orders of magnitude, from approximately $10^4$ $M^{-1}$ to $10^{11}$ $M^{-1}$. In general, values for the apparent rate constant for antibody-antigen bond formation was relatively constant, ranging from $10^4$ $M^{-1}$ $s^{-1}$ to $10^8$ $M^{-1}$ $s^{-1}$. The apparent rate constant for breaking of the bound pair varied over the range of $10^3$ $s^{-1}$ to $10^5$ $s^{-1}$. The calculation of intrinsic rate constants from these apparent rate constants is also discussed.

To evaluate the intrinsic rate constants or the reaction equilibrium constants, the rate constants for encounter complex must be estimated. This requires estimates of the diffusion coefficient of the ligand in solution and the radius of the encounter complex. Various methods are available to estimate this diffusion coefficient. Several estimates of encounter complex radius are available. Bell (1) proposed the radius was 0.75 $\eta$m. Pecht (12) gives a range of 0.3 $\eta$m to 1.5 $\eta$m, calculated from molecular radii. Based on the area of receptor exposed on the cell surface (39), an encounter complex radius of 1.1–1.4 $\eta$m is obtained.

## 2.4 Estimating Equilibrium and Rate Constants for Porins

In this section, methods for calculating the apparent rate constants for porins are illustrated. The particular porin in question is LamB, the maltoporin of *E. coli*. This outer membrane protein is responsible for the transport of malto-oligosaccharides across the cell membrane and has also been used to mediate the specific adhesion of *E. coli*. The difference between methods of calculating equilibrium and apparent rate constants for the porins and receptors is that often the ligand quickly passes through the porin while it remains attached to the receptor. Different experimental techniques and analysis are therefore employed.

### 2.4.1 *In Vitro Methods*

Porins that act as ion channels can be studied in vitro using a lipid bilayer containing the porin of interest. The apparent equilibrium constant for a ligand and porin is determined by measuring the conductance of a membrane containing the porin as a function of ligand concentration. The ligands that bind to the porins block these, and the membrane conductance therefore decreases as the porins are blocked. In this analysis, it is assumed that the porin possesses a single binding site accessible from both ends and each end displays the same apparent rate constants. The method has also been used to investigate how different mutations in the porin affect the affinity for different sugars (57).

In the experiment where soluble ligand, $L$, is initially present in the same concentration, $\tilde{C}_L$, on each side of the membrane, identical ligand concentrations will be found in the reservoirs at equilibrium. The fraction of porins that are not occupied can then be expressed as

$$\frac{C_{RF}}{C_R} = \frac{1}{1 + K\tilde{C}_L} \tag{18}$$

where $C_R$ is the initial or total porin density, $C_{RF}$ is the density of free or unoccupied porin, and $K$ is the equilibrium binding constant.

Equilibrium binding to porins can also be investigated by placing the ligand on only one side of the membrane. In this case, the conductance of the membrane quickly drops and then levels off. When this experiment is repeated with different concentrations of ligand, an "equilibrium" constant is calculated. This constant is determined at a true equilibrium condition for ligands that do not pass through the porin. But for ligands that permeate a porin, the point of stable conductance probably does not occur at equilibrium. This point more likely occurs when the binding of ligand to porin is at a quasi-steady state.

The relationship between this "equilibrium" constant and the one defined in Eq. (18) is determined by assuming the conductance measurement is made at a quasi-steady state. The ligand concentration on one side of the membrane is assumed to be zero and the concentration on the other side is at its original value. The fraction of unoccupied porin as a function of the equilibrium constant is

$$\frac{C_{RF}}{C_R} = \frac{1}{1 + \frac{K}{2}\tilde{C}_L} \tag{19}$$

The "equilibrium" constant determined from membrane conductance after adding ligand to one reservoir is half the true equilibrium constant. This is consistent with the results of Benz et al. (58).

To determine the relative apparent rate constants of porin-ligand attachment, the liposome swelling assay has been used (59–61). In this procedure, liposomes, containing the porin, are formed in a solution of dextran or stachyose. The liposomes are then transferred to an isotonic solution of the ligand to be studied. The ligand binds to the porins and enters the liposome while the dextran or stachyose is retained. As ligands enter, the liposome swells owing to the entrance of water driven by osmotic pressure. The initial rate of liposome swelling, measured as a decrease in optical density, is assumed proportional to the flux of the ligand into the liposome. From this relative flux, the relative apparent rate constants for ligand porin interaction are calculated (58).

The molecular flux of ligand into the liposomes, which occupy a fraction, $f$, of the total volume, $V_t$, is

$$J = -V_t \frac{d((1-f)\tilde{C}'_L)}{dt} = V_t \frac{d(f\tilde{C}''_L)}{dt} \tag{20}$$

where $\tilde{C}'_L$ and $\tilde{C}''_L$ are the concentration of ligand outside and inside the liposome, respectively. It is assumed in this equality that the rate of ligand accumulation in the porins is small compared to the rate at which the ligand number inside or outside the liposome changes.

From the data of Luckey and Nikaido (59), it is observed that the increase in the flux into the liposome is proportional to the porin number. This would occur under two conditions, if the binding is reaction limited or $N_{RF}s_c$ is much less than $\pi r$. Here, $r$ is the characteristic radius of a liposome. In either case, the simplification of the apparent rate constants [Eqs. (14a) and (14b) or (15a) and (15b)] allows Eq. (20) to be written as

$$\begin{aligned}\frac{d((1-f)\tilde{C}'_L)}{dt} &= -k_f(1-f)(\tilde{C}'_L)\tilde{C}_{RF} + k_r\tilde{C}_{LR} \\ &= k_f f(\tilde{C}''_L)\tilde{C}_{RF} - k_r\tilde{C}_{LR}\end{aligned} \tag{21}$$

where $k_f$ and $k_r$ are independent of $N_{RF}$. The form of the apparent rate constants depend on whether binding is reaction limited or $(N_{RF}s_c) \ll \pi r$. Equation 21 can be solved for the number of occupied porins and combined with the total receptor balance to yield an expression for the concentration of free porins:

$$\tilde{C}_{RF} = \frac{\tilde{C}_R}{1 + \dfrac{k_f}{2k_r}((1-f)(\tilde{C}'_L) + f(\tilde{C}''_L))} \tag{22}$$

Using these expressions, the ligand flux into the liposome becomes

$$J = V_t k_f \tilde{C}_R \cdot \frac{(1-f)\tilde{C}'_L - f\tilde{C}''_L}{2 + K((1-f)\tilde{C}'_L + f\tilde{C}''_L)} \tag{23}$$

The initial flux into the liposome is taken to be proportional to the initial rate of liposome swelling, $Y$. During this period of the experiment, the concentration of ligand in the liposome is close to zero. If the total volume of the liposomes is small compared to the total volume of the system, $f \ll 1$, the relative apparent rate constant, $k_f^r$ is calculated to be

$$k_f^r = \frac{Y}{V_t \tilde{C}_R} \cdot \frac{2 + K\tilde{C}_L'}{\tilde{C}_L'} \tag{24}$$

and the relative rate constant for a ligand vacating a porin is

$$k_r^r = \frac{k_f^r}{K} \tag{25}$$

Thus, for the same liposome preparations, relative rate constants for different ligands or various operating conditions can be determined.

### 2.4.2 *In Vivo Methods*

Using radioactive or fluorescently labeled ligands that will not cross the cell membrane, equilibrium concentrations of bound versus free ligand can be measured in vivo (62). From such experiments, equilibrium constants and the mean number of binding sites on a cell are determined.

The apparent association and disassociation rate constants for ligands that do not permeate the porins can be measured in vivo from ligand accumulation on the cell. The analysis of the intrinsic rate constant is similar to that discussed for ligand-receptor systems.

For ligands that permeate porins, if the apparent rate constants for porin interacting with ligand in the bulk solution and in the periplasmic space are taken to be equal, the initial flux of ligand into the cell can be used to estimate the apparent rate constants. During this initial period, the concentration of ligand in the periplasmic space is assumed to be zero, and Eq. (23) is solved for the apparent rate constant.

### 2.4.3 *Apparent and Intrinsic Rate Constants for Maltoporin*

Apparent rate constants for malto-oligosaccharides can be determined using the maltose flux into cells (62), the relative flux into liposomes (59,61), and Eqs. (23) and (24) (Table 1). The flux measurements reported by Ferenci et al. (62) were taken as representative values. The time period of data collection was short, and the rate constants determined in the presence of competitors agreed quite well.

Also presented in Table 1 are the equilibrium constants for the compounds binding to the maltoporin reported by Benz et al. (58) and several estimates of apparent rate constants for ligand binding determined from in vivo ligand accumulation data. All constants were calculated assuming that there was $1.55 \times 10^5$ maltoporin per cell. This value was calculated from the amylopectin binding data presented by Ferenci et al. (62). It agrees well with previously published values of $3 \times 10^4$ to $10^5$ (63–65).

**Table 1** Apparent Rate Constants for the Maltoporin

| | K($M^{-1}$) | $k_f$ $(Ms)^{-1}(\times 10^5)$ | | |
|---|---|---|---|---|
| Ligand | Benz[c] | Luckey[a] | Nakae[a] | in vivo accumulation[b] |
| Maltose | 100 | 8.20 | 8.20 | 8.20[d] |
| Maltotriose | 2,500 | 92.0 | 87.8 | |
| Maltotetraose | 10,000 | 104.4 | 197.8 | 4.60[e] |
| Maltopentaose | 17,000 | | 83.9 | 4.00[e] |
| Maltohexaose | 15,000 | | 107 | 2.31[e] |
| Maltoheptaose | 15,000 | 20.57 | <49 | 0.77[e] |
| Trehalose | 46 | 3.99 | 4.30 | |
| Lactose | 18 | 0.33 | 1.23 | |
| Sucrose | 67 | 0.16 | <0.38 | |
| Gentibiose | 250 | 6.89 | 3.44 | |
| Melibiose | 180 | 4.15 | 8.05 | |
| Celliobiose | 6.7 | 0.40 | 1.13 | |
| D-Glucose | 9.5 | 9.43 | 3.64 | |
| D-Galactose | 24 | 9.1 | 4.49 | |
| D-Fructose | 1.7 | 3.81 | | |
| D-Mannose | 6.3 | 4.92 | 3.26 | |

[a]Calculated using relative permeabilities reported by Luckey and Nikaido (59) or Nakae et al. (61) with maltose flux as described in text as basis.
[b]Apparent rate constants calculated from in vivo accumulation rates using Eq. (23).
[c]From Ref. 58.
[d]From Ref. 62.
[e]From Ref. 64.

To estimate the intrinsic rate constants, the binding to the maltoporin was assumed to be reaction limited. This assumption appears reasonable in light of the results of Luckey and Nikaido (59).

The encounter complex radius was estimated to be 0.4 $\eta$m, the diameter of the pore (58). Note that under reaction limitation the apparent rate constants are independent of the membrane diffusion coefficient. The intrinsic reaction rate constants and equilibrium constants estimated for the maltoporin are presented in Table 2.

## 3 DEPENDENCE OF CELL ADHESION ON LIGAND-RECEPTOR BOND FORMATION

The ability to form the receptor-ligand bond is essential in the process of specific cell adhesion. The success of the cell adhesion event is therefore

**Table 2** Intrinsic Reaction Rate Constants

| Ligand | $K^a$ ($M^{-1}$) | $K_{rxn}$ ($\times 10^3$) | $k_{+1}$ ($s^{-1}$) ($\times 10^7$) | $k_{-1}$ ($s^{-1}$) ($\times 10^3$) |
|---|---|---|---|---|
| Maltose | 100 | 1.95 | 1.60 | 8.2 |
| Maltotriose | 2,500 | 48.6 | 17.9 | 3.68 |
| Maltotetraose | 10,000 | 195.2 | 20.3 | 1.04 |
| Maltoheptaose | 15,000 | 285.7 | 4.0 | 0.14 |
| Trehalose | 46 | 0.90 | 0.78 | 8.67 |
| Lactose | 18 | 0.35 | 0.065 | 1.83 |
| Sucrose | 67 | 1.29 | 0.031 | 0.24 |
| Gentibiose | 250 | 4.87 | 1.34 | 2.75 |
| Melibiose | 180 | 3.52 | 0.81 | 2.30 |
| Celliobiose | 6.7 | 0.13 | 0.078 | 5.97 |
| D-Glucose | 9.5 | 0.18 | 1.83 | 99.3 |
| D-Galactose | 24 | 0.47 | 1.77 | 37.9 |
| D-Fructose | 1.7 | 0.03 | 0.74 | 224.1 |
| D-Mannose | 6.3 | 0.12 | 0.96 | 78.0 |

[a]From Ref. 58.

dictated to some degree by the kinetics of the bond formation. Experimental investigation of parameters that alter bond formation and their role on adhesion is often complicated. In many cases analysis of adhesion experiments is difficult, owing to the interaction between the variables or lack of fundamental information about the system. The framework reviewed in the previous section can help to clarify the interaction of the parameters affecting ligand-receptor bond formation and specific cell adhesion. In the following discussion, selected studies of specific cell adhesion are viewed within the framework of ligand-receptor bond formation. The effects of ligand-receptor binding kinetics, ligand-receptor density, and mobility of membrane components on specific cell adhesion are considered. Even within this framework, the interaction between various parameters complicates the evaluation of adhesion effects. This is particularly true when considering the role of membrane properties on receptor or ligand mobility.

## 3.1 Effects of Ligand-Receptor Equilibrium Binding

To accurately describe specific cell adhesion as a function of ligand receptor binding kinetics, a well defined experimental program is required. Unfortunately, such a program is not readily available for many ligand-receptor systems of interest. In many cases, the information on the kinetics of bond formation or the adhesion of cells to surfaces is incomplete or nonexistent.

One ligand-receptor system that mediates specific cell adhesion and has been studied in some detail is the carbohydrate binding of the *E. coli* maltoporin.

The maltoporin is the *lam*B gene protein product in *E. coli*. It plays a role in the transport of malto-oligosaccharides across the cell outer membrane (66). The maltoporin displays binding specificity for a range of malto-oligosaccharides, amylopectin, amylose, and other carbohydrates, and the transport rates and equilibrium binding constants for many of these compounds have been determined (38,58–60,62,64,67–69). These data have been used to calculate apparent and intrinsic rate constants (see above). Numerous investigations have focused on the structure of the maltoporin (38,70–73), its mode of expression (66,74), and its integration into the outer membrane (75–77). The interaction of the maltoporin with starch has also been used to mediated the specific adhesion of *E. coli* (38,78–81). Within this background lies much of the information necessary for studying maltoporin-mediated adhesion to immobilized carbohydrates.

Several reports of adhesion using cells that express mutated maltoporin provide good examples of the effect of ligand-receptor affinity on adhesion (38,78,79). In the work reported by Charbit et al. (38), maltoporins containing a single amino acid substitution were studied. The maltose transport rate through the maltoporin, equilibrium binding of amylopectin, and cell adhesion to immobilized starch were determined for 21 substitutions in the maltoporin. Cell adhesion was measured as the fraction retained in a packed bed of starch-Sepharose.

In this case, it is possible to estimate equilibrium constant for amylopectin binding. The equilibrium binding is reported as the percent binding observed relative to the wild-type maltoporin. Using $\beta$, the fraction of amylopectin bound by cells expressing the mutated maltoporin compared to binding by cells expressing the wild-type maltoporin, and the equilibrium constant by wild-type binding, $K^{wt}$, the equilibrium binding constant for the mutated maltoporin is

$$K^{m} = \frac{K^{wt}\beta}{1 + K^{wt}\tilde{C}_{L0}(1 - \beta)} \tag{26}$$

The initial ligand concentration, $\tilde{C}_{L0}$, is 40 $\mu$g/ml and is essentially constant under the experimental conditions. The equilibrium binding constant for wild-type maltoporin is estimated to be 2.9 ml/mg (62). The equilibrium binding constants for the mutated maltoporin are calculated to range from 2.9 ml/mg to $<0.03$ ml/mg. This is the apparent equilibrium constant for soluble ligand binding to the cell surface receptor.

The adhesion of cells expressing the mutant maltoporin was a function of maltoporin affinity. A strong correlation between the estimated equilibrium constant and cell adhesion was found. The rank correlation gave a Spearman's rho of 0.84.

The amino acid substitution affects both starch binding and maltose transport. The relative rate of maltose transport also correlates strongly with the equilibrium binding of amylopectin (Spearman's rho = 0.88) or cell adhesion (Spearman's rho = 0.84). It is unclear, however, exactly how these alterations occur. It is suggested that in some cases the amino acid substitution alters the binding site while in others the accessibility of the binding site is changed. For one mutant, at least, access to the binding site is restricted from one side of the porin (72).

Cell adhesion has been used to isolate *E. coli* mutants based on differential binding characteristics of the maltoporin (78,79). Clune et al. (79) report isolation of a population that displayed greater adherence to a starch-sepharose support than a wild-type population under suboptimal conditions. The equilibrium constant for amylopectin binding to the mutant maltoporin was found to be about 6 times that of the wild-type maltoporin. For the mutant maltoporin, the equilibrium constant for binding of maltose also increased. Using slightly different conditions for selection, another population was isolated that expressed increased binding of amylopectin but decreased binding of maltose. Compared to the wild type, this strain also displayed increased adhesion under suboptimal conditions and was more difficult to elute from the packed bed using maltose.

## 3.2 Ligand-Receptor Density Effects

Cell adhesion mediated by receptor-ligand interaction is expected to be strongly influenced by receptor and ligand density. The density would influence both the rate at which receptor-ligand bonds form and the final density of receptor-ligands bonds between the cell and adhesion surface. The number of bonds between the cell and surface directly influences the strength of cell adhesion (82).

One property of the ligand-receptor density that has been observed is, discounting the effects of nonspecific adhesion, that a threshold density is required before adhesion occurs (9,13,14,83–85).

A good illustration of the threshold ligand density was reported by Weigel et al. (83). Cell adhesion was first observed at the threshold ligand density. The fraction of cell that attached increased with ligand density until maximum adhesion occurred at densities that were 10–20% above the threshold value. Further increasing the ligand density did not lead to further increase in cell attachment. The rate of attachment also increased with ligand density, while the lag period before cell attachment decreased as the ligand density increased. The increase in the cell adhesion with ligand concentration has been reported for a large variety of cells and ligands (5,78,84,86–88).

It has been reported that the ability of cells to adhere or remain attached under the influence of removal forces depends on ligand and receptor density (5,15,89–92). The fraction of glioma cells that remain attached to a surface coated with fibronectin or tenascin under a normal force increased with the amount of ligand on the surface (87). The force required to separate cells agglutinated by Con A was found to increase sharply with the Con A concentration and then level off (5). Increased antibody concentration resulted in an increase in the shear and normal forces required to separate agglutinated cells (90).

Increasing the shear forces on a cell (83) or receptor-coated bead (9) resulted in an increase in the ligand density required for attachment. The studies with the receptor-coated beads suggested that there was a linear relationship between the two factors over a limited range of ligand and receptor densities.

Wattenburger et al. (85) used liposomes containing receptors to study attachment to ligand-coated surfaces. The probability of adhesion was found to be a function of receptor density, increasing with increasing receptor number. The liposome behavior before attachment was dependent on receptor density. Liposomes with high receptor density were observed to attach immediately upon contacting the ligand-coated surface. If the liposomes contained a low amount of receptor, they would continually adhere to the surface and then release and not become permanently attached. Liposomes with intermediate levels of receptor would adhere and detach from the surface several times before becoming permanently attached. The extent of adhesion and detachment before permanent adhesion also appeared to be a function of the forces on the liposome.

## 3.3 Mobility of Membrane Components and Membrane Deformation

The enclosure around a cell consists of a set of membranes and related structural components. Because of this makeup, the cells display elasticity and deform when subjected to external physical forces (93). The cells themselves can also initiate deformation through such activities as spreading and villi extension. Hochmuth and Waugh (94) classified any deformation of the membrane as being the function of three basic types of deformation. The basic deformation types were represented by either the shear modulus, the area expansion modulus, or a bending modulus.

The membranes that enclose the cells are described using the generally accepted fluid mosaic model (7). In this model the components of the membrane are free to diffuse in the plane of the membrane. In reality, free diffusion of these components is rarely observed. Movement is often hindered by interaction with other membrane components or with cell structures on either side of the membrane (15,17,31,32).

Both the deformation of the cell and the mobility of the membrane components are expected to affect the absolute rate at which bonds between the cell and an adhesion surface form, the number of bonds, and the probability of cell adhesion. For the bond formation to occur the cell and the surface must approach within a short distance of each other. This distance has been estimated to be less than 15 $\eta$m (87,95). The elasticity of the cell membrane allows the membrane to deform to fit the contours of an adhesion surface it encounters. This brings a greater number of receptors into range for immediate bond formation with ligand without the time lag associated with diffusion into the area of contact.

During the collision of a cell with an adhesion surface, the deformation of the cell membrane may provide a local environment in which bonds can form in an unstressed state. After this short period, forces exerted on the cell that tend to move it away from the adhesion surface would exert stress on the bond. It has been noted that a single bond could interrupt the thermal movement of the cell for a period (5) and temporarily halt cell movement under conditions of low hydrodynamic stress (96). In this way the membrane properties directly influence the rate and strength of cell adhesion.

The degree of cell deformation is dependent on the membrane and surface elastic properties. When rat basophilic leukemia cells were brought into contact with a surface containing a ligand for a specific cell surface receptor, membrane deformation upon adhesion was dependent on the nature of the adhesion surface (18). Use of a polyacrylamide bead or a solid-phase lipid vesicle as the adhesion surface caused the cell to deform and wrap around the spherical surface. The cell did not deform upon adhesion to a fluid-phase vesicle. However, the vesicle distorted to partially engulf the cell.

A cell adhering in a specific manner to a surface is reported to have a greater area in surface contact than a similar cell adhering in a nonspecific manner (16). The area in contact has been reported to increase with ligand mobility (92) and ligand concentration (5,91). However, at least at the levels studied by Sung et al. (91), the initial rate of increase in the contact area was the same for high and low ligand density. A larger steady-state contact area was achieved at the higher ligand density.

The time-dependent increase of the cell area in close contact with the surface has been reported to be a function of the particular ligand-receptor pair (87). In this report, glioma cell attachment to fibronectin and tenascin was studied. The area of contact was initially the same for cells adhering to a surface coated with either ligand. However, after incubation at 37°C for 15 min, the area in close contact increased about 125-fold for cells on fibronectin while approximately doubling for cells adhering to tenascin. This change in surface contact area also correlated with an increase in adhesion strength as determined by a centrifugal force assay. The adhesion strength for fibroblasts on fibronectin did not increase when actin polymerization was inhibited. Lotz et al. (87) suggest that the strengthening of adhesion over the time scale of 15 min was the result of interaction between the receptor and cytoskeleton. However, the molar density of fibronectin was several times that of tenascin.

The relationship between adhesion and fibronectin receptor interaction with the cytoskeleton is consistent with the observations of Duband et al. (17). These investigators found that fibronectin receptors where concentrated around fibrillar streaks and in focal contact areas on stationary cells. Of the receptors concentrated in these areas, only a small fraction displayed mobility. For this mobile fraction, the lateral diffusion coefficient was approximately the same as for receptors in membrane areas outside fibrillar streaks or focal contact areas.

The strength of cell adhesion, evaluated by the force required to remove the cell, is dependent on mobility of the ligands and receptors and their density. Chan et al. (15) studied the removal of Jurkat cells adhering to a membrane containing mobile or immobile ligands. The ligands were isoforms of lymphocyte function-associated antigen 3 in a planar membrane. At all but the highest ligand densities, after a contact time of 20 min, much higher shear stress was required to remove cells from membranes containing the mobile ligands compared to cells adhering to membranes containing immobile ligands. The shear stress for removal was about the same at the highest ligand densities. However, the development of strong adhesion was much faster for cells adhering to the membrane with mobile ligand. The strength of adhesion was also found to be dependent on the mobility of the cell surface receptor.

The increase of adhesion strength over time has been correlated with the accumulation of receptors in the area of cell contact with the adhesion surface (18). It is unclear whether the increased adhesion strength was solely due to increased receptor-ligand bond density. The redistribution of receptors into the area in contact with the adhesion surface was demonstrated to be independent of cytoskeletal involvement. Receptors accumulated in membrane blebs when the bleb was brought into contact with a ligand-bearing vesicle. In the bleb, the membrane is separated from the cytoskeleton and the receptor complex is reported to possess a higher lateral diffusion coefficient than found in the normal membrane (32). This finding is noted by McCloskey and Poo (18) to support the observation that receptor accumulation in the contact region occurred more quickly in the membrane blebs than in a normal membrane.

The effects of altering membrane rigidity and the mobility of membrane components on specific cell adhesion have been reported (5,13,14,97,98). In these investigations, cell-cell adhesion was promoted by a lectin and the membrane properties altered by gluteraldehyde treatment. Treating the cells with gluteraldehyde crosslinks proteins in the cell membrane (99). This is expected to increase membrane rigidity and decrease mobility of the membrane components. Cells treated in this fashion displayed very low tendency to adhere in a specific manner. Significant adhesion was reported when one of the cells in these cell-cell adhesion experiments was treated with gluteraldehyde.

Adhering cells that were treated with gluteraldehyde were found to be more easily disrupted by shear forces than expected from observations using untreated cells (5,89,97,98). The area of cell membranes in contact with the adhesion surface was found to be much less for the gluteraldehyde-treated cells compared to the untreated cells (5,98). Also, the lectin was found to accumulate in the contact area between cells that were not treated with gluteraldehyde, depleting the rest of the surface, while the treated cells displayed no sign of lectin redistribution on adhesion.

Clustering of receptors (16–18) and other membrane components (16) in the area in contact with the adhesion surface increases the local concentration over time. This accumulation is expected to alter the rate constants for encounter complex formation through both the protein density effect on the local translational diffusion coefficient and the decreased mean separation distance between unbound receptors. Interactions of the membrane protein with other structures that affect protein mobility would similarly alter the rate constants for encounter complex formation.

## 4 NONSPECIFIC REPULSIVE FORCES AND HYDRODYNAMIC FORCES ON ADHERED CELLS

The fate of a cell that is approaching an adhesion surface or one that is already attached depends to a great extent on the external forces on the cell. As mentioned in previous sections, many parameters of cell adhesion affect the retention of cells on a surface in the face of removal forces. The concentration of an agglutinating compound, the kinetics of ligand-receptor bond formation, receptor and ligand density, contact time, and cell surface contact area alter the ability of the cells to adhere under conditions where removal forces are present. All these factors can influence the number of ligand-receptor bonds and presumably the strength of adhesion. In the following discussion the effects of removal forces on cell adhesion and the role of increased bond number or cytoskeletal involvement on cell adhesion are outlined.

Any force that acts to disrupt cell adhesion mediated by specific ligand-receptor bonds must break these bonds. Although the ligand-receptor bonds are reversible, the probability that several bonds holding a cell to a surface are all broken at a given time is normally quite low. The stresses on the bonds act to hasten the approach to the state where all ligand-receptor bonds are broken. Bell (1) used this reasoning and the theory of strength of solids to express the rate constant for breakup, $k_-$, of the stressed ligand-receptor bond as increasing exponentially with the tensile force on the bond, $F/N_B$.

$$k_- = k_-^0 \exp\left(\frac{\delta F}{k_B T N_B}\right) \tag{27}$$

where $F$ is the total removal force on the cell, $k_B$ is the Boltzmann constant, $N_B$ is the number of ligand-receptor bonds, $T$ is temperature, and $\delta$ is a characteristic length of the free energy minimum of the bond, estimated by Bell to be on the order of 0.5 $\eta$m (1).

The force required to remove a cell from a surface, assuming that all bonds are equally stressed, was then calculated as a function of ligand and receptor number and the bond formation kinetics. It was noted that while removal forces exerted on a cell are often shearing, mobile ligands and receptors could orient themselves such that stress on the individual bonds is tensile. This approach has been employed in several models to express the effect of cell removal forces on breakup of ligand-receptor bonds (6,100) although it may greatly overestimate bond lifetime (96).

Nonspecific repulsive forces between a cell and an adhesion surface stress bonds or discourage the close approach of cells to the surface. Bell et al. (95) developed an approach for calculating the competing forces for bond formation and cell adhesion in the presence of nonspecific repulsive forces.

Often in a cell adhesion assay, an adhering cell is subjected to a removal force to indicate the strength of adhesion. An adhesion assay that uses centrifugal force to stress the bonds of adhering cells has been employed (87). Micropipettes have been used to capture adhering cells and pull them away from the adhesion surface (18,93,101). In these experiments the separation force is normal to the plane of adhesion. The general behavior of adhering cells was to remain attached when subjected to low removal force. As the force was increased, a range was reached where the probabilistic nature of the adhesion was displayed. In this range, a fraction of the cells remained adhered to the surface while the rest detached. At a large enough removal force, all adhered cells were found to release. This behavior was found to be a function of adhesion time, the bond density, and the size of the contact area (18,87).

Tozeren et al. (101) used micromanipulation to pull apart two adhering cells and analyze the effect of the separation. Analysis of these experiments indicated that the contact area decreased as the cells separated, but the energy density function increased. These observations were explained in context of the view that the bond receptor-ligand pair moves along the separating edge into the contact area (93,101).

A cell in a fluid environment is subject to the hydrodynamic forces of the bulk fluid. Two cells adhering together in a moving fluid experience hydrodynamic forces that can cause breaking of the adhesive bonds. In a study of the forces acting on agglutinated cells, estimates of the hydrodynamic forces were coupled with observations of breakup of a cell doublet (89,90,102). The shear rate required for separating the cells decreased with an increase in fluid viscosity. The shear force and normal force required for separating the cells increased with the concentration of agglutinating antibody, indicating that doublets with increased bond number were more resistant to breakup.

A cell adhering to a surface experiences removal forces as the fluid moves past. Hydrodynamic forces on the cell, due to directional fluid flow, tend to induce a torque on an adhered cell which stresses ligand-receptor bonds. The distribution of forces on the bonds is expected to be a function of the bond position relative to the direction of the flow. Several stress distribution models have been introduced to estimate stress as a function of bond position in the contact area (1,100,103–105).

The effects of fluid forces on specific cell adhesion and detachment were investigated by Lauffenburger and co-workers using several well-defined model systems (6,9,10,85). Using the antibody-coated latex beads, the removal of adhering cells as a function of shear stress was evaluated (6). There existed a transition region of shear stress over which partial removal of adhering beads was observed. Shear stresses in this region resulted in the removal of a fraction of the beads while the rest remained attached to the surface. The fraction adhering decreased sharply as shear increased. At shear stresses above those of the transition region, all beads were removed.

The behavior of individual cells interacting with a ligand-coated surface yields more insight into adhesion behavior in a moving fluid. Wattenburger et al. (85) studied the interaction of receptor-bearing liposomes in a moving fluid with a ligand-coated surface. The liposomes displayed various types of behavior. Those that contained a high density of receptors adhered immediately to the surface. Liposomes with intermediate to low receptor density would contact the surface a number of times. The final fate of these liposome, whether they adhered or not, was dependent on the shear rate. The number of collisions before adhesion increased with the shear rate.

## 5 NOMENCLATURE

Not surprisingly, there is little agreement on the use of symbols in this area of research. We have tried to use a consistent nomenclature in this chapter and to avoid dual use of symbols. To achieve this, we have often had to alter the nomenclature that was used in the original references. This should be kept in mind when consulting the original references. The balance equations on the number of bonds in the contact area that appear in adhesion models can be written using either numbers of bonds or concentration (density) of bonds. Here we have used the convention that numbers are indicated by $N$, surface concentrations are indicated by $C$, and volume concentrations are indicated by $\tilde{C}$.

$a$ Radius of protein.
$C_B$ Density or surface concentration of ligand-receptor bonds.
$C_L$ Total ligand density or concentration.
$C_{LF}$ Density of free or onbound ligands.
$C_{L0}$ Initial ligand concentration.
$\tilde{C}'_L, \tilde{C}''_L$ Ligand volume concentration outside and inside of liposomes, respectively.
$C_R$ Total receptor or porin density.
$C_{RF}$ Density of free or unbound receptors or porins.
$d_+$ Rate constant for formation of encounter complex.
$d_-$ Rate constant for breakup of encounter complex.
$D$ Diffusion coefficient of ligand in solution, the sum of the translational and rotational diffusion coefficients in solution (area/time).
$D_m$ Receptor diffusion coefficient in the membrane.
$D_r$ Rotational diffusion coefficient in the membrane.
$D_T$ Translational diffusion coefficient in the membrane.
$f$ Fraction of volume occupied by liposomes.
$F$ Tensile force on ligand-receptor bonds.
$h$ Membrane thickness.
$J$ Ligand flux.
$k_B$ Boltzmann's constant, $1.380622\ 10^{-23}$ J $K^{-1}$.
$k_f$ Apparent rate constant for ligand-receptor bond formation.
$k_r$ Apparent rate constant for ligand-receptor bond breakup.
$k_+$ Rate constant for ligand-receptor bond formation.

$k_-$ Rate constant for ligand-receptor bond breakup.
$K$ Equilibrium binding constants.
$K_{\text{diff}}$ Equilibrium constant for encounter complex formation. Equal to $d_+/d_-$
$N_B$ Number of ligand-receptor bonds on an adhered cell.
$N_{RF}$ Number of unbound receptors on a cell or liposome.
$N_R$ Total number of receptors available for binding.
$r$ Cell radius.
$s_c$ Radius of encounter complex.
$s_l$ Mean separation distance between unbound ligands.
$t_+$ Mean encounter time for encounter complex formation.
$t_-$ Mean time for a receptor to exit an encounter complex.
$T$ Absolute temperature.
$V_t$ Total volume of liposome suspension.
$\gamma$ Euler's constant, $\gamma = 0.5772156649\ldots$
$\delta$ Characteristic length of the free-energy minimum of a ligand-receptor bond.
$\mu$ Viscosity of fluid surrounding a cell.
$\mu_m$ Viscosity of cell membrane.
$Y$ Initial rate of liposome swelling.

## REFERENCES


1. Bell, G.I. (1978). Models for the specific adhesion of cells to cells, *Science*, *200*: 618.
2. Eigen, M. (1974). Diffusion control in biochemical reactions, in *Quantum Statistical Mechanics in the Natural Sciences* (S.L. Mintz and S.H. Widmanger, eds.), Plenum Press, New York, p. 37.
3. DeLisi, C. (1980). The biophysics of ligand-receptor interactions, *Q. Rev. Biophys.*, *13*: 201.
4. Lauffenburger, D., and DeLisi, C. (1983). Cell surface receptors: physical chemistry and cellular regulation, *Int. Rev. Cyt.*, *84*: 269.
5. Capo, C., Garrouste, Benoliel, A-M., Bongrand, P., Ryter, A., and Bell, G.I. (1982). Concanavalin-A-mediated thymocyte agglutination: a model for a quantitative study of cell adhesion, *J. Cell Sci.*, *56*: 21.
6. Cozens-Roberts, C., Lauffenburger, D.A., and Quinn, J.A. (1990). Receptor-mediated cell attachment and detachment kinetics. I. Probabilistic model and analysis, *Biophys. J.*, *58*: 841.
7. Singer, S.J., and Nicolson, G.L. (1972). The fluid mosaic model of the structure of cell membranes, *Science*, *175*: 720.
8. Kotyk, A., Janacek, K., and Koryta, J. (1988). *Biophysical Chemistry of Membrane Functions*, Wiley, Chichester, p. 89.
9. Cozens-Roberts, C., Quinn, J.A., and Lauffenburger, D.A. (1990). Recep-

tor-mediated adhesion phenomena, model studies with the radial-flow detachment assay, *Biophys. J.*, *58*: 107.

10. Cozens-Roberts, C., Quinn, J.A., and Lauffenburger, D.A. (1990). Receptor-mediated cell attachment and detachment kinetics. II. Experimental model studies with the radial flow detachment assay, *Biophys. J.*, *58*: 857.
11. Akiyama, S.K., Nagata, K., and Yamada, K.M. (1990). Cell surface receptors for extracellular matrix components, *Biochim. Biophys. Acta*, *1031*: 91.
12. Pecht, I. (1982). Dynamic aspects of antibody function, in *The Antigens*, Vol. 6 (M. Sela, ed.), Academic Press, New York, p. 1.
13. Rutishauser, U., and Sachs, L. (1975). Cell-to-cell binding induced by different lectins, *J. Cell Biol.*, *65*: 247.
14. Rutishauser, U., and Sachs, L. (1975). Receptor mobility and the binding of cells to lectin-coated fibers, *J. Cell Biol.*, *66*: 76.
15. Chan, P-Y., Lawrence, M.B., Dustin, M.L., Ferguson, L.M., Golan, D.E., and Springer, T.A. (1991). Influence of receptor lateral mobility on adhesion strengthening between membranes containing LFA-3 and CD2, *J. Cell Biol.*, *115*: 245.
16. Andre, P., Benoliel, A-M., Capo, C., Foa, C., Buferne, M., Boyer, C., Schmitt-Verhulst A-M., and Bongrand, P. (1990). Use of conjugates made between T cell clone and target cells to study the redistribution of membrane molecules in cell contact areas, *J. Cell Sci.*, *97*: 335.
17. Duband, J-L., Nuckolls, G.H., Ishihara, A., Hasegawa, T., Yamada, K.M., Thiery, J.P., and Jacobson, K. (1988). Fibronectin receptor exhibits high lateral mobility in embryonic locomoting cells but is immobile in focal contacts and fibrillar streaks in stationary cells, *J. Cell Biol.*, 107: 1385.
18. McCloskey, M.A., and Poo, M. (1986). Contact redistribution of specific membrane components: local accumulation and development of adhesion, *J. Cell Biol.*, *102*: 2185.
19. Berg, H.C., and Purcell, E.M. (1977). Physics of chemoreception, *Biophys. J.*, *20*: 193.
20. Peters, R., and Cherry, R.J. (1982). Lateral and rotational diffusion of bacteriorhodopsin in lipid bilayers: experimental test of the Saffman-Delbruck equations, *Proc. Natl. Acad. Sci. USA*, *79*: 4317.
21. Saffman, P.G., and Delbruck, M. (1975). Brownian motion in biological membranes, *Proc. Natl. Acad. Sci. USA*, *72*: 3111.
22. Hughes, B.D., Pailthorpe, B.A., White, L.R., and Sawyer, W.H. (1982). Extraction of membrane microviscosity from transitional and rotational diffusion coefficients, *Biophys. J.*, *37*: 673.
23. Tank, D.W., Wu, E.S., Meers, P.R., and Webb, W.W. (1982b). Lateral diffusion of gramicidin C in phospholipid multilayers; effects of cholesterol and high gramicidin concentration, *Biophys. J.*, *40*: 129.
24. Saxton, M.J. (1987). Lateral diffusion in an archipelago; the effect of mobile obstacles, *Biophys. J.*, *52*: 989.
25. Saxton, M.J. (1992). Lateral diffusion and aggregation; a Monte Carlo study, *Biophys. J.*, *61*: 119.

26. Saxton, M.J. (1993). Lateral diffusion in an archipelago; dependence on tracer size, *Biophys. J.*, *64*: 1053.
27. Scalettar, B.A., Abney, J.R., and Owicki, J.C. (1988). Theoretical comparison of the self diffusion and mutual diffusion of interacting membrane proteins, *Proc. Natl. Acad. Sci. USA*, *85*: 6726.
28. Abney, J.R., Scalettar, B.A., and Owicki, J.C. (1989). Self diffusion of interacting membrane proteins, *Biophys. J.*, *55*: 817.
29. Minton, A.P. (1989). Lateral diffusion of membrane proteins in protein-rich membranes, *Biophys. J.*, *55*: 805.
30. Jacobson, K., Ishihara, A., and Inman, R. (1987). Lateral diffusion of proteins in membranes, *Annu. Rev. Physiol.*, *49*: 163.
31. Wier, M.L., and Edidin, M. (1986). Effects of cell density and extracellular matrix on the lateral diffusion of major histocompatibility antigens in cultured fibroblasts, *J. Cell Biol.*, *103*: 215.
32. Tank, D.W., Wu, E-S., and Webb, W.W. (1982). Enhanced molecular diffusibility in muscle membrane blebs: release of lateral constraints, *J. Cell Biol.*, *92*: 207.
33. Gumbiner, B., and Louvard, D. (1985). Localized barriers in the plasma membrane: a common way to form domains, *TIBS*, November, 435.
34. Kusumi, A., Sako, Y., and Yamamoto, M. (1993). Confined lateral diffusion of membrane receptors as studied by single particle tracking (nanovid microscopy). Effects of calcium-induced differentiation in cultured epithelial cells, *Biophys. J.*, *65*: 2021.
35. Eisinger, J., and Halperin, B.I. (1986). Effects of spatial variation in membrane diffusibility and solubility on lateral transport of membrane components, *Biophys. J.*, *50*: 513.
36. de Beus, A., and Eisenger, A. (1992). Modulation of lateral transport of membrane components by spatial variations in diffusivity and solubility, *Biophys. J.*, *63*: 607.
37. van Oss, C.J., Absolom, D.R., and Neumann, A.W. (1980). The "hydrophobic effect": essentially a van der Waals interaction, *Colloid Polymer Sci.*, *258*: 424.
38. Charbit, A., Gehring, K., Nikaido, H., Ferenci, T., and Hofnung, M. (1988). Maltose transport and starch binding in phage-resistant point mutants of maltoporin, *J. Mol. Biol.*, *201*: 487.
39. van Oss, C.J., and Absolom, D.R. (1984). Nature and thermodynamics of antigen-antibody interactions, in *Molecular Immunology, A Textbook* (M.Z. Atassi, C.J. van Oss, and D.R. Absolom, eds.), Marcel Dekker, New York, p. 337.
40. Levitzki, A. (1984). *Receptors, a Quantitative Approach*, Benjamin/Cummings, Menlo Park, CA.
41. Ruzgas, T.A., Razumas, V.I., and Kulis, Y.Y. (1992). Ellipsometric study of antigen-antibody interactions in the solid-solution interphase region, *Biophysics*, *37*: 46.
42. Goldstein, B., and Wolfsy, C. (1980). Theory of equilibrium binding of a

bivalent ligand to cell surface antibody: the effect of antibody heterogeneity on cross-linking, *J. Math. Biol.*, *10*: 347.
43. Perelson, A.S., and DeLisi, C. (1980). Receptor clustering on a cell surface. I. Theory of receptor cross-linking by ligands bearing two chemically identical functional groups, *Math. Biosci.*, *48*: 71.
44. Perelson, A.S. (1980). Receptor clustering on a cell surface. II. Theory of receptor cross-linking by ligands bearing two chemically distinct functional groups, *Math. Biosci.*, *49*: 87.
45. Perelson, A.S. (1980). Receptor clustering on a cell surface. III. Theory of receptor cross-linking by multivalent ligands: description by ligand states, *Math. Biosci.*, *53*: 1.
46. Dower, S.K., Titus, J.A., and Segal, D.M. (1984). The binding of multivalent ligands to cell surface receptors, in *Cell Surface Dynamics: Concepts and Models* (A.S. Perelson, C. Delis, and F.W. Wiegel, eds.), Marcel Dekker, New York, p. 277.
47. Klotz, I.M. (1985). Ligand-receptor interactions: facts and fantasies, *Q. Rev. Biophys.*, *18*: 227.
48. Almagor, H., and Levitzki (1990). Analytical determination of receptor-ligand dissociation constants of two populations of receptors from displacement curves, *Proc. Natl. Acad. Sci. USA*, *87*: 6482.
49. Hill, T.L. (1975). Effect of rotation on the diffusion-controlled rate of ligand-protein association, *Proc. Natl. Acad. Sci. USA*, *72*: 4918.
50. Schmitz, K.S., and Schurr, J.M. (1972). The role of orientation constraints and rotational diffusion in bimolecular solution kinetics, *J. Phys. Chem.*, *76*: 534.
51. Baldo, M., Grassi, A., and Raudino, A. (1989). Modeling the mechanism of enzyme reactivity by the rototranslation diffusion equation, *Phys. Rev. A*, *39*: 3700.
52. Baldo, M., Grassi, A., and Raudino, A. (1989). Effect of the internal rotations of the reactants in diffusion-controlled chemical reactions: an application to the enzyme kinetic problems, *Phys. Rev. A*, *40*: 1017.
53. DeLisi, C., and Wiegel, F.W. (1981). Effect of nonspecific forces and finite receptor number on rate constants of ligand-cell bound-receptor interactions, *Proc. Natl. Acad. Sci. USA*, *78*: 5569.
54. Shoup, D., and Szabo, A. (1982). Role of diffusion in ligand binding to macromolecules and cell-bound receptors, *Biophys. J.*, *40*: 33.
55. Erickson, J., Goldstein, B., Holowka, D., and Baird, B. (1987). The effect of receptor density on the forward rate constant for binding of ligands to cell surface receptors, *Biophys. J.*, 52: 657.
56. Behling, R.W., and Jelinski, L.W. (1990). Importance of the membrane in ligand-receptor interactions, *Biochem. Pharm.*, *40*: 49.
57. Benz, R., Francis, G., Nakae, T., and Ferenci, T. (1992). Investigation of the selectivity of maltoporin channels using mutant LamB proteins: mutations changing the maltodextrin binding site. *Biochim. Biophys. Acta*, *1104*: 299.
58. Benz, R., Schmid, A., and Vos-Scheperkeuter, G.H. (1987). Mechanism of

sugar transport through the sugar-specific LamB channel of *Escherichia coli* outer membrane, *J. Membrane Biol.*, *100*: 21.

59. Luckey, M., and Nikaido, H. (1980). Specificity of diffusion channels produced by λ phage receptor protein of *Escherichia coli*, *Proc. Natl. Acad. Sci. USA*, *77*: 167.
60. Luckey, M., and Nikaido, H. (1980). Diffusion of solutes through channels produced by phage lambda receptor protein of *Escherichia coli*: inhibition by higher oligosaccharides of maltose series, *Biochem. Biophys. Res. Commun.*, *93*: 166.
61. Nakae, T., Ishii, J., and Ferenci, T. (1986). The role of the maltodextrin-binding site in determining the transport properties of the LamB protein, *J. Biol. Chem.*, *261*: 622.
62. Ferenci, T., Schwentorat, M., Ullrich, S., and Vilmart, J. (1980). Lambda receptor in the outer membrane of *Escherichia coli* as a binding protein for maltodextrins and starch polysaccharides, *J. Bacteriol.*, *142*: 521.
63. Braun, V., and Krieger-Brauer, H.J. (1977). Interrelationship of the phage λ receptor protein and maltose transport in mutants of *Escherichia coli* K12, *Biochim. Biophys. Acta*, *469*: 89.
64. Ferenci, T. (1980). The recognition of maltodextrins by *Escherichia coli*, *Eur. J. Biochem.*, *108*: 631.
65. Schwartz, M. (1983). Phage λ receptor (LamB protein) in *Escherichia coli*, *Meth. Enzymol.*, *97*: 100.
66. Shuman, H.A., and Treptow, N.A. (1985). The maltose-maltodextrin transport system of *Escherichia coli* K-12, in *The Enzymes of Biological Membranes*, 2nd ed., *Vol. 3. Membrane Transport* (A.N. Mortonosi, ed.), Plenum Press, New York, p. 561.
67. Benz, R., Schmid, A., Nakae, T., and Vos-Scheperkeuter, G.H. (1986). Pore formation by LamB of *Escherichia coli* in lipid bilayer membranes, *J. Bacteriol.*, *165*: 978.
68. Szmelcman, S., Schwartz, M., Silhavy, T.J., and Boos, W. (1976). Maltose transport in *Escherichia coli* K12, *Eur. J. Biochem.*, *65*: 13.
69. Ferenci, T., Muir, M., Lee, K-S., and Maris, D. (1986). Substrate specificity of the *Escherichia coli* maltodextrin transport system and its component proteins, *Biochim. Biophys. Acta*, *860*: 44.
70. Clement, J.M., and Hofnung, M. (1981). Gene sequence of the λ receptor, an outer membrane protein of *E. coli* K12, *Cell*, *27*: 507.
71. Gehring, K., Charbit, A., Brissaud, E., and Hofnung, M. (1987). Bacteriophage λ receptor site on the *Escherichia coli* K-12 LamB protein, *J. Bact.*, *169*: 2103.
72. Dargent, B., Charbit, A., Hofnung, M., and Pattus, F. (1988). Effect of point mutations on the in vitro pore properties of maltoporin, a membrane of *Escherichia coli* outer membrane, *J. Mol. Biol.*, *201*: 497.
73. Ferenci, T., Saurin, W., and Hofnung, M. (1988). Folding of maltoporin with respect to the outer membrane, *J. Mol. Biol.*, *201*: 493.
74. Raibaud, O., and Richet, E. (1987). Maltotriose is the inducer of the maltose regulon of *Escherichia coli*, *J. Bacteriol.*, *169*: 3059.

75. Benson, S., and Silhavy, T.J. (1983). Information within the mature LamB protein necessary for localization to the outer membrane of *E. coli* K12, *Cell*, *32*: 1325.
76. Nakae, T., and Ishii, J.N. (1982). Molecular weights and subunit structure of LamB proteins, *Ann. Microbiol.* (*Inst. Pasteur*), *133 A*: 21.
77. Lepault, J., Dargent, B., Tichelaar, W., Rosenbusch, J.P., Leonard, K., and Pattus, F. (1988). Three-dimensional reconstruction of maltoporin from electron microscopy and image processing, *EMBO J.*, *7*: 261.
78. Ferenci, T., and Lee, K.S. (1982). Directed evolution of the Lambda receptor of *Escherichia coli* through affinity chromatographic selection, *Mol. Biol.*, *160*: 431.
79. Clune, A., Lee, K-S., and Ferenci, T. (1984). Affinity engineering of maltoporin: variants with enhanced affinity for particular ligands, *Biochem. Biophys. Ress. Commun.*, *121*: 34.
80. Roos, J.W., and Hjortso, M.A. (1989). Determination of population balances in a mixed culture by specific cell adhesion, *Biotech. Technol.*, *3*: 7.
81. Roos, J.W., and Hjortso, M.A. (1991). Control of an *Escherichia coli* mixed culture via affinity binding, *Biotech. Bioeng.*, *38*: 380–388.
82. Tozeren, A. (1990). Cell-cell, cell-substrate adhesion: theoretical and experimental considerations. *J. Biomech. Eng.*, *112*: 311.
83. Weigel, P.H., Schnaar, R.L., Kuhlenschmidt, M.S., Schmell, E., Lee, R.T., Lee, Y.C., and Roseman, S. (1979). Adhesion of hepatocytes to immobilized sugars, *J. Biol. Chem.*, *254*: 10830.
84. Jones, G.E., Arumugham, R.G., and Tanzer, M.L. (1986). Fibronectin glycosylation modulates fibroblast adhesion and spreading, *J. Cell Biol.*, *103*: 1663.
85. Wattenburger, M.R., Graves, D.J., and Lauffenburger, D.A. (1990). Specific adhesion of glycophorin liposomes to a lectin surface in shear flow, *Biophys. J.*, *57*: 765.
86. Aumailley, M., Mann, K., von der Mark, H., and Timpl, R. (1989). Cell attachment properties of collagen type VI and Arg-Gly-Asp dependent binding to its $\alpha$2(VI) and $\alpha$3(VI) chains, *Exp. Cell Res.*, *181*: 463.
87. Lotz, M.M., Burdsal, C.A., Erickson, H.P., and McClay, D.R. (1989). Cell adhesion to fibronectic and tenascin: quantitative measurements of initial binding and subsequent strengthening response, *J. Cell Biol.*, *109*: 1795.
88. Xia, Z., Goldsmith, H.L., and van de Ven, T.G.M. (1993). Kinetics of specific and nonspecific adhesion of red blood cells on glass, *Biophys. J.*, *65*: 1073.
89. Tha, S.P., and Goldsmith, H.L. (1986). Interaction forces between red cells agglutinated by antibody. I. Theoretical, *Biophys. J.*, *50*: 1109.
90. Tha, S.P., Shuster, J., and Goldsmith, H.L. (1986). Interaction forces between red cells agglutinated by antibody. II. Measurement of hydrodynamic force of breakup, *Biophys. J.*, *50*: 1117.
91. Sung, K-L., Kuhlman, P., Maldonado, F., Lollo, B.A., Chien, S., and Brian, A.A. (1992). Force contribution of the LFA-1/ICAM-1 complex to T cell adhesion, *J. Cell Sci.*, *103*: 259.

92. Tozeren, A., Sung, K-L.P., Sung, L.A., Dustin, M.L., Chan, P.Y., Springer, T.A., and Chien, S. (1992). Micromanipulation of adhesion of a Jurkat cell to a planar bilayer membrane containing lymphocyte function-associated antigen 3 molecules, *J. Cell Biol.*, *116*: 997.
93. Evans, E., and Leung, A. (1984). Adhesivity and rigidity of erythrocyte membrane in relation to wheat germ agglutin binding, *J. Cell Biol.*, *98*: 1201.
94. Hochmuth, R.M., and Waugh, R.E. (1987). Erythrocyte membrane elasticity and viscosity, *Annu. Rev. Physiol.*, *49*: 209.
95. Bell, G.I., Dembo, M., and Bongrand, P. (1984). Cell adhesion competition between nonspecific repulsion and specific bonding, *Biophys. J.*, *45*: 1051.
96. Kaplanski, G., Farnarier, C., Tissot, O., Pierres, A., Benoliel, A-M., Alessi, M-C., Kaplanski, S., and Bongrand, P. (1993). Granulocyte-endothelium initial adhesion: analysis of transient binding events mediated by E-selectin in a laminar shear flow, *Biophys. J.*, *64*: 1922.
97. Marquardt, M.D., and Gordon, J.A. (1975). Glutaraldehyde fixation and the mechanism of erythrocyte agglutination by concanavalin A and soybean agglutinin, *Exp. Cell Res.*, *91*: 310.
98. Van Blitterswijk, W.J., Walborg, E.F., Feltkamp, C.A., Hilkmann, H.A.M., and Emmelot, P. (1976). Effect of glutaraldehyde fixation on lectin-mediated agglutination of mouse leukaemia cells, *J. Cell Sci.*, *21*: 579.
99. Peters, K., and Richards, F.M. (1977). Chemical cross-linking reagents and problems in studies of membrane structures, *Annu. Rev. Biochem.*, *46*: 523.
100. Hammer, D.A., and Lauffenburger, D.A. (1987). A dynamical model for receptor-mediated cell adhesion to surfaces, *Biophys. J.*, *52*: 475.
101. Tozeren, A., Sung, K-L.P., and Chien, S. (1989). Theoretical and experimental studies on cross-bridge migration during cell disaggregation, *Biophys. J.*, *55*: 479.
102. Tees, D.F.J., Coenen, O., and Goldsmith, H.L. (1993). Interaction forces between red cells agglutinated by antibody. IV. Time and force dependence of break-up, *Biophys. J.*, *65*: 1318.
103. Schmid-Schoenbein, G.W., Fung, Y., and Zweifach, B.W. (1975), Vascular endothelium-leukocyte interaction, *Circ. Res.*, *36*: 173.
104. Evans, E.A. (1985). Detailed mechanics of membrane-membrane adhesion and separation. I. Continuum of molecular cross-bridges, *Biophys. J.*, *48*: 175.
105. Evans, E.A. (1985). Detailed mechanics of membrane-membrane associated adhesion and separation. II. Discrete kinetically trapped molecular cross-bridges, *Biophys. J.*, *48*: 185.

# 2

# Mathematical Models of Specific Cell Adhesion Phenomena

**Martin A. Hjortso**
*Louisiana State University, Baton Rouge, Louisiana*

**Joseph W. Roos**
*Ethyl Corporation, Richmond, Virginia*

## 1 INTRODUCTION

The expressions for the kinetics of ligand-receptor bond formation, described in Chapter 1, can be used to develop models of specific cell adhesion phenomena. Such models must necessarily embrace much more than just the kinetics of bond formation. To be complete, they must also model the transport of the cells from the bulk fluid to the vicinity of the adhesion surface, the deformation of the cells to form the contact area, the diffusion of receptors into the contact area and the effect of fluid drag forces on the cell and the ligand-receptor bonds. The models presented in the literature so far are still quite primitive and describe highly idealized situations in which many of the complicating factors have been ignored or simplified away.

Before a cell can adhere to a solid support it must be transported from the bulk solution to the vicinity of the adhesion surface, close enough for the receptors on the cell surface to make contact with the immobilized ligands on the adhesion surface. In general, this transport process is a very complex combination of fluid mechanical and diffusional mechanisms. In particular in systems with complex geometrics, such as cell affinity chromatographic columns, an exact analysis of the transport process is difficult, if not impossible. Some simplification of the fluid mechanical problem is possible because the small size of cells makes it valid to assume Stoke's flow. Yet, the transport problem remains difficult and is usually ignored or drastically simplified in analysis of cell adhesion. A rare case for which an analysis is possible is for spherical cells that reach the adhesion surface via sedimentation. In such cases, the time required for contact is the time needed for the sedimentation followed by drainage of the liquid film between the cell and the solid surface. This would show up as a delay or lag before cell adhesion begins. Such adhesion delays have in fact been observed experimentally (1,2).

The cell adhesion models that have been published in the literature can roughly be divided into two classes, which, by analogy with microbial growth models, may be termed segregated and nonsegregated models. In nonsegregated models, no distinction is made between different cells in the population and a cell is assumed to have a well-defined, deterministic number of ligand-receptor bonds holding it to the adhesion surface. Segregated models, on the other hand, acknowledge that in a population of otherwise identical cells, different cells may be in different adhesion states, i.e., may have a different number of ligand-receptor bonds. Nonsegregated models are effectively models of a single cell that either does or does not adhere. Segregated models are models of cell populations that predict the fraction of cells in a given adhesion state or the probability of finding a cell in a given adhesion state. However, both modeling approaches make use of the rate expressions described in Chapter 1 and build on the same physical picture of the cell adhesion process.

The events that can take place during specific cell adhesion are illustrated in Figure 1. Fluid forces may cause the cell to roll or slide across the ligand-coated surface until enough bonds have formed to hold the cell stationary (a). The domain in which cell receptors are close enough to the adhesion surface for ligand-receptor bonds to form (b) is known as the contact area. This area may change in size and shape continuously during the adhesion process. Receptors can enter and leave the contact area either by diffusion within the cell membrane or by convection due to the rolling of the cell (c). Finally, ligand-receptor bonds become stressed if they leave the contact area due to the rolling of the cell or cell removal forces (d). Such stress on the bonds increases the rate at which they break.

## 2 FORCES ON MEMBRANES

The bonds between receptors on a cell surface and complementary ligands on a solid or another cell will exert a force that deforms the cell away from its relaxed shape. The deformed cell shape determines the size and shape of

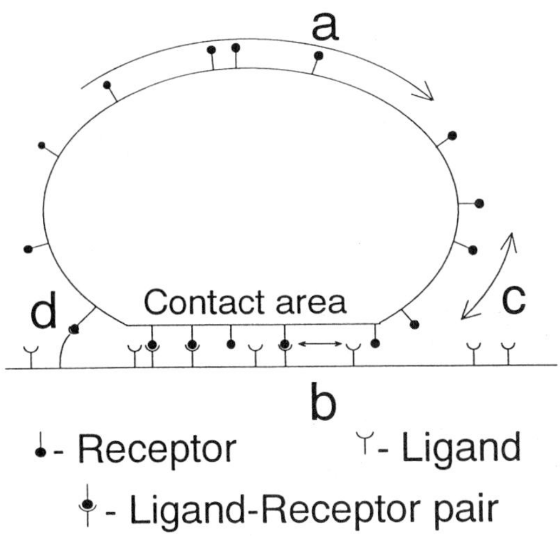


**Figure 1** Important events that take place during specific cell adhesion. Fluid or body forces may cause the cell to slide or roll across the adhesion surface (a). Only if enough bonds are formed to balance the fluid forces will the cell come to a rest. The domain over which the cell surface is close enough to the adhesion surface for the receptors and ligands to react (b) is called the contact area. The shape and size of this area may change continuously during the adhesion process. Cell receptors outside the contact area can enter the area by diffusion in the membrane (c). Similarly, unbound receptors in the contact area can leave by diffusion. External forces on the cell may work to peel the membrane away from the adhesion surface, thus stressing ligand-receptor bonds and increasing their rate of breakage. For simplicity, the ligand-receptor bonds are drawn as vertical, although this is not necessarily the case.

the contact area, the domain in which receptors and ligands are close enough to interact and form ligand-receptor bonds. Models of this process are still far from predicting the shape of the deformed cell from the properties of the membrane and the ligand-receptor pair. Instead, the simpler problem of peeling of a hypothetical strip of the membrane from a flat, ligand-coated surface has been considered. The physical situation addressed in this problem is illustrated in Figure 2. A membrane strip is pulled off an adhesion surface with a tension $T_0$ applied to the free end of the strip in a direction given by the angle $\Theta$, the macroscopic contact angle. Following an initial transient after application of the tension, the membrane strip will pull off or readhere with a fixed peeling velocity $V_p$. The central problem is to determine this steady-state peeling velocity for a given ligand-receptor pair.

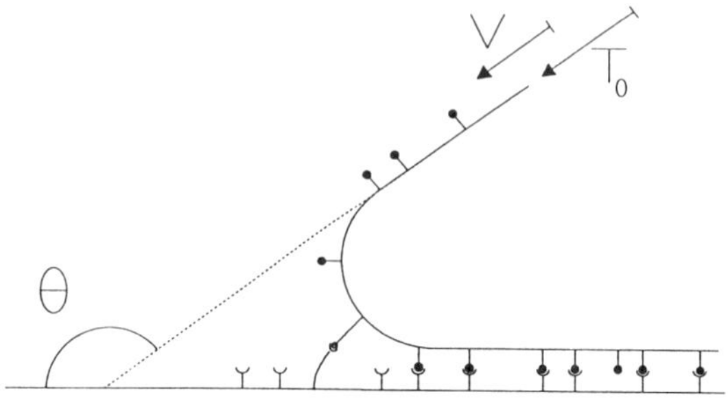


**Figure 2** Geometry of the membrane-peeling problem. The free end of the membrane is being pulled away from the adhesion surface with a tension $T_0$ in a direction given by Θ. At steady state, a constant peeling velocity $V$ is attained.

An obvious approach to this problem would be to use a continuum approximation to model the effect of the discrete ligand-receptor bonds. Such an analysis has been carried out by Evans (3). In this continuous model, the peeling process is reversible in the sense that the minimum tension required to peel off the membrane equals the tension needed to prevent readhesion of the membrane. In this case, one can define an adhesive energy density, the free energy reduction per area of contact formation, Γ. This quantity is then found to be related to the tension and the macroscopic contact angle through the Young equation,

$$\Gamma = T_0 \cdot (1 - \cos\Theta) \tag{1}$$

Dembo et al. (4) expanded on this result and found that when the ligand-receptor bonds behave as Hookean springs, the adhesive energy density per area of contact is given by

$$\Gamma = k_B T C_R \ln(1 + K) \tag{2}$$

where $k_B$ is Boltzmann's constant, $T$ the absolute temperature, $C_R$ the total receptor density (both bound and unbound receptors), and $K$ the equilibrium constant for unstressed bond formation. A similar result, except for a factor of two, has been shown to hold for cell-cell adhesion (5).

From the analysis, Evans also obtained an expression for the membrane contour in the contact area. The result shows that peeling the membrane away from the surface does not necessarily stress all the bonds. Close to the edge of the contact area, the stiffness of the membrane forces it to overshoot in its approach toward the surface. In this region the separation between the membrane and the surface is less than elsewhere in the contact area and bonds are slightly compressed. A similar result was obtained by Dembo et al. (4) from numerical solution of their model. Determination of the membrane contour is a valuable result of this type of analysis. The membrane contour determines the stresses on the ligand-receptor bonds in the contact area and thus appears indirectly in force and torque balances on adhered cells. Therefore, good models that predict the membrane contour are important in formulation of accurate models of cell adhesion.

Evans (6) also analyzed a peeling model that retained the assumption of discrete ligand-receptor bond forces. In this discrete model it was assumed that the receptors were "kinetically trapped," i.e., effectively immobile due to their diffusion rates being much lower than the rates at which they form bonds with immobilized ligands. The model was solved by numerically determining the membrane contour that minimized the sum of the membrane elastic energy and the free energy of the ligand-receptor bonds. As in the continuous model, some of the contours obtained showed a region where the ligand-receptor bonds were compressed. Significantly, however, the analysis showed that the continuum approximation is only valid if the ligand-receptor bonds are so densely spaced that the distance between them is smaller than the range of the ligand-receptor bond forces. While the continuum model predicted the existence of a critical tension value below which the membrane would spread and above which it would peel away, the discrete model predicted two critical tension values. The larger of these is the minimum tension required to separate the membranes, while the smaller is the maximum tension that will permit the membrane to readhere.

According to the discrete model, cells can thus exhibit a type of irreversible adhesion behavior. Adhered cells may show little tendency to increase the size of their contact area while still requiring a large force to be removed from the adhesion surface. The reason for this behavior is that the adhesion force is concentrated into discrete, sparsely distributed ligand-receptor bonds. If there are no ligand-receptor bonds close to the edge of the contact area, the membrane can be peeled off with a tension only slightly larger than the tension required to prevent readhesion. Eventually, however, further peeling would require breaking of a ligand-receptor bond. The tension would have to be increased to accomplish this. Thus, the larger of the critical tensions corresponds to the tension required to break one ligand-receptor bond while the smaller critical tension corresponds to the tension that barely allows formation of one more ligand-receptor bond.

An alternative explanation of irreversible adhesion phenomena is the existence of so-called "catch bonds" (4). These are ligand-receptor bonds that decrease, rather than increase, their rate of breakage when stressed.

Evidently, such bonds completely prevent the removal of the cell from the adhesion surface except by pulling the receptor out of the cell membrane, an event that may require only a modest amount of energy (7).

In the case of cell-cell adhesion, both membranes deform. The equivalent peeling problem corresponds to two membranes that are peeled away from each other. A generalization of the Young equation has been derived for this case (8), which can be used if the bonds are immobile and the continuum approximation is valid.

However, as mentioned in Chapter 1, experiments have indicated (9) that, in cell-cell adhesion, the ligand-receptor bonds may be mobile and this mobility strongly affects the peeling process. As the membranes are pulled away from each other, the bond stress exerts a force on the receptors that pulls them back toward the adhered part of the membranes and, thus, concentrates bonds in the boundary of the contact area.

Tozeren (5,10) developed a model of peeling of two identical membranes, taking into account bond mobility and diffusivity. Numerical solutions for peeling at 90 degrees showed that the speed of peeling strongly influenced the separation. At low speeds, diffusion effectively counteracts the concentration effect of the peeling, while at high speeds, the bond density in the separation zone becomes more than triple the density of the interior of the contact area. Likewise, the tension and the adhesive energy density were found to increase with the speed of peeling.

Widely applicable solutions for the peeling velocity as a function of the membrane properties and the ligand-receptor kinetics remain an elusive goal, but solutions for special cases have been found. Substantial simplifications of the peeling problem are possible if the mechanical properties of the membrane are ignored and the attached and free parts of the membrane are simply assumed to follow straight lines, intersecting at a some angle. Using this assumption, together with simple kinetic expressions for ligand-receptor formation and breakup, Tozeren (10) derived an equation relating the steady-state membrane tension and peeling velocity for 90 degree peeling. However, this analysis does not depend on the assumed reaction kinetics and could be used to derive similar equations for various ligand-receptor binding kinetics and peeling at any angle.

Dembo et al. (4) proposed a somewhat complex expression for the peeling velocity. The expression was not rigorously derived from the model equations but was suggested from studies of numerical solutions of the model. The numerical solutions also provided results for the initial transient that follows the onset of peeling. Interestingly, the authors found that the transient dynamics can be both quite complex and long-lasting and suggests that some adhesion phenomena may best be understood in terms in this transient peeling dynamics.

## 3 MODELS OF SINGLE CELLS

The kinetics of ligand-receptor bond formation, as described in Chapter 1, has been used as a basis for several models of different cell adhesion phenomena. Bell (7) proposed a model for cell-cell adhesion between identical cells. In this model, the contact area is assumed to be of constant size and is essentially treated as a two-dimensional, well-mixed reactor in which the ligand-receptor bond formation takes place following elemental kinetics. Neglecting diffusion of receptors into the contact area, the balance equation on the concentration of bonds, $C_B$, becomes

$$\frac{dC_B}{dt} = k_f(C_{R1} - C_B)(C_{R2} - C_B) - k_r C_B \tag{3}$$

where $C_{Ri}$ is the total number of receptors available for binding per unit area on cell $i$ and $k_f$ and $k_r$ are the forward and reverse rate constants for bond formation, respectively. The steady-state bond density is easily found as

$$C_B = \frac{1}{2}\left(C_{R1} + C_{R2} + \frac{k_r}{k_f}\right) - \frac{1}{2}\sqrt{\left(C_{R1} + C_{R2} + \frac{k_r}{k_f}\right)^2 - 4C_{R1}C_{R2}} \tag{4}$$

Cozens-Roberts et al. (11) extended this type of model to cell-surface attachment in a quiescent fluid. Assuming an immobilized ligand concentration much larger than the receptor concentration, such that the ligand concentration is effectively constant, they obtained a balance equation on ligand-receptor as

$$\frac{dN_B}{dt} = k_f C_L (N_R - N_B) - k_r N_B \tag{5}$$

where $N_B$ is the number of ligand-receptor bonds, $N_R$ the total number of receptors available for binding, and $C_L$ the immobilized ligand density.

This equation was modified slightly to obtain a model for detachment of cells subjected to a fluid drag force. The fluid drag pulls the adhered cell away from the surface and, thus, stresses the ligand-receptor bonds, which is assumed to cause an increase in the rate of bond breakage. This type of bond has been termed a "slip bond" (4) and the effect is modeled by applying the expression first proposed by Bell (7) to the apparent rate constant,

$$k_r = k_r^0 e^{\left(\frac{\delta F}{k_B T N_B}\right)} \tag{6}$$

where $F$ is the total removal force on the cell, $N_B$ the number of ligand-receptor bonds, and $\delta$ a characteristic length of the free energy minimum of the bond, estimated by Bell to be on the order of 0.5 $\eta$m.

The solution to the attachment model is easily obtained analytically, while numerical integration is needed for the detachment model.

This simple model of cell attachment illustrates some of the problems with the kind of models termed nonsegregated. The model is completely deterministic and predicts that all cells will have the same number of ligand-receptor bonds. The number of bonds can be found from the steady-state solution to the detachment equation and must satisfy

$$N_B = \frac{N_R}{1 + \dfrac{k_r^0}{k_f C_L} e^{\delta F / N_B k_B T}} \tag{7}$$

The existence of a physically meaningful root ($0 < N_B < N_R$) to this equation indicates that the cells do not detach, but that an adhesion steady state exists. However, the physical picture of the process leads us to a different conclusion. Due to random fluctuations, some cells in the adhered population will have a different number of ligand-receptor bonds than that predicted by the steady-state solution. Specifically, there is some nonzero probability that there will be cells without ligand-receptor bonds. In a true detachment process, these cells will be able to escape to the bulk fluid and will not be replaced. One would therefore expect the eventual detachment and removal of all cells. If the number of bonds predicted by the steady-state solution above is large, then the fluctuations around this value are relatively small and the probability of a cell occupying a state with no bonds will be low. Consequently, the rate of detachment would be very low, possibly so low as to be experimentally undetectable. On the other hand, if the steady-state solution predicts only a few bonds, the fluctuations around this value are expected to significantly affect the dynamics of the system. Thus, a better interpretation of the number of ligand-receptor bonds, $N_B$ in the model above, may be as the average number of bonds per cell in the adhered population.

Single-cell adhesion models can easily be applied to populations of cells with a distribution of receptor numbers. The receptor numbers that give rise to adhesion or detachment of a cell are found by solving the attachment or detachment equations for all values of receptor number found in the population. This determines which cells in the population attach or detach, and the fraction of attached or detached cells then equals the integral of the distribution of receptor numbers over the attached or detached states. Results of such model calculations have been presented by Saterbak et al. (12), who base their calculations on the single-cell model proposed by Cozens-Roberts et al. (11). It is obvious that this modeling approach, based on the assumption that the population contains a distribution of receptor numbers, can predict that a fraction of the cells detach rather than all or none. However, the model retains some of the all-or-nothing character of single-cell models in that it will predict total cell detachment if the conditions are such that the cells with the largest number of receptors detach or no detachment if cells with the lowest number of receptors do not detach.

## 4 CELL ATTACHMENT IN UNIFORM SHEAR FLOW

Cell attachment in shear flow occurs in stirred fluids and in situations where the fluid moves past the solid support. This situation arises in such cases as blood flow or cell affinity chromatography. In most cases, it is probably reasonable to assume that the fluid gradient does not change significantly over the length scale of the cells. This simplifies the analysis, because the attaching cells effectively experience a uniform velocity gradient. Similarly, the reasonable assumption of Stoke's flow can be used to simplify the fluid mechanical problem. A Stoke's flow solution is known for the dynamics of a neutrally buoyant sphere in a shear flow (13), but it has been questioned if this solution is a good representation of genuine cells. Tissot and co-workers (14) compared the solution to rotational and translational measurements of lymphoid cells in a flow chamber and found that the solution did not compare well with experimental values. However, by including a term for solid friction between cell wall and adhesion surface in the mathematical model, good agreement was obtained. The analysis also indicated that the gap between a steady rolling cell and the adhesion surface was approximately one-quarter of the cell radius. This high value was attributed to the fact that the cells were not perfect spheres but had various projections on the cell surface that forced the cell away from the wall of the flow chamber. The dynamics of cells in shear flow has also been studied in experiments by Tempelman and co-workers (15), who found that leukocytes traveled faster than spherical polystyrene beads, an observation that can be interpreted as a large separation distance between the cells and the solid surface. However, the authors dismiss this possibility because the implied cell–solid surface distance would be too large to permit cell adhesion and instead suggest that two distances need to be considered to understand the motion and adhesion dynamics. One is the minimum distance between the adhesion surface and protrusions from the cell surface, such as the tips of microvilli; the other is the much larger hydrodynamic separation distance.

Using a dynamically similar physical model to measure the fluid forces on attached cells, Schmid-Schoenbein et al. (16) determined the shear force on a white blood cell in a vein to be in the range $4*10^{-5}$ dynes to $234*10^{-5}$ dynes. As expected, the exact value depended on the size of cell and vein and on the magnitude of the fluid velocity. However, also affecting the value was the volume fraction of cellular content in the blood. This last result indicates that cell adhesion can be affected by the concentration of other cells in the surrounding fluid.

A model study, such as the one used by Schmid-Schoenbein et al., is of somewhat limited use in practical applications because it does not provide a mathematical model that quantitatively determines the effect of bond strength, receptor diffusivity, and so forth on the probability of cell adhesion. To derive such a mathematical model, Hammer and Lauffenburger (17,18) used Goldman's solution (13) for a solid sphere in a shear flow to estimate the fluid force on a rolling or attached cell. With this estimate, they were able to formulate a force balance on the cell and determine whether adhesion would occur. The complete model and model analysis are too involved to permit a detailed review, but the essential model assumptions and simplifications as well as the more useful results are outlined below.

The model, termed the "point attachment model," consists of two balance equations on free and ligand-bound receptor density in the contact area. In our notation the balance equations are

$$\frac{dC_B}{dt} = k_f \cdot C_L \cdot C_{RF} - k_r \cdot C_B \tag{8a}$$

$$\frac{dC_{RF}}{dt} = -k_f \cdot C_L \cdot C_{RF} + k_r \cdot C_B + \Delta \cdot (C_{c_0} - C_{RF}) \tag{8b}$$

where $C_{c_0}$ is the receptor density on the cell outside of the contact area and $\Delta$ can be thought of as a type of mass transfer coefficient that accounts for the flux of unbound receptors into the contact area. Implicit in the model equations are the assumptions that the density of immobilized ligands is constant, i.e., that the ligands are present in much higher numbers in the contact area than the receptors and that the receptor density outside of the contact area is constant, which would be valid if the contact area is a small fraction of the total cell surface area.

The main source of complexity in the model arises from the effect of fluid shear forces on the bond stress, and thus on the value of $k_r$. Initially, as the cell rolls over the ligand-coated surface, the bonds are all unstressed. The rolling motion of the cell causes receptors on the front of the cell to move into the contact area where they can interact with immobilized ligands. Once a ligand-receptor bond is formed, the forward rolling of the cell causes the bond to move toward the rear of the contact area. During this movement, the bond is assumed to be unstressed, but is instantaneously stressed once it reaches the rear of the contact area. The magnitude of the stress is dependent on the fluid force on the cell and the number of bonds that have formed. The characteristic time for bond formation followed by transport of the bond to the rear edge of the contact area is the contact time, $T_c$, for the cell. Thus the value of $k_r$ changes discontinuously at a time equal to the contact time. The cell will then adhere only if a sufficiently large number of bonds form during this contact time to stop the cell from rolling. In the model, the stress on the bond is calculated from a force balance on the cell with the added assumption that all bonds are stressed equally. The effect of bond stress on the reverse rate constant $k_r$ is calculated as described previously [Eq. (6)].

For the two limiting cases of high and low ligand-receptor affinity, analytical results can be obtained. In the case of very high ligand-receptor affinity, one single ligand-receptor bond is sufficient to immobilize the cell. Thus, adhesion depends solely on whether or not a ligand-receptor bond forms during the contact time and is therefore dependent on the rate of bond formation. In this regime, the critical number of cell receptors required for adhesion of a single cell, $N_C$, is given by

$$N_C = \frac{4 \cdot \left(\dfrac{r}{r_a}\right)^2}{1 - e^{-k_f \cdot C_L \cdot T_c}} \tag{9}$$

where $r_a$ is the radius of the contact area and other symbols are as previously defined.

In the low-affinity regime, attachment occurs if enough ligand-receptor bonds form to overcome the removal forces on the cell, i.e., if a stable solution exists for Eqs. (8a) and (8b). It is found that

$$N_C = \frac{k_r^0}{k_f C_L} \cdot \frac{4 \cdot \left(\dfrac{r}{r_a}\right)^2 \cdot e \cdot \delta \cdot F}{k_B \cdot T} \tag{10}$$

where all symbols have their previous definition. Not surprisingly, the critical number of receptors required for adhesion is inversely proportional to the ligand-receptor equilibrium constant, $k_r^0/k_f$.

A complete analysis of the model was performed using phase plane analysis (19). The interested reader is referred to the original papers, where the authors present the results of the analysis in terms of a plot of the number of receptors required for adhesion versus dimensionless contact time and ligand-receptor dissociation constant as well as several plots illustrating the effects of changes in contact area, reaction rate constants, and receptor mobility.

A radically different stochastic modeling approach to cell adhesion in shear flow was used in the model presented by Hammer (20) and by Hammer and Apte (21). This model considers a rigid cell, approaching a uniformly reactive surface due to gravity, van der Waals, and electrostatic forces. Adhesion is assumed to be mediated by immobile receptors, randomly distributed over tips of microvilli. The use of the contact area concept is completely avoided in the model because any receptor is assumed capable of forming a ligand-receptor bond. The reaction rate however, is strongly dependent on the distance from the receptor to the adhesion surface. Similarly, the rate of breakup of ligand-receptor bonds is dependent on the distance between the adhesion surface and the receptor site on the cell surface. Simulations of the model required substantial computing time and were carried out by considering small time steps over which ligand-receptor bonds form or break at random with probabilities that depended on the kinetics of bond formation and breakup. The bonds, considered to act as springs, then determine the force on the cell from which its new translational and rotational velocity can be found. The model permits simulations with either slip bonds, which weaken when stretched, or catch bonds, which strengthen when stretched, though results are not presented for catch bonds.

The model contains a fairly large number of parameters. Of these the so-called fractional spring slippage was found to be particularly important in determining the dynamics of the cell. This dimensionless parameter is positive for slip bonds and negative for catch bonds and its magnitude is a measure of how strongly the reaction rate for ligand-receptor bond breakup is affected by stress on the bond. Surprisingly, the affinity of the ligand-receptor bond did not strongly influence the dynamics of adhesion. From the simulations, five different dynamic regimes were identified and statistical tools for identifying these regimes were presented. In the unbound and rolling regimes, the cell never attaches and becomes immobile. In tumbling, the cell rolls over the surface, interrupted by brief periods of adhesion, while in transient adhesion, the cell is largely immobile except for short periods of rolling or tumbling. Finally, in adhesion, the cell is permanently immobilized.

A numerical study of adhesion of cells of various different morphologies was done by Olivier and Truskey (22). In their model, they assumed the rate of cell deformation to be small compared to the rate of detachment and the cells could therefore be treated as rigid bodies. Two-dimensional, steady-state flows of the complete Navier-Stoke's equation around four different cell morphologies, obtained form microscopy measurements of bovine aortic endothelial cells, were found by finite element analysis. The morphologies considered ranged from a spherical cell touching the adhesion surface to an almost flat cell, completely spread over the surface. It was found that the drag force on the cell decreased by a factor of 2 between the spherical cell and the fully spread cell. A much more dramatic effect was seen for the torque, which decreased by a factor of 20. Using the numerically found values of the fluid drag and torque, a force balance was formulated and from this balance the force on individual ligand-receptor bonds was found. The number of receptors available for the different cell morphologies was determined from a model of receptor transport into the contact area. To set up the balance, the ligand-receptor bonds were assumed to be at a quasi-steady state and bond stress distributions were assumed known. Two different stress distributions were used in the calculations, a uniform distribution and a distribution in which the bond stress was assumed to decrease exponentially with distance from the edge of the contact area. The first distribution reflects a cell that only detaches when all bonds break simultaneously, while the second distribution is more representative of the already described gradual peeling process in which a cell detaches when ligand-receptor bonds at the peeling edge are stressed beyond their breaking point. The uniform stress model was found to predict stronger adhesion than the peeling model by factors ranging from 2.4 to more than 15, depending on the cell morphology.

A model of ligand-receptor-mediated cell aggregation in a shear flow has been developed by Bell (23). The physical situation considered in this model is superficially very similar to the case considered in the "point attachment model." Both models analyze ligand-receptor-mediated adhesion of cells in a fluid velocity gradient, and one might expect that the models would predict qualitatively similar behavior. Unfortunately, the assumptions in the two models are so different that a comparison of them would not be meaningful. As opposed to the "point attachment model," Bell did not consider the details of the fluid flow and forces around the cells, but assumed that cells move with a velocity equal to that of the surrounding fluid. Colliding cells then stick if enough bonds form during the collision to hold the cells together against a Stoke's drag. The number of bonds formed depends on the size of the contact area and the duration of the collision, and both of these in turn are functions of geometry of the cell-cell impact. To quantify the geometry of the impact, Bell introduced the impact parameter, defined as the distance of closest approach between the centers of the colliding cells if the cells did not interact. Thus, the impact parameter equals zero in a head-on collision and equals the sum of the two cell radii when the cells pass as close as possible without touching. Both the contact time and the contact area are assumed to be functions of this parameter as follows:

$$\text{Contact time} = 2\sqrt{\mathrm{D}^2 - b^2}/V_r \tag{11}$$

$$\text{Contact area} = \eta(\mathrm{D}^2 - b^2) \tag{12}$$

where $b$ is the impact parameter, D the cell diameters, $V_r$ the relative velocities of the cells, and $\eta$ a constant that depends on how easily the cells are deformed. Although these two expressions yield predictions that are intuitively appealing, they are not derived from basic principles and there is unfortunately no obvious, simple way that the parameter $\eta$ can be calculated for a given type of cells.

The remainder of the model development does not rely on any new concepts, but makes use of rate expressions already derived. The specific rate of bond formation is modeled by an equation similar to Eq. (7a) of Chapter 1. The total number of bonds formed in a collision is found as the product of the contact area, the contact time, the specific rate of bond formation, and the receptor-ligand surface concentrations. Finally, if the total number of bonds formed in a collision exceeds the critical number of bonds required to hold the cells together against a Stoke's drag, the two colliding cells will adhere. Thus, the model is deterministic in the sense that for given collision conditions, it predicts that cells either attach or do not attach.

The probability of cells sticking after an arbitrary collision is calculated as the ratio of collisions that result in sticking over the total number of collisions. The result can be expressed as

$$P = 1 - 3 \int_{\beta_m}^{1} \sqrt{1 - A\beta^{-2}(1 - \beta^2)^{3/2}}\, \beta^2 d\beta \tag{13}$$

The parameter $A$ is given by

$$A = \frac{4\eta\epsilon C_{R1} C_{R2} (D_1 + D_2) f_c}{3\mu G^2} \tag{14}$$

where $\epsilon$ is a constant indicating how much the ligand-receptor reaction falls short of its diffusion limit, $C_{Ri}$ and $D_i$ are the receptor surface concentration and receptor diffusion coefficient on cell $i$. The critical force required to break a ligand-receptor bond is represented by $f_c$, $\mu$ is the fluid viscosity, and $G$ is the shear rate. Notice, that since neither $\eta$ nor $\epsilon$ is given in terms of more fundamental parameters, their product may conveniently be combined into a single adjustable constant.

The parameter $\beta_m$ is the root, in the interval from 0 to 1, of,

$$\beta_m^2 = A(1 - \beta_m^2)^{3/2} \tag{15}$$

This equation is simply a qubic in $\beta_m^2$ and an analytical, albeit cumbersome, closed-form solution is possible for $\beta_m$.

Alternatively, one might want to know the rate at which cells stick together. This quantity is the product of the sticking probability and the rate of collisions and is found to be

$$R_{\text{stick}} = \frac{32}{3} PGC_{\text{cell}} r^3 \tag{16}$$

where $r$ is the cell radius and $C_{\text{cell}}$ the cell concentration.

The dimensionless group $A$ features prominently in the solution and contains all the parameters associated with the kinetics of ligand-receptor bond formation. Thus, the dependence of the rate of sticking on the parameters associated with ligand-receptor kinetics is solely through $A$. Small values of $A$ occur for high values of the shear rate $G$, i.e., high rates of collision or for small values of the parameters $\eta$, $\epsilon$, $C_{Ri}$, and $D_i$. In all cases, small values of these parameters result in low rates of ligand-receptor bond formation. Thus, small values of $A$ correspond to conditions with frequent cell-cell collisions but a low probability of adhesion between colliding cells. Large values of $A$, on the other hand, correspond to conditions in which the cells collide infrequently, but have a high probability of sticking if they do collide. A maximum in the adhesion rate is found to occur for a value of $A$ close to unity.

## 5 SEGREGATED MODELS OR ADHESION OF CELL POPULATIONS

The immobile, adhered cell does not reach a true steady state as ligand-receptor bonds continuously form and break, thus changing the number of ligand-receptor bonds that hold the cell in place. However, in a population of adhered cells, a distribution of adhesion states can be defined by specifying the adhesion states of a cell as the number of ligand-receptor bonds on the cell. Given an unchanging environment and sufficient time, this distribution will reach a steady state.

The distribution of adhesion states is a useful concept when trying to make sense of cell adhesion data. The reaction processes that are involved in specific cell adhesion are all reversible, yet experimental observations have indicated otherwise. For instance, it has been reported that under conditions that severely limit cell adhesion, already adhering cells are not necessarily released (24) and once cells have adhered, very few are released unless steps are taken to promote release (1,25,26). This could be interpreted as irreversible adhesion but can also be explained in terms of the distribution of adhesion states. Cells in a state with zero ligand-receptor bonds may be removed to the bulk fluid as soon as they appear. However, if the distribution of adhesion states is shifted away from the state with zero ligand-receptor bonds, the probability that a cell resides in this state is extremely low. Thus, the rate at which adhered cells enter the zero-bond state will be low and will be the rate-limiting step in cell removal. If all cells in a zero-bond state are removed from the solid, all reversibly adhered cells will eventually elute, but the elution rate may be so low that it will be impossible to detect experimentally.

The cell state with few ligand-receptor bonds is analogous to a high-energy transition state through which adhered cells must pass before being transported to the bulk fluid. Consequently, cell adhesion and elution are often carried out under different conditions. Cell adhesion is stimulated by minimization of the fluid drag forces on the cells and by the absence of soluble ligands or receptors that can block formation of immobilized ligand-cell receptor bonds. Elution, on the other hand, is enhanced by high fluid shear rates and high concentrations of soluble, competing ligands or receptors. Because of the different conditions under which cell adhesion and cell elution experiments are carried out, the two processes may best be modeled separately.

Specific adhesion of cell populations is further complicated by the fact that cell populations rarely consist of identical cells. Unless the cells have undergone terminal differentiation or are synchronized, cells will be in different parts of the cell cycle and will have slightly different adhesion properties. At cell division, the total number of receptors is conserved, while the total cell surface area is increased. Thus, younger cells have a lower surface concentration of receptors than older cells. The smaller size of younger cells will also result in a lower possible contact area for adhesion and a different value of the fluid drag forces.

Further complications in determining adhesion states may arise from particular environmental conditions. For example, in geometrically complex systems, such as a cell affinity chromatograph column, fluid forces on an attached cell vary considerably with its position on the support. In this situation, there will be a distribution of the magnitude of fluid force experienced by adhering cells. Another case is the adhesion of cells in the presence of a soluble competing ligand. The distribution of adhesion states will be two-dimensional, requiring specification of both the number of immobilized ligand-receptor bonds and the number of soluble ligand-receptor bonds. The distribution of cells throughout adhesion states would have to account for competition for cell surface receptors between the immobilized and soluble ligand.

Thus, the distribution of adhesion states in a cell population is a multidimensional frequency function, the shape of which depends on the fluid dynamics of the system, the ligand-receptor pair, the concentrations of immobilized and soluble ligand, and the distribution of receptor numbers within the cell population.

A simple segregated model of identical cells, adhering under identical conditions to a surface with a ligand density much greater than the cell receptor density, has been proposed by Cozens-Roberts et al. (11). The model consists of a set of coupled ordinary differential equations, one equation for the probability of each adhesion state, where the *i*th adhesion state is defined as a cell with *i* ligand-receptor bonds. The equation for the fraction of cells in the *i*th state or, equivalently, the probability that a cell is in the *i*th adhesion state is

$$\begin{aligned}\frac{dP_i}{dt} = {} & k_f C_L (N_R - (N_B - 1))P_{i-1} \\ & - k_f C_L (N_R - N_B)P_i \\ & - k_r N_B P_i \\ & k_r (N_B + 1)P_{i+1} \end{aligned} \tag{17}$$

The terms on the right-hand side represent the rate at which cells in state $i$ appear or disappear through formation of one more bond from cells in state $i - 1$, formation or breakage of one more bonds, and breakage of one bond from cells in state $i + 1$. Notice that the set of equations is homogeneous in $P_i$, a common property of many kinds of segregated models. Because of this homogeneity, multiplying a solution by any constant will give another solution. The problem statement is therefore complete only if a scaling condition of some kind is specified with the governing equations. For the model above, the condition is clearly that the sum of the probabilities over all adhesion states must equal 1. However, one could equally well require that the sum equals the total number of cells in the adhered population or the total surface concentration of adhered cells. The $P_i$'s would then represent the number of cells or the concentration of cells in adhesion state $i$.

No radically new modeling concepts are used in deriving this segregated model. Like the nonsegregated models described above, the model basically regards the contact area as a well-mixed reactor in which ligand-receptor bonds form and break according to elemental kinetics. The main challenge of models such as this arises from their complexity. Methods of analysis and solution must be developed that are efficient enough to render the models practical yet accurate. When searching for analysis methods, it is worth noting the similarity between this cell adhesion problem and problems in polymer kinetics. In both cases, one deals with systems in which a large number of comparable bonds can form and where the state of the system is given by the distribution of bond numbers among either cells or polymer molecules. Thus, one can hope that the methods that are used in analysis of polymer kinetics may prove useful in analysis of cell adhesion problems. In fact, the methods used in solution of the model above, solution by generating functions and by continuum approximation, are already well known from polymer kinetics (27).

The authors solved the model for cell attachment in the absence of fluid drag using generating functions. This solution technique is virtually identical to the use of z-transforms which may be more familiar to some readers. Adapting the solution of McQuarrie (28), they obtained the following expression for the transient mean bond number,

$$<N_B> = N_R k_f C_L \frac{1 - \exp(-(k_f C_L + k_r)t)}{k_f C_L + k_r} \tag{18}$$

The expression on the right-hand side is also the solution to the nonsegregated model given in Eq. (5), indicating again that nonsegregated models can at best predict average properties.

The case of detachment of cells subjected to a fluid drag force was solved by converting the model equations to a single partial differential equation using a continuous variable approximation. The partial differential equation was then solved numerically and the authors present a series of plots of the distribution of adhesion states and the fraction of adhered cells following the onset of fluid drag. Unfortunately, an incorrect initial condition, a distribution of adhesion states with a variance larger than the variance of the steady-state distribution, was used in the solutions and the plots therefore represent a somewhat artificial detachment scenario (29). Plots obtained with the correct initial condition have since been published (12). In this paper, the model analysis was also extended to included populations of cells with different total receptor numbers. As for single-cell models, the analysis was done by solving separately for all possible total receptor numbers and integrating the product of the fraction of attached or detached cells and the distribution of receptor numbers over all possible receptor numbers. The results show that the qualitative shape of the detachment curves after the onset of fluid drag depends strongly on whether the total receptor number is the same for all cells or not. For the case where all cells have the same total receptor number, the fraction of adhered cells pass through an initial lag period before decreasing to zero in a manner qualitatively similar to an exponential decay. For a population of cells with different receptor numbers, the curves show little lag and the fraction of adhered cells does not decrease to zero but reaches some plateau value greater than zero. This dynamics is quite reasonable in light of the physical picture we have of cell adhesion. A population of adhered cells will respond to the onset of a fluid drag force by shifting the distribution of states toward lower values of the number of ligand-receptor bonds, which in turn will increase the probability that a given cell will be in a state with no ligand-receptor bonds and thus detach. When all cells have the same number of receptors, the distribution of adhesion states is so narrow that significant cell detachment is not observed until the whole distribution has shifted sufficiently far toward low values of ligand-receptor bonds. The lag observed in this case can therefore be interpreted as the time required for the distribution of states to attain the shape valid under conditions of fluid drag. The lag is not expected if the variation in total receptor number among the cells is sufficiently large because there will then always be subpopulations of cells with so few receptors that they will detach almost as soon as the fluid drag is applied. The plateau observed when the total receptor number is not identical for all cells indicates that only the cells with the largest numbers of receptors remain attached. It does not imply that cell detachment has ceased. The remaining attached cells will eventually detach, but their rate of detachment is so slow that relative to the time scale of the solution, they appear almost as irreversibly attached. Only solutions over much longer times will reveal the detachment of these cells.

It should be pointed out that while the generating function approach that was used to solve the model for cell attachment is exact, the continuous variable approximation is just that, an approximation. Even if the partial differential equation that comes out of the analysis can be solved analytically, several approximations are used in deriving the equation. The distribution of adhesion states is inherently discrete and the continuous approximation is obtained by the following trick: The number of bonds, $i$, is treated as if it were a continuous variable. The dependent variables $P_{i+1}$ and $P_{i-1}$ are then expanded in Taylor series around $P_i$ and the expansions are substituted into the discrete model in such a way that one is left with derivatives of $P(t,i)$ with respect to $t$ and $i$. The accuracy of this approximation method for adhesion models is not known. However, results from analysis of analogous polymeric systems (30) suggest a strong dependence on both the total number of bonds and the number of terms retained from the Taylor series expansion.

Significant enhancements in rates of cell detachment can be obtained when a soluble compound that binds to the receptor or ligand is present while the fluid drag force is applied (26,31–33). The reason for this is clear: The soluble compound can replace either the immobilized ligand or the receptor in the ligand-receptor bond thus driving the cell to an adhesion state with fewer immobilized ligand-receptor bonds. This effect has been explored in an elution model using soluble ligands (34). The elution model, like several of the previous models, treats the contact area as a constant-volume, well-mixed reactor with some fixed concentration of receptors and soluble and immobilized ligands. However, because the receptors can react with either free or immobilized ligands, the state of a cell is given by two integer parameters, the number of free ligand-receptor bonds and the number of immobilized ligand-receptor bonds. The distribution of states in the elution model is thus a discrete, two-dimensional distribution. The balance equation on the total number of cells, $N_{i,j}$, in a state with $i$ immobilized ligand bonds and $j$ soluble ligand bonds is written as

$$\frac{dN_{i,j}}{dt} = r_1 \cdot N_{i-1,j} + r_2 \cdot N_{i+1,j} + r_3 \cdot N_{i,j+1} + r_4 \cdot N_{i,j-1} - (r_5 + r_6 + r_7 + r_8) \cdot N_{i,j} \tag{19}$$

where the $r$'s represent the specific rates of the bond formation or breakup reactions that cause a cell to move to a new adhesion state. These rates are modeled in detail in the paper using basic concepts and ideas already described. The steady-state model equations are coupled, linear equations that were easily solved, while the dynamic equations were integrated numerically using as initial condition the steady-state distribution of adhesion states for a system without removal forces or soluble ligands. The use of a numerical solution method voids the need for some commonly used simplifying assumptions such as the frequently used assumption of a constant concentration of unbound immobilized ligands.

Solutions of the elution model for the steady-state distribution of states show that a small increase in the receptor number can significantly decrease the number of cells in adhesion states with no or few immobilized ligand-receptor bonds. Also shown in the paper are detachment curves, which again exhibit a strong dependence on receptor number. The receptor number therefore has a significant effect on the adhesion dynamics of the population, an effect that has also been observed experimentally (1,25). Because of this fact, selective elution of cell subpopulations based on receptor number can be used as a rapid cell separation tool in such diverse situations as processing of bone marrow for transplants (32), isolation of recombinant cells (35), and control of mixed cultures (36). Figure 3 shows an application of the elution model to this process when the distributions of receptor numbers in the two populations overlap, i.e., if the two populations express the same receptor but at different levels. The figure shows the distribution of receptor numbers of the released cells (not the distribution of adhesion states) for two populations at various times after the start of the elution process. At early times, the cells released are almost entirely those belonging to the population with the lowest number of receptors in the contact area. At later times, this population is almost completely eluded while a substantial fraction of the other population remains attached. Clearly, in this sepa-

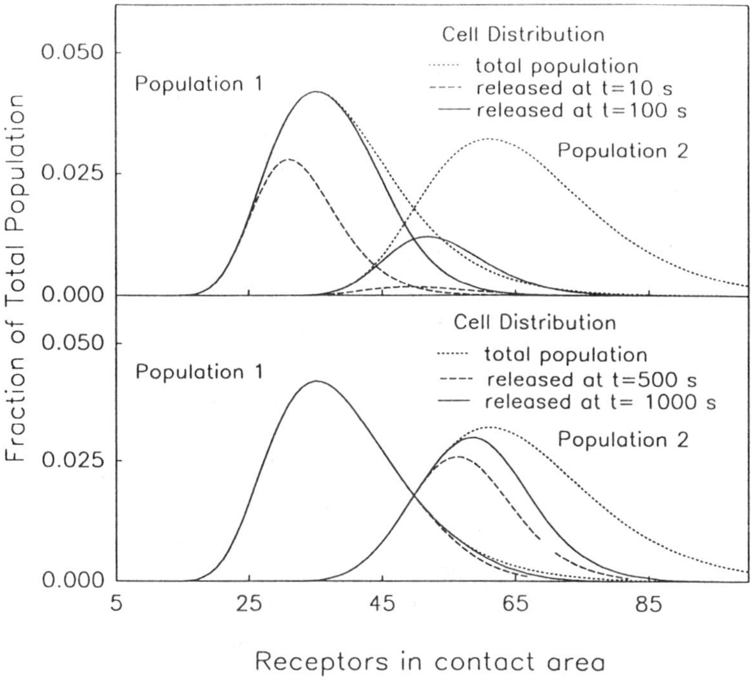


**Figure 3** Separation of two specifically adhered cell populations based on elution with a soluble ligand. The receptor number distributions of the released populations are plotted at four different times together with the initial distributions of the adhering populations. (Top) early times, (bottom) later times (37).

ration process, as in most others, there is a tradeoff between purity and yield.

## 6 NOMENCLATURE

Not surprisingly, there is little agreement on the use of symbols in this area of research. We have tried to use a consistent nomenclature in this chapter and to avoid dual use of symbols. To achieve this, we have often had to alter the nomenclature that was used in the original references. This should be kept in mind when consulting the original reference. The balance equations on the number of bonds in the contact area that appear in adhesion models can be written using either numbers of bonds or concentration (density) of bonds. Here we have used the convention that numbers are indicated by $N$, surface concentrations are indicated by $C$, and volume concentrations are indicated by $\tilde{C}$.

| | |
|---|---|
| $b$ | Impact parameter. |
| $C_B$ | Density or surface concentration of ligand-receptor bonds. |
| $C_{c_0}$ | Density of receptors outside the contact area. |
| $C_{Ri}$ | Total receptor density on cell $i$. |
| $C_L$ | Total ligand density or concentration. |
| $C_{RF}$ | Density of free or unbound receptors or porins. |
| D | Cell diameter. |
| $D_i$ | Receptor diffusion coefficient on cell $i$. |
| $f_c$ | Critical force required to break ligand-receptor bond. |
| $F$ | Tensile force on ligand-receptor bonds. |
| $G$ | Shear rate. |
| $k_B$ | Boltzmann's constant, $1.380622\ 10^{-23}$ J $K^{-1}$. |
| $k_f$ | Apparent rate constant for ligand-receptor bond formation. |
| $k_r$ | Apparent rate constant for ligand-receptor bond breakup. |
| $K$ | Equilibrium binding constants. |
| $N_B$ | Number of ligand-receptor bonds on an adhered cell. |
| $N_C$ | Critical number of receptors required for adhesion. |
| $N_R$ | Total number of receptors available for binding. |
| $N_{i,j}$ | Total number of cells, $N_{i,j}$, in a state with $i$ immobilized ligand bonds and $j$ soluble ligand bonds. |
| $P$ | Probability of cells sticking in a collision. |
| $P_i$ | Probability that a cell is in an adhesion state with $i$ ligand-receptor bonds. |
| $r$ | Cell radius. |
| $r_a$ | Radius of contact area. |
| $t$ | Time. |
| $T$ | Absolute temperature. |
| $T_c$ | Contact time in point attachment model. |
| $T_0$ | Tension in peeling problems. |
| $V_p$ | Peeling velocity. |
| $V_r$ | Relative velocity of cells. |
| $\Gamma$ | Adhesive energy density. |
| $\delta$ | Characteristic length of the free energy minimum of a ligand-receptor bond. |
| $\Delta$ | Mass transfer coefficient for flux of free receptors into contact area. |
| $\epsilon$ | Parameter indicating how much a ligand-receptor reaction rate falls short of the diffusion limit. |
| $\eta$ | Parameter for cell deformability. |

$\mu$ Viscosity of fluid surrounding a cell.
$\Theta$ Macroscopic contact angle between adhesion surface and membrane.

## REFERENCES


1. Weigel, P.H., Schnaar, R.L., Kuhlenschmidt, M.S., Schmell, E., Lee, R.T., Lee, Y.C., and Roseman, S. (1979). Adhesion of hepatocytes to immobilized sugars, *J. Biol. Chem.*, *254*: 10830.
2. Liao, N., St.John, J., Du, Z.J., and Cheung, H.T. (1987). Adhesion of lymphoid cell lines to fibronectin-coated substratum: biochemical and physiological characterization and the identification of a 140-kDa fibronectin receptor, *Exp. Cell Res.*, *171*: 306.
3. Evans, E.A. (1985). Detailed mechanics of membrane-membrane adhesion and separation. I. Continuum of molecular cross-bridges, *Biophys. J.*, *48*: 175.
4. Dembo, M., Torney, D.C., Saxman, K., and Hammer, D. (1988). The reaction-limited kinetics of membrane-to-surface adhesion and detachment, *Proc. R. Soc. Lond. B*, *234*: 55.
5. Tozeren, A. (1990). Cell-cell, cell-substrate adhesion: theoretical and experimental considerations. *J. Biomech. Eng.*, *112*: 311.
6. Evans, E.A. (1985). Detailed mechanics of membrane-membrane adhesion and separation. II. Discrete kinetically trapped molecular cross-bridges, *Biophys. J.*, *48*: 185.
7. Bell, G.I. (1978). Models for the specific adhesion of cells to cells, *Science*, *200*: 618.
8. Tozeren, A., Sung, K-L.P., and Chien, S. (1989). Theoretical and experimental studies on cross-bridge migration during cell disaggregation, *Biophys. J.*, *55*: 479.
9. Evans, E., and Leung, A. (1984). Adhesivity and rigidity of erythrocyte membrane in relation to wheat germ agglutinin binding, *J. Cell Biol.*, *98*: 1201.
10. Tozeren, A. (1989). Adhesion induced by mobile cross-bridges: steady state peeling of conjugated cell pairs, *J. Theor. Biol.*, *140*: 1.
11. Cozens-Roberts, C., Lauffenburger, D.A., and Quinn, J.A. (1990). Receptor-mediated cell attachment and detachment kinetics. I. Probabilistic model and analysis, *Biophys. J.*, *58*: 841.
12. Saterbak, A., Kuo, S.C., and Lauffenburger, D.A. (1993). Heterogeneity and probabilistic binding contributions to receptor-mediated cell detachment kinetics, *Biophys. J.*, *65*: 243.
13. Goldman, A.J., Cox, R.G., and Brenner, H. (1967). Slow viscous motion of a sphere parallel to a plane wall. II. Couette flow, *Chem. Eng. Sci.*, *22*: 653.
14. Tissot, O., Pierres, A., Foa, C., Delaage, M., and Bongrand, P. (1992). Motion of cells sedimenting on a solid surface in a laminar shear flow, *Biophys. J.*, *61*: 204.
15. Tempelman, L.A., Park, S., and Hammer, D.A. (1994). Motion of model leukocytes near a wall in simple shear flow, *Biotechnol. Prog.*, *10*: 97.

16. Schmid-Schoenbein, G.W., Fung, Y., and Zweifach, B.W. (1975). Vascular endothelium-leukocyte interaction, *Circ. Res.*, *36*: 173.
17. Hammer, D.A., and Lauffenburger, D.A. (1987). A dynamical model for receptor-mediated cell adhesion to surfaces, *Biophys. J.*, *52*: 475.
18. Hammer, D.A., and Lauffenburger, D.A. (1989). A dynamical model for receptor-mediated cell adhesion to surfaces in viscous shear flow, *Cell Biophys.*, *14*: 139.
19. Hammer, D.A. (1987). An analysis of receptor-mediated cell adhesion under conditions of flow, Ph.D. thesis. University of Pennsylvania, Philadelphia.
20. Hammer, D.A. (1992). Simulation of cell rolling and adhesion on surfaces in shear flow—microvilli-coated hard spheres with adhesive springs, *Cell Biophys.*, *2*: 145.
21. Hammer, D.A., and Apte, S.M. (1992). Simulation of cell rolling and adhesion on surfaces in shear flow: general results and analysis of selectin-mediated neutrophil adhesion, *Biophys. J.*, *63*: 35.
22. Olivier, L.A., and Truskey, G.A. (1993). A numerical analysis of forces exerted by laminar flow on spreading cells in a parallel plate flow chamber assay., *Biotechnol. Bioeng.*, *42*: 963.
23. Bell, G.I. (1981). Estimate of the sticking probability for cells in uniform shear flow with adhesion caused by specific bonds, *Cell Biophys.*, *3*: 289.
24. Rutishauser, U., and Sachs, L. (1975). Cell-to-cell binding induced by different lectins, *J. Cell Biol.*, *65*: 247.
25. Edelman, G.M., and Rutishauser, U. (1974). Specific fractionation and manipulation of cells with chemically derivatized fibers and surfaces, *Meth. Enzymol.*, *34*: 195.
26. Ferenci, T., and Lee, K.S. (1982). Directed evolution of the lambda receptor of *Escherichia coli* through affinity chromatographic selection, *Mol. Biol.*, *160*: 431.
27. Ray, H.W. (1972). On the mathematical modeling of polymerization reactors, *J. Macromol. Sci.-Rev.*, *C8*: 1.
28. McQuarrie, D.A. (1963). Kinetics of small systems. I, *J. Chem. Phys.*, *38*: 433.
29. Hwang, Y. (1991). Specific cell adhesion and its application in mixed culture fermentations, Ph.D. thesis. Louisiana State University, Baton Rouge.
30. Zeman, R.J., and Amundson, N.R. (1965). Continuous polymerization models. I. Polymerization in continuous stirred tank reactors, *Chem. Eng. Sci.*, *20*: 331.
31. Clune, A., Lee, K-S., and Ferenci, T. (1984). Affinity engineering of maltoporin: variants with enhanced affinity for particular ligands, *Biochem. Biophys. Res. Commun.*, *121*: 34.
32. Hertz, C.M., Graves, D.J., Lauffenburger, D.A., and Serota, F.T. (1985). Use of cell affinity chromatography of separation of lymphocyte subpopulations, *Biotechnol. Bioeng.*, *27*: 603.
33. Roos, J.W., and Hjortso, M.A. (1989). Determination of population balances in mixed culture by specific cell adhesion, *Biotechnol. Tech.*, *3*: 7.

34. Roos, J.W., and Hjortso, M.A. (1993). A mathematical model of specific cell adhesion: release of specifically adhering populations by soluble ligands, *Biofouling*, *6*: 381.
35. Roberts, P. (1991). Mathematical modeling and experimental analysis of cell affinity chromatography, M.S. thesis, Louisiana State University, Baton Rouge.
36. Roos, J.W., and Hjortso, M.A. (1991). Control of an *Escherichia coli* mixed culture via affinity binding, *Biotechnol. Bioeng.*, *38*: 380–388.
37. Roos, J.W. (1990). Specific cell adhesion and its application to monitoring and control of mixed culture bioreactors, Ph.D. thesis, Louisiana State University, Baton Rouge.

# 3

# Cell Adhesion in Animal Cell Culture: Physiological and Fluid-Mechanical Implications

**Manfred R. Koller**
*Aastrom Biosciences, Inc., Ann Arbor, Michigan*

**Eleftherios T. Papoutsakis**
*Northwestern University, Evanston, Illinois*

## 1 INTRODUCTION

Cell adhesion to other cells and to various substrata, such as the extracellular matrix or other artificial supports, is a critical process in the formation of tissues, cell differentiation, and morphogenesis (1,2). In this chapter we describe how cell adhesion affects cultured cells with respect to their function and productivity. The importance of cell adhesion in animal cell biotechnology has not been explicitly acknowledged, probably because in terms of sophistication, the technology is still in its infancy. A number of key steps and cellular processes are, however, critically dependent on cell adhesion. We aim to make the biological implications of cell adhesion better known to the community of biotechnologists with the expectation that more attention to this key aspect of cell life and function will result in better processes and expanded cell culture possibilities. The general and specific biological mechanism of cell adhesion will be examined in considerable detail. We will briefly review the nature of cell culture substrata and then discuss the molecular and cellular implications of cell adhesion. Subsequent discussion will focus on biotechnological applications, but no attempt will be made to cover all aspects of such applications. The emphasis will be on cell culture in bioreactors for the production of more cells and their glycoprotein products. In that case, the fluid-mechanical aspects of cell injury are intricately related to the adhesion process, so the current understanding of the fluid-mechanical mechanisms through which cells are damaged in microcarrier bioreactors will be examined in some detail. Aspects of cell adhesion that pertain to artificial organs and cell interactions with biocompatible surfaces, such as those encountered in prostheses, will not be discussed. This is a rather separate field of investigation, despite the commonality of the underlying biology. The importance of cell adhesion will, however, be discussed as it applies to the emerging field of tissue engineering. As an example of tissue engineering, the long-term cultivation of bone marrow cells will be discussed in some detail.

## 2 ADHESION MECHANISMS

The attachment and subsequent spreading of a cell on an appropriate substratum is the result of the action of many different related and unrelated forces. These forces may be divided into nonspecific physical forces and the forces due to specific adhesion molecules present on the cell surface. Since, at the atomic level, specific adhesion interactions also act through nonspecific physical forces, the distinction between specific and nonspecific interactions is made on the basis of the need for specific molecules to be present between the cell and substratum. When discussing specific adhesion, it is also necessary to introduce the components of the extracellular matrix that are critical in mediating specific adhesion. A good understanding of the nonspecific and specific interactions that allow cells to attach and spread is important for the development of improved anchorage-dependent animal cell culture media and bioreactor systems.

### 2.1 Nonspecific Physical Forces

Several types of nonspecific physical forces may act between a cell and the substratum. Among these forces are: coulombic (or electrostatic) forces, which exist between any charged bodies, van der Waals (or electrodynamic) forces, which may arise between electrically neutral bodies, and steric stabilization forces, which result from the hydrated layers of long-chain polymer molecules that exist on cell membranes (i.e., the glycocalyx) (3).

Coulombic interactions are repulsive between bodies of like charge and are attractive between bodies of opposite charge. The magnitude of the force is directly proportional to the product of the charges on the two bodies, and it decays exponentially with the distance of separation between the two objects. Electrostatic forces are weakened by increasing the concentration of salt in the intervening medium. In biological systems, electric charges arise from mobile ions (principally $Na^+$, $K^+$, $Cl^-$, $Ca^{2+}$, $Mg^{2+}$, $H^+$, and $OH^-$) and from fixed chemical moieties that are covalently bound to macromolecules (such as $PO_4^{3-}$, $COO^-$, and $NH_3^+$) and undergo acidic or basic dissociation or association in aqueous solution. Therefore, the total force will depend on the pH and ionic strength of the medium. Under most conditions, intact cells have an overall net negative charge (4) and are therefore repelled from most substrata (i.e., tissue culture plastic, glass) since they also carry a net negative charge. The magnitude of this repulsive electrostatic force must not be very large since it is known that cells do in fact adhere to negatively charged substrata. Positively charged counterions in the medium probably shield the negative charges and reduce the repulsive force (5). In contrast, cells would be attracted to positively charged substrata such as DEAE-dextran and polylysine. But since most cells are grown in medium containing serum, the positively charged surface of these substrata would be substantially altered by the adsorption of negatively charged serum proteins. Theoretical calculations have shown that the electrostatic force is quite small in comparison to the other physical forces (6,7) and is therefore not the major force involved in the attachment of cells to substrata. This conclusion has been demonstrated experimentally as well, since negatively charged BHK cells were unable to attach to positively charged polylysine under serum-free conditions in vitro (8).

A nonspecific attractive force known as the van der Waals force may exist between two bodies. The strength of interaction decreases relatively slowly as a power of the distance of separation, as compared with coulombic forces, which decay exponentially. The van der Waals force is attractive until the bodies become very close ($<3$–$5$ Å) at which point there is a very strong repulsive force due to the overlapping of outer electron clouds of the atoms in each body. The van der Waals force arises as a consequence of the spontaneous transient fluctuations of charge distribution which occur in a body. At any given instant of time, the charge distribution in a body is asymmetric, thereby causing an electrical interaction with other parts of the body as well as other nearby bodies that are also experiencing this phenomenon. These charge fluctuations arise from thermal motion and changes in the positions and momenta of electrons and atomic nuclei in the bodies. The energy of van der Waals interaction between two atoms is very small (about 1 kcal/mol) and is much less than the energy of electrostatic interaction between two charged atoms. There are, however, a vast number of atoms on the cell and substratum which may interact in this manner. This fact, along with the fact that electrostatic forces decay much more rapidly with distance, causes the van der Waals force to be more important than electrostatic forces in the initial attachment of cells to a substratum (7).

The steric stabilization force is well known in the theory of colloid suspensions (9). This force is also significant in the approach of a cell to the substratum or another cell. Because many animal cell membranes are extensively coated with a hydrated layer of long-chain polymer molecules (proteins and carbohydrates), a repulsive force is generated as the cell approaches the substratum (3). During this approach, the polymer molecules on the cell surface are compressed between the rigid surfaces of the substratum and cell membrane. The steric compression of the polymer molecules results in a force that opposes further cell approach to the surface. In addition, water of the hydration layer is squeezed out of the membrane-substratum space, and an osmotic pressure develops as the solvent attempts to return into the gap (see Fig. 1). The magnitude of these steric and os-

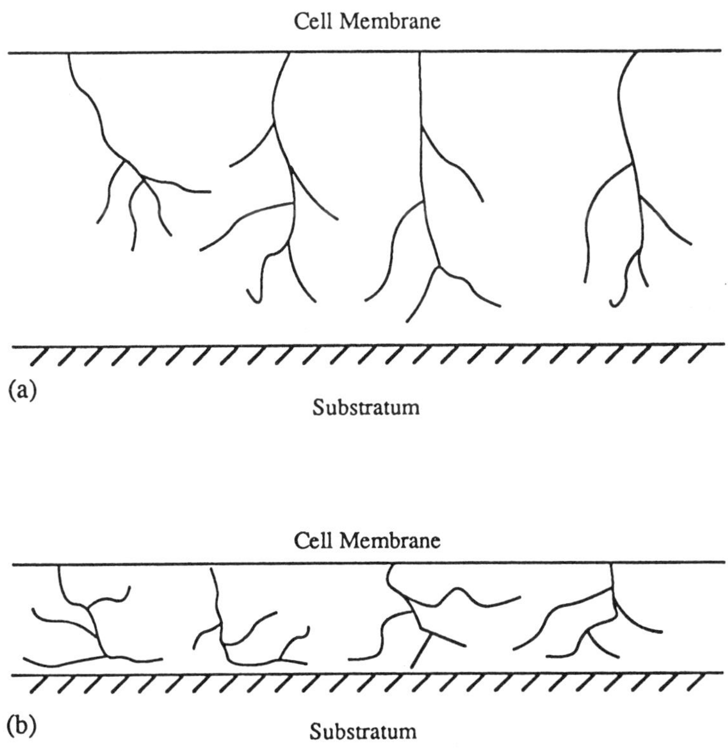


**Figure 1** Long-chain polymer molecules on the cell membrane are not hindered in (a), and there is water of hydration associated with these moieties. As the cell approaches the substratum in (b), two forces develop which repel the cell from the surface: (1) the steric compression of the polymer molecules into a smaller space, and (2) the osmotic pressure of the water of hydration, which tends to flow back into the gap which has a very high local polymer concentration. These forces combine to generate the repulsive steric stabilization force. (Adapted from Ref. 3.)

**Table 1** Nonspecific Physical Forces

| Type | Comments |
|---|---|
| Coulombic (or electrostatic) | Relatively weak forces that are shielded by intervening medium |
| van der Waals (or electrodynamic) | Causes numerous interactions that allow initial attachment |
| Steric stabilization | A steric and osmotic effect that resists close contact |

motic forces will depend on the number and stiffness of the polymers present on the cell membrane. There may also be a 2-5-nm layer of serum proteins adsorbed onto the substratum (10), but this would not greatly affect the steric stabilization force (3). These forces may be quite large, and it has been reported that they are a more significant source of cell-substratum repulsion than electrostatic repulsion between a cell and a negatively charged substratum (6).

Van der Waals forces may be significant in the initial attachment of a cell to the substratum, but, as will be discussed later, other specific forces dominate in maintaining cell adhesion and promoting cell spreading. Coulombic forces are shielded by ions and proteins that are present in any cell culture medium. These forces are therefore not significant with the distances involved in cell-substratum interactions. Finally, a significant amount of cell-substratum repulsion may be developed upon close interaction due to the steric stabilization force, which varies with the number and type of polymer molecules present on the cell membrane. The effects of these nonspecific physical forces are summarized in Table 1.

## 2.2 Specific Cell Adhesion and the Extracellular Matrix

The physical forces discussed in the previous section are nonspecific and therefore do not require the presence of any specific molecules. The remarkable specificity with which most biological reactions occur tends to suggest that these nonspecific physical forces are only part of the picture. The elegantly specific binding that occurs throughout biological systems, such as between enzyme and substrate, antibody and antigen, and hormone and receptor, must also have a counterpart in the arena of cell adhesion. Specific interactions are of major importance in cell adhesion, and nonspecific interactions, which act over the long range of up to 100 nm (7), are only important in the initial approach and binding of a cell to a substratum.

Many anchorage-dependent cells exhibit a dependence on serum in vitro

(10,11). It is known that the substratum absorbs serum components very rapidly, so that a multilayered heterogeneous mosaic of adsorbed proteins is formed (10). Although cells attach to the substratum more rapidly under serum-free conditions, the final extent of cell attachment in serum-containing media reaches the level obtained in serum-free media within 30 min. In the absence of serum, however, the cells do not flatten or spread out, as almost all the cells do within 1 h in media containing ≥1% serum and 1 mM $Ca^{2+}$ (10). This spreading of the cell onto the substratum is accompanied by a cytoskeletal rearrangement, which is absolutely necessary for the survival and proliferation of most normal mammalian cells (see Sec. 4). The components of serum that are responsible for this effect were identified by several investigators using fractionated serum and/or purified proteins from other sources. Proteins that are capable of mediating cell attachment and spreading include: concanavalin A (10), fibronectin (12), chondronectin (13), ricin (8), laminin (14), and vitronectin (15). Fibronectin, chondronectin, laminin, and vitronectin, along with collagen and proteoglycans, make up the extracellular matrix (ECM) that fills the extracellular spaces between cells both in vivo and in vitro. While these molecules are the mediators of cell attachment from the extracellular side, intracellular molecules such as α-actinin (16), vinculin (17), and talin (18) have been isolated and implicated in the formation of cell-substratum and cell-cell contacts.

Further insight into the nature and molecular composition of contact sites between cells and the substratum, and other cells, has been obtained through immunolabelled electron microscopy of fibroblasts (19). In this technique, proteins thought to be present at contact sites were specifically

**Table 2** Structural Proteins Implicated in Cell Adhesion

| Extracellular | Transmembrane | Cytoplasmic |
|---|---|---|
| Fibronectin (a 550-kDa dimer) | Integrins: α-subunits (~110–155 kDa) | Actin (42 kDa) |
| Vitronectin (65- and 80-kDa forms) | β-subunits (~140 kDa) | α-Actinin (100 kDa) |
| Laminin (400 kDa) | Cell surface proteoglycans (~400 kDa) | Fimbrin (68 kDa) |
| Collagen | | Vinculin (115 kDa) |
| | | Talin (a 550-kDa dimer) |
| | | Tensin (150 and 200-kDa forms) |

labeled by antibodies directed at them. Electron micrographs of these labeled cell preparations revealed the spatial arrangement of proteins that are involved in contact site formations. In these studies, three types of cell-substratum (or cell-cell) associations were elucidated: (1) ECM contacts, (2) close contacts, and (3) focal adhesions.

In ECM contacts, the cell surface is relatively far removed from the substratum (50–100 nm) and is sporadically connected to the substratum and to other cells by large and cable-like filamentous aggregates of ECM components. ECM contacts stain densely for fibronectin extracellularly and moderately for $\alpha$-actinin intracellularly. Some ECM contacts (~20%) stain densely for vinculin intracellularly, and it was suggested that these sites exhibit end-to-end binding of extracellular filaments to intracellular actin microfilaments (through an unidentified integral membrane protein), while the sites negative for vinculin are characterized by lateral association of the two filament structures. The number of ECM contacts, as a percentage of total cell-substratum contacts, increased dramatically from 28% in early culture (6–12 h) to 84% in late culture (24–36 h). In late culture, it was found that ECM contacts constituted 30% of the cell-cell contacts in the culture.

Close contacts are characterized by a 30–50-nm spacing between the cell surface and the substratum and are broad areas that are often found surrounding sites of focal adhesion. A meshwork of actin microfilaments that stains positively for $\alpha$-actinin is often seen subtending the membrane at these sites. All close-contact sites stain moderately for fibronectin, but are negative for vinculin, suggesting that they are similar to the more predominant type of ECM contact discussed above. In early culture, close contacts were numerous in forming 50% of the cell-substratum contacts, but this number declined sharply to 10% in late culture. Close contacts were found to comprise 40% of the cell-cell contacts in late culture, making close contacts the most abundant type of cell-cell contact between fibroblasts.

Even though they occupy only a small fraction of the interface, focal adhesions (also known as focal contacts or adhesion plaques) are the strongest sites of cell-substratum and cell-cell adhesion. Focal adhesions are formed between many cells and the appropriate substrata, and with other cells as well. Although focal adhesions begin to form in some transformed cells (see Sec. 5.4), they do not mature and acquire attached stress fibers as do normal focal adhesions, which correlates with the observed growth of these cells in suspension (20). Stress fibers contain parallel bundles of actin microfilaments and are responsible for generating the stress or tension that determines the flattened shape of many cells (21). These stress fibers are anchored to the plasma membrane at focal adhesion sites. In addition, focal adhesions probably also form the nucleation sites needed to regulate the assembly of the stress fibers (22), which occurs as the cell attaches and flattens on the substratum. In fibroblast cultures, these sites made up 22% of the total cell-substratum contacts in early culture, with the number falling to 6% in late culture. Focal adhesions were seen in 28% of the cell-cell contacts in late culture (19).

Focal adhesions are characterized by a 10–15 nm spacing between the lipid bilayer membrane and the substratum and are relatively confined to areas 2–10 $\mu$m long and 0.25–0.5 $\mu$m wide found mostly near the perimeter of cells (22). Actin microfilament bundles are known to terminate at the cytoplasmic side of the plasma membrane of these sites (23), with all filaments being aligned in the same orientation (24). Focal adhesions label intracellularly for vinculin and $\alpha$-actinin, with vinculin being situated closer to the membrane than $\alpha$-actinin (which extends up to 60 nm from the membrane) (19). In addition, a flexible, elongated protein called talin (25) and a protein named tensin (26) are also found intracellularly at these sites, and both have been shown to bind vinculin with high affinity in vitro (27). The only intracellular focal adhesion proteins that have been shown to bind to actin directly are fimbrin (28) and $\alpha$-actinin (29). In addition, $\alpha$-actinin has been shown to bind to vinculin in vitro (26).

The major extracellular protein of focal adhesions is vitronectin. Although both fibronectin and vitronectin are capable of forming focal adhesions independently (30), fibronectin is specifically cleared from these sites over time and is replaced by vitronectin (31). The majority of adhesion-promoting activity in serum is accounted for by vitronectin (30), which is present at a concentration of ~300 $\mu$g/ml in serum. Without these serum proteins, or in the presence of nonspecific proteins such as BSA, cells will not form focal adhesions, although eventually some cells will synthesize and secrete enough fibronectin onto the surrounding substratum to allow their attachment (32). Fibronectin and vitronectin have several properties that allow them to act as attachment proteins. For instance, fibronectin is a very sticky molecule which may bind to fibrin, heparan, collagen, cell receptors, DNA, IgG, plasminogen, and even to another fibronectin molecule (33). The essential features of these molecules in the formation of focal adhesions are the heparan sulfate proteoglycan binding and the cell receptor binding domains (22). Heparan sulfate is a glycosaminoglycan, which is a repeating polymer of -(*N*-acetylglucosamine-uronic acid)$_n$-disaccharide units. The sugars in the polymer chain are sulfated to varying degrees, and many heparan sulfate chains bind to a core protein to form a heparan sulfate proteoglycan. Proteoglycans are a very diverse group of large macromolecules that fill up much extracellular space and interact with many other molecules through their charged moieties. Proteoglycans also exist on the surface of cells, and it is probably these cell surface proteoglycans that play an augmenting role in the attachment of cells to fibronectin and vitronectin (34). The segment of fibronectin and vitronectin that binds to cell receptors contains the tripeptide sequence Arg-Gly-Asp (RGD), which is responsible for recognition (35). When coated onto a surface, short synthetic polypeptides containing the RGD sequence have been shown to promote cell attachment, whereas in solution they competitively inhibit cell attachment to a surface coated with either fibronectin or the polypeptides themselves (33). The RGD sequence is the cell recognition site of a number of extracellular matrix proteins, including: fibronectin, vitronectin, collagen types I, III, IV, V, and VI, and laminin (36).

Cells are able to interact with the RGD sequence through a number of integral transmembrane protein receptors that belong to the integrin superfamily of proteins (36). Each integrin is a heterodimer of one of several distinct $\beta$-subunits ($\sim 140$ kDa) noncovalently associated with one of several distinct $\alpha$-subunits ($\sim 110$–$155$ kDa) (37). Different combinations of the $\alpha$- and $\beta$-subunits allow many integrins to be formed, and each one has a different specificity for extracellular matrix components (36). For example, four distinct integrins are known to bind fibronectin, and some of them will bind to other ECM components while others do not (33). The sequencing of $\alpha$-subunits has shown that they contain two sequences that are homologous to a known $Ca^{2+}$ binding site (33). This might explain the need for $Ca^{2+}$ in medium in order to form focal adhesions. It is believed that both subunits contribute to the binding of RGD sequences, and the $\beta$-subunit has been shown to bind talin as well (38). This allows the integrins to act as a transmembrane link between the intracellular cytoskeletal network and the extracellular matrix components.

With the information presented above regarding the locations and interactions of proteins known to be present at focal adhesions, it is possible to propose a model for the molecular structure of a focal adhesion. The model is presented in Figure 2 and is described in the accompanying legend. This model is incomplete and will remain so until all the details of focal adhesions have been elucidated.

Aside from the relatively well-investigated mechanisms of cell attachment described above, many other cell surface receptors have been described in the literature. Many of these receptors are specific to certain types of cells and therefore do not lend themselves to a general discussion of cell attachment. Some examples of receptors for extracellular matrix components include: the 67-kDa laminin and elastin receptor (39), a different 67-kDa muscle-cell-specific laminin receptor (40), and chondrocyte-specific anchorin, which binds to collagen type II (41). Another type of molecule that is important in specific cell-cell interactions is the cell adhesion molecule (CAM). Many of these have been described in the literature,

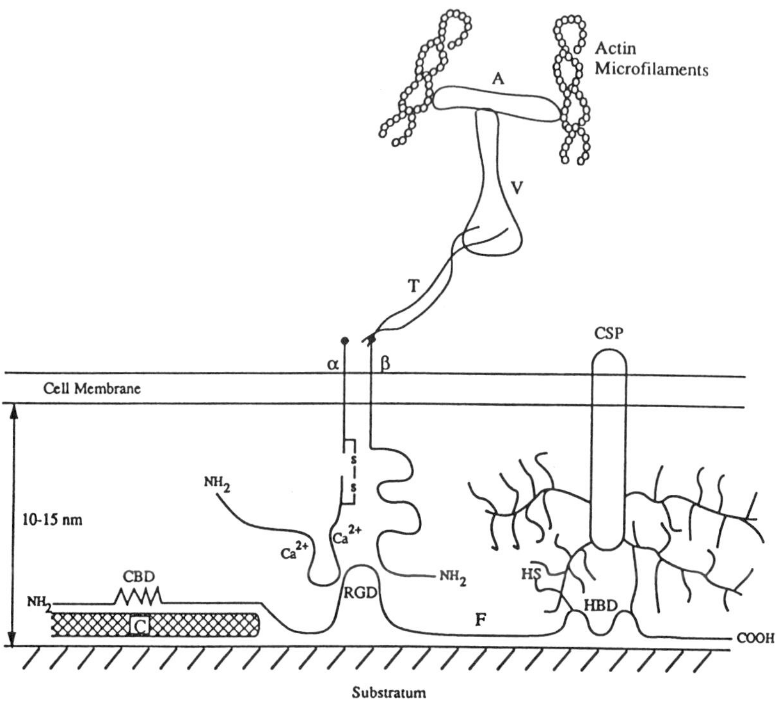


**Figure 2** Known interactions of focal adhesion components. Components with unknown functions, such as fimbrin and tensin, are not shown. Fibronectin (F) may bind to collagen (C) through its collagen-binding domain (CBD) or may simply become adsorbed to the substratum. The $\alpha$- and $\beta$-integrin subunits ($\alpha$ and $\beta$) bind to fibronectin through RGD sequence recognition. Talin (T) binds to the $\beta$-integrin subunit and to vinculin (V). Vinculin in turn binds to $\alpha$-actinin (A), which is associated with the actin microfilaments. The interaction of cell and substratum may also be stabilized by cell surface proteoglycans (CSP), which have heparan sulfate moieties (HS). Fibronectin has a heparan sulfate binding domain (HBD), which completes this interaction. (Adapted from Refs. 32 and 33.)

including the neural cell adhesion molecule (N-CAM) and the liver cell adhesion molecule (L-CAM). The CAMs are thought to play a major role in development by allowing cells to communicate with their neighbors, and they have been implicated in the induction of differentiation (for review see Ref. 42). Finally, one of the most important cell-cell interactions that has been studied is the tight junction that forms between epithelial cells. Tight junctions are restricted to these types of cells that are responsible for sealing body cavities. For example, epithelial cells of the intestinal lumen form tight junctions that completely encircle the cell and thereby form an impermeable seal between the intestinal lumen, which contains bacteria, and the internal abdominal cavity. These cell-cell interactions may be mimicked in vitro, and a variety of interesting culture systems, such as the growth of artificial skin, may result from this technology.

The mechanism of cell adhesion and spreading involves initial nonspecific physical forces and subsequent specific adhesion interactions of the cell with the substratum and extracellular matrix. The requirement for serum in the growth media of anchorage-dependent cells is in part due to the need for attachment and spreading factors such as fibronectin and vitronectin. The formation of ECM contacts, close contacts, and focal adhesions allows a cell to adhere to the substratum and to adjacent cells. The design of effective serum-free media and culture systems should account for the requirement of adhesion which is common to all anchorage-dependent mammalian cells.

## 3 CELL CULTURE SUBSTRATA

Many natural and artificial substrata have been used for the culture of anchorage-dependent cells in a variety of geometric configurations. Both chemical and physical factors have been implicated in the influence of substratum on cell adhesion. An overview of popular substratum materials will give an insight into the requirements for good cell adhesion. This will be followed by a discussion of different configurations that are employed in the culture of anchorage-dependent cells.

### 3.1 Natural Substrata

Natural substrata are derived from materials that are part of the extracellular matrix, to which many types of cells are attached in vivo. These materials are all, or in part, proteins that may influence the growth and differentiation of certain cell types. Many of these materials have been discussed in Sec. 2.2 and will be further discussed in Sec. 4. Some examples include collagen, proteoglycans, fibronectin, laminin, and elastin. These materials are usually used to derivatize artificial substrata to enhance cell attachment and growth. For example, the use of either native or denatured collagen-coated (43) and laminin-coated (14) surfaces has been well established in the culture of various cells. A complete natural substrata called the biomatrix may be produced by solubilizing away all the other components of a tissue (i.e., non-ECM proteins, lipids, and nucleic acids) by treatment with NaCl and detergents. The result is an intact extracellular matrix that may be used as a substratum for the growth of any desired cell that will adhere (44). Although these natural substrata allow the study of cell growth and differentiation in a physiologically relevant system, these components are difficult and expensive to produce for large-scale culture. Artificial substrata have therefore been used more frequently in the culture of anchorage-dependent cells.

## 3.2 Artificial Substrata

The first (one of the most widely used) substratum for cell culture is glass. Glass is nontoxic, easy to clean and sterilize, and inexpensive. In addition, the amount of negative charge present can be manipulated by alkali treatment and thereby adjusted to optimize the attachment and growth of a specific cell type (45). Aluminum borosilicate glass is preferred over soda lime glass because of its very low content of alkaline oxides. These alkaline oxides may be released into the medium during culture, resulting in poor cell growth (46). In addition to being economically favorable, the reuse of glass cultureware is desirable for improving cell attachment and spreading. For unknown reasons, new glass that is marginally satisfactory may give improved results after two or three cycles of use (46).

Although cells will adhere to many plastic surfaces, by far the most widely used plastic substratum is "tissue-culture-treated" polystyrene. Regular polystyrene dishes (bacterial dishes) are not suitable for anchorage-dependent cell growth. Tissue culture treatment of polystyrene is accomplished by increasing the charge density of the surface through chemical or physical methods. Chemical processes consist of treating the surface with oxidizing agents (such as $KMnO_4$) or strong acids (such as $H_2SO_4$), while physical processes depend on ultraviolet light exposure (47) or glow discharge between two electrodes at a higher voltage (48). These treatments result in a high density of negatively charged surface groups on the polystyrene. Although tissue-culture-treated polystyrene is not reusable or autoclavable, many cells will adhere better to this surface than to glass (49).

Other novel substrata that have been developed for the culture of anchorage-dependent cells include gas-permeable FEP-Teflon (50) and various metal films (51). Positively charged polymers such as DEAE-dextran (52), polylysine, polyarginine, polyhistidine, and polyornithine (53) have also been shown to support cell attachment and spreading.

## 3.3 Characteristics of Adhesive Substrata

Although each cell type will probably vary in its requirements for an adhesive substratum, the three most important parameters of a surface are solidity, charge density, and wettability. These factors must be considered when developing a suitable substratum for the growth of anchorage-dependent cells.

As will be discussed in Sec. 4, anchorage-dependent cells will not grow in semisolid matrices such as methylcellulose or agar. A rigid surface must be provided for the cells to spread out upon. The ability of a substratum to resist tensile forces generated by a spreading cell is important for the regulation of proliferation and differentiation. The signal for growth of a cell upon spreading may be, in part, transduced to the nucleus by tension that is generated within the intracellular cytoskeleton (54). In the case of microcarrier cultures, it was initially thought that a less rigid surface would be required to reduce cell damage due to bead-bead collisions (55). It appears, however, that this problem is not any worse when rigid microcarriers are used, and that rigid surfaces provide the best substrata for anchorage-dependent cell growth.

Pure surfaces of metals, metallic oxides, and glasses have high charge density and organic surfaces have low charge density (56). As discussed above, the surface charge density can be modified by either chemical or physical treatments. Cells adhere better to surfaces with a high charge density, and studies of sulfonated polystyrene have shown that 2–5 negatively charged groups/$nm^2$ promote maximum cell attachment and spreading (57). It is generally accepted that cells can grow on surfaces that are either positively or negatively charged, and that the proper charge density is more important than the polarity (46).

The substratum wettability appears to affect protein adsorption, which in turn can affect cell adhesion (58). In addition, surfaces become even more wettable following protein adsorption, which further affects cell adhesion (59). Hydrophilic substrata are generally more suitable for anchorage-dependent cell growth with the notable exceptions of agar and poly-(HEMA), which are both wettable and nonadhesive (see Sec. 4).

## 3.4 Cell Culture Substrata Configuration

The substrata discussed above may be arranged in a myriad of ways for the growth of anchorage-dependent cells. The most common small-scale stationary technique involves the use of tissue culture flasks with a usable surface area ranging from 25 to 150 $cm^2$ each. Because of the high cost of cell culture medium (particularly serum), it is advantageous to increase the cell number per unit volume of medium. High cell densities will also allow a higher concentration of cell product in the medium, thereby simplifying the purification process. For anchorage-dependent cells, increasing cell density requires the maximization of the surface area to volume ratio. Although roller bottles allow more usable area than stationary flasks, the cells are not always covered by medium and a large number of bottles and the associated rolling racks are required for large-scale production.

An alternative technique that allows high-density cell growth is the use of microcarrier culture. Microcarriers are small, usually spherical particles that are suspended in the growth medium by gentle agitation. The cells are initially allowed to attach to the microcarriers under stationary conditions, and once agitation begins, the cells attached to each microcarrier will grow to confluence. Advantages of this technology include: high surface area to volume ratio, relatively homogeneous conditions with all cells in one vessel as opposed to many individual bottles or flasks, representative sampling is easily performed by removing some of the microcarriers, and the technique is easily amenable to scale-up (60). We have demonstrated the ability to produce more concentrated recombinant protein product using microcarrier culture in two systems (61,62).

In addition to the requirements for substrata presented in Sec. 3.3, there are additional considerations in the design of microcarriers. To keep the beads in suspension with low agitation, their specific gravity should be between 1.02 and 1.04 (55). With respect to microcarrier size, two factors must be considered: (1) smaller beads will give more surface area per gram of microcarriers, and (2) larger beads will present a flatter surface to the cells and allow more cells to grow on each bed. A bead diameter of roughly 200 $\mu$m is optimal for the balance of these two factors (55). The beads should be made of a material that is relatively nonporous, as this will prevent the loss of adsorbed proteins inside the beads. Finally, the beads should be transparent to allow easy microscopic examination. All types of materials have been used for the formulation of microcarriers, and a comprehensive list of these substrata along with some important properties is presented in Table 3.

## 4 BIOLOGICAL IMPLICATIONS OF ADHESION

The survival and proliferation of many cells is absolutely dependent on the attachment of the cell to an appropriate surface. This phenomenon, known as anchorage dependence (63), has not been as extensively studied as some other growth regulatory mechanisms. As a result, the specific mechanisms responsible for the coupling of cell attachment to cell growth and differentiation have yet to be elucidated. The studies that have been performed reveal that many molecular processes in the normal cell are dependent on adhesion and the subsequent spreading of the cell on the substratum. A thorough understanding of the cellular processes that are disrupted upon cell detachment and the time course of cellular recovery upon reattachment is important for any work involving the culture of anchorage-dependent cells.

**Table 3** Commercially Available Microcarriers

| Name and manufacturer | Chemical composition | Size ($\mu$m) | Culture area ($cm^2/g$) |
|---|---|---|---|
| Cytodex 1 Pharmacia | DEAE positively charged groups distributed throughout a crosslinked dextran matrix | 131–210 | 6000 |
| Cytodex 2 Pharmacia | DEAE positively charged groups distributed in outer layer of a crosslinked dextran matrix | 114–198 | 5500 |
| Cytodex 3 Pharmacia | Denatured collagen type I covalently linked to the surface of a crosslinked dextran matrix | 133–215 | 4600 |
| Superbeads Flow Labs | DEAE positively charged groups distributed throughout a crosslinked dextran matrix | 135–205 | 5500 |
| Dormacell Pfeifer | DEAE positively charged groups distributed throughout a crosslinked dextran matrix | 140–240 | 7000 |
| Gelibeads KC Biological | Crosslinked gelatin (denatured collagen) | 115–235 | 3800 |
| Biosilon Nunc | Tissue-culture-treated polystyrene | 160–300 | 250 |
| Cytospheres Lux | Tissue-culture-treated polystyrene | 160–300 | 250 |
| Bioplas SoloHill Engineering | Tissue-culture-treated polystyrene | 90–150<br>150–210 | 325<br>475 |
| Collagen SoloHill Engineering | Collagen-coated polystyrene | 90–150<br>150–210 | 325<br>475 |
| Bioglas SoloHill Engineering | Glass | 90–150<br>150–210 | 325<br>475 |
| Bio-Carriers Biorad | Polyacrylamide | 120–180 | 4700 |

## 4.1 Transformed Cells

The importance of anchorage dependence in the control of cellular growth may be seen in the study of malignantly transformed cells. In these cells, the requirement for cell attachment has been lost, or can be easily overcome, and the cells proliferate rapidly in suspension (64). Cell transformation also results in a loss of the other normal constraints on cell proliferation such as a decreased requirement for growth factors and a loss of capacity for growth arrest. These changes are often accompanied by a change in cellular morphology and the development of in vivo tumor-forming capacity (malignancy). Transformation may be induced by several factors, including viral infection, irradiation, and exposure to carcinogenic chemicals. Different degrees of cellular transformation are possible, with a complete loss of anchorage dependence occurring only in the extreme, malignantly transformed cell. Unfortunately, the term "transformed" is also loosely applied to spontaneously immortalized cell lines, such as 3T3 cells, although these immortalized cells quite often retain nearly normal growth regulation. These types of immortalized cell lines and primary cell strains that exhibit anchorage dependence are the focus of the following discussion.

## 4.2 Effect of Cell Adhesion on Cellular Processes and Growth

Cells that are not allowed to attach to an appropriate substratum are spherical, whereas attached anchorage-dependent cells begin attaining a very flat and spread-out morphology within 30 min of attachment (65). This flattening of the cell increases its surface area approximately 10-fold, and the cell membrane also appears to be more permeable in its spread-out configuration (66). It has also long been known that normal anchorage-dependent cells exhibit a density-dependent inhibition of growth (63). When cells grown in culture reach a confluent monolayer, cell growth stops and the number of viable cells may remain constant for long periods of time without further cell division. Only when a wound is made in a confluent monolayer do the neighboring cells begin to divide, but once the gap is filled, the cells return to their quiescent state (67).

The first quantitative data obtained relating the degree of cell spreading to DNA synthesis involved experiments utilizing modified substrata with varying adhesiveness (65). By using increasing thicknesses of poly(2-hydroxyethyl methacrylate) (poly(HEMA)) polymer films on tissue culture plastic, cells were maintained in different shapes ranging from very flat to spherical for up to 7 days. DNA synthesis rates were found to be inversely proportional to the height of the cells, even though all cells were plated at the same subconfluent density. When compared with cells in confluent monolayers, sparsely plated cells held at the same height by the poly-(HEMA) dishes exhibited similar rates of DNA synthesis. The conclusion made was that cells in confluent monolayers are forced into a more rounded morphology due to crowding by their neighbors, and it is this rounded morphology that results in decreased DNA synthesis, or the well-known density-dependent inhibition of growth. The cells do not appear to have a shape threshold that must be reached to promote growth. When DNA synthesis is plotted versus the degree of spreading, a classic dose-response curve is obtained.

To further study the effects of suspension on anchorage-dependent cells, many investigators have utilized 3T3 and 3T6 fibroblasts in semisolid methylcellulose culture in which the cells could not attach or flatten (68–72). The viability of anchorage-dependent cells maintained in these suspension cultures may remain relatively unchanged for up to 72 h, but the cells become arrested in $G_1$ phase and are unable to initiate DNA synthesis (68). After this 72-h period, viability and plating efficiency decline rapidly until the 8th or 9th day, at which time almost all the cells are dead (65). Following suspension, the production of mRNA is reduced fivefold within a few hours, but the total amount of mRNA within the cell remains constant (69). This is achieved by a concomitant stabilization of the mRNA against breakdown, thereby balancing the rates of production and decay of mRNA within the cell. The rate of protein synthesis declines much more slowly in these cells, but after 72 h it falls to only 15% of control level protein synthesis. Although mRNA levels remain stable, the mRNA is somehow reversibly modified, making it unavailable for translation (70). When this modified mRNA is translated by in vitro reticulocyte or wheat germ lysate translation systems, the polypeptides produced are neither full-sized nor homogeneous. The cytoskeletal framework is also affected by the suspension of these cells. It appears that although the actin filaments and other associated proteins of the cytoskeletal framework undergo a reorganization, there is no significant depolymerization of these filaments upon suspension (73). This change in cytoskeletal organization is highly significant, because it has been shown that mRNA molecules bind polyribosomes and are translated only after the mRNA is attached to the cytoskeletal framework (74). Upon reattachment of the cells to a solid substratum, the previously modified mRNA is again made available for translation, and protein synthesis increases sevenfold within 4 h. The recovery in protein synthesis requires only attachment to a surface and will proceed even if cell spreading is inhibited by concanavalin A. The recovery of new mRNA and DNA synthesis occurs slowly, requires cell spreading, and does not fully recover until 18 h after reattachment (71). Cells that are recovering after reattachment synthesize large amounts of a few major structural proteins. For example, even though mRNA production is very low during the first 6 h of reattachment, significant amounts of new actin mRNA are being produced, and actin synthesis accounts for 25% of the total protein synthesis (72). It has also been shown that c-*fos* and c-*myc*, two growth-associated proto-oncogenes, are induced upon reattachment of the cell to tissue culture plastic (75). In addition, when the cells were allowed to attach to a fibronectin matrix, the expression of c-*fos* was further enhanced and the cells were more responsive to EGF and insulin. This demonstrates clearly that cell attachment and cell shape can profoundly influence the expression of genes and the translation of mRNA, although the mechanisms by which these processes occur remain unknown. There have been studies implicating many factors including cell volume (66), membrane surface area (76), and nuclear shape (77) in the morphological control of cell growth. Another important factor in anchorage-dependent growth is the binding of ECM proteins (e.g., fibronectin and vitronectin) by transmembrane receptors (e.g., integrins). Phosphorylation events at focal adhesions (78) and cytoskeletal reorganizations (72) that occur upon attachment may be involved in the transduction of mitogenic signals to the nucleus.

In some cases, both the adhesion of a cell and its consequent stimulation may be mediated by the same membrane-bound factor. Several growth and morphogenesis factors are known to be produced as membrane-anchored precursors. These include factors that have the consensus EGF-like structure, such as EGF itself, and transforming growth factor-$\alpha$ (TGF-$\alpha$) (79). Once the TGF-$\alpha$ precursor becomes anchored to the membrane, the actual TGF-$\alpha$ portion is cleaved off to give the soluble growth factor. Apparently, however, this cleavage process is relatively inefficient, and the membrane-bound growth factor precursor may remain stable for several hours (80). This membrane-anchored growth factor is free to bind to its receptor during this time, which in effect will bind two cells together through a growth factor-receptor complex. The same interaction will also result in stimulation of the cell that is expressing the growth factor receptor. This dual interaction has been demonstrated for the interaction between the TGF-$\alpha$ precursor and the EGF receptor. The TGF-$\alpha$ was expressed in a bone marrow stromal cell line, which was then able to bind and stimulate proliferation of a hematopoietic progenitor cell line that does not bind to cells without the TGF-$\alpha$ (see Sec. 6 for a discussion of the hematopoietic system) (80). The term "juxtacrine" was coined to describe this type of intercellular stimulation mediated by the binding of a membrane-bound growth factor to its receptor on an adjacent cell. This interaction may facilitate cell-cell adhesion and also mediate cell homing to appropriate cells that allow their stimulation and proliferation. This type of cellular homing is important in bone marrow cultures and is discussed further in Sec. 6.

## 4.3 Cell Adhesion and Differentiation

Another interesting cellular response that may occur as a result of cell attachment or shape change is differentiation. As has been discussed, the attachment and/or spreading of a cell results in the activation of genes associated with growth and structure. In some types of cells, the change in cell shape due to an alteration of the degree or type of attachment serves as a signal for the differentiation of the cell into a different phenotype. For example, human epidermal keratinocytes grown in culture will stop proliferating and undergo terminal differentiation (as evidenced by involucrin protein synthesis) when forced into a rounded morphology (81). As was seen for DNA synthesis previously, there appears to be no shape threshold for triggering differentiation, because a plot of involucrin synthesis versus area of substratum contact again yields a classic dose-response curve. The program of terminal differentiation of these cells in vivo also appears to be related to the degree of both cell-cell and cell-ECM adhesion, although the precise mechanism remains unknown. There is sufficient evidence that the composition and organization of the extracellular matrix may play a role in influencing the shape of a cell (82). For example, chondrocytes attain a polygonal morphology in response to plasma fibronectin whereas they assume a flattened, elongated morphology on cellular fibronectin (83). This change in morphology may also be accompanied by a profound change in sensitivity to soluble growth factors as well (84). When corneal epithelial cells are grown on tissue culture plastic, they assume a flattened fibroblastoid morphology and are responsive to fibroblast growth factor (FGF) but not epidermal growth factor (EGF). If these cells are grown instead on collagen, the cells attain their in vivo cuboidal morphology and sensitivity to EGF is restored while sensitivity to FGF is lost. It is therefore critical to maintain an exogenous extracellular matrix similar to that found in vivo for a particular cell type if the shape and characteristics of a cell are to be duplicated in vitro.

## 4.4 Cell Adhesion and Morphogenesis

Perhaps one of the most striking transformations that may occur in cellular systems is that of morphogenesis (spatial arrangement and tissue formation). Angiogenesis, the formation of capillary tubes by vascular endothelial cells, is one model system that has been used to study this phenomenon in vitro by several investigators (85,86). Although these cells require soluble factors such as FGF and transforming growth factor-$\beta$ in order to form their three-dimensional tubular network, it is the insoluble factors of the extracellular matrix that dominate the process by controlling cell shape and responsiveness to soluble factors (54). It appears that two distinct domains in laminin bind to the cell through two distinct receptors, the first allowing initial attachment and the second mediating the cellular reorganization into a capillary-like structures (85). In addition to the chemical signals that may occur upon cell attachment, such as phosphorylation of integrins (87) and initiation of the phosphatidylinositol (PI) breakdown cycle (88), it appears that tensile forces generated in the cytoskeleton may also play a role in the signal transduction of ECM binding (54). Because the ECM components are insoluble, they are capable of resisting cell-generated tensile forces, thereby increasing tension and transmitting mechanical stresses throughout the cell. The tension generated by cell-ECM contacts may affect stretch activated membrane channels (89), polymerization of actin and tubulin (90), and activate differentiation specific genes (91). These findings suggest that the effect of extracellular matrix components may be due in part to their insolubility, and that these signals are transduced in a mechanochemical manner.

### 4.5 Summary of the Biological Implications of Cell Adhesion

The ability of anchorage-dependent cells to attach and spread on an appropriate substratum has profound implications for the cell and the cell culturist. DNA, RNA, and protein synthesis are adversely affected by the suspension of these cells. Resumption of these molecular processes may be significantly delayed upon reattachment, and a large amount of cellular energy is spent on the synthesis of structural components. The time course of these events is summarized in Figure 3. The degree of cell attachment and spreading allowed may also affect the differentiated characteristics of a cell, but it does so in a dose-response manner as opposed to a threshold effect. The different components of the extracellular matrix may affect different cell types in different ways, mediating both cell shape and responsiveness to growth factors. In addition, the insolubility of the ECM components allows them to impart mechanical stresses to the cell upon cell-ECM binding, thereby signaling the cell in ways that soluble factors cannot. There is a very complex relationship between the extracellular matrix and cell shape, and this can affect all aspects of cellular function from simple proliferation to differentiation and tissue formation.

## 5 CELL ATTACHMENT AND MICROCARRIER CULTURES: FLUID FORCES, AGGREGATION, AND CELL INJURY

Most normal and many transformed animal cells are anchorage-dependent for growth; i.e., they require a surface for attachment. One of the most efficient means to provide a surface for cell growth is the use of microcarriers (small polymeric or glass beads of 180–250 $\mu$m average diameter). The surface properties and other characteristics of microcarriers have been extensively reviewed (55,92,93). In this section, we shall first briefly review the literature that establishes the importance of cell attachment, of cell type, and of the extracellular matrix on the ability of cells to sense and

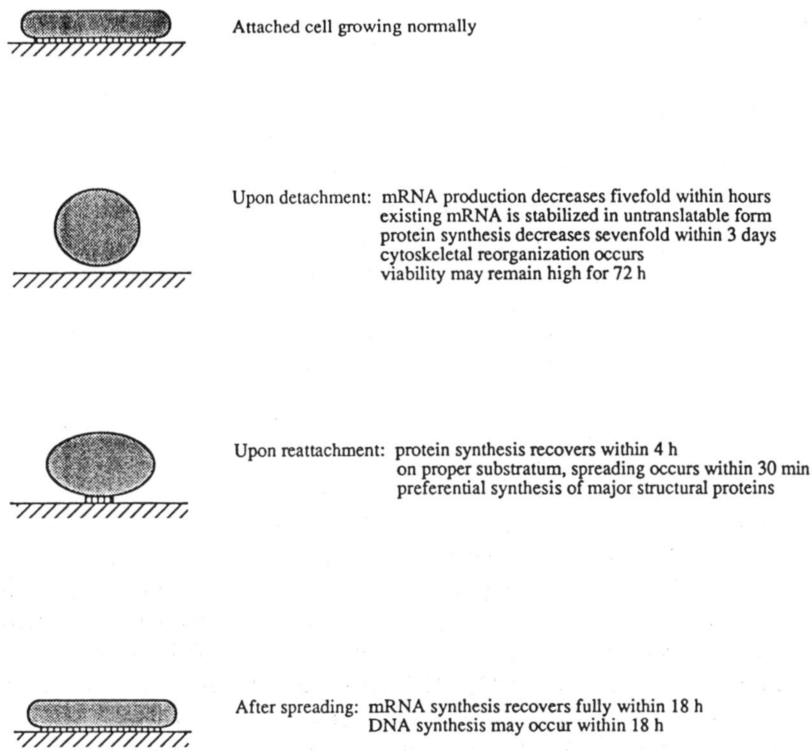


**Figure 3** Molecular processes that occur upon detachment and reattachment of a cell to the substratum.

respond to fluid forces. These factors are also important when considering the way in which shear stresses affect cell metabolism. Then we shall review what is known about the types of forces that affect cells in microcarrier cultures and how these effects are presently modeled at a macroscopic level. A discussion on cell and microcarrier aggregation and how they are affected by fluid forces will then follow.

## 5.1 Fluid Effects on Cells Exposed to Well-Defined Shear

Anchorage-dependent cells exposed to laminar shear in defined-flow chambers for periods of 3–24 h have exhibited a range of responses. The most evident response of a monolayer of (endothelial, at least) cells to a steady shear stress is a change of morphology from the characteristic cobblestone, random, polygonal pattern to one in which the cells are elongated in the direction of flow (94–99). This change is accompanied by a cytoskeletal restructuring, specifically the alignment of microtubules (96) and the formation of actin stress fibers (100,101). The rate and amount of elongation is an increasing function of shear stress, with changes becoming apparent in as little as 1 h or at shear stresses as low as 4 dyne/cm$^2$. Similar elongation and orientation results have been reported for human embryonic kidney (HEK) cells (an epithelial cell) with effects becoming evident at shear stresses at or above 6.5 dyne/cm$^2$ (99).

The differences among cells and the importance of cell attachment on the ability of cells to resist fluid forces have been demonstrated by studies on the shear stress levels that are necessary to detach various cells from various substrata. Although there are numerous studies of a single cell type, few studies have been performed with more than one cell type under the same conditions (same flow device). At relatively higher levels of shear stress (26 and 54 dyne/cm$^2$), the HEK cells detach from the surface and exhibit reduced viability (99). In contrast, no cell loss was reported for bovine aorta endothelial cells at shear stress levels of 85 dyne/cm$^2$ (97). This is a demonstration of the large variability in the response of various cells to fluid forces. It has also been found that the nature of the surface to which the cells are adhering, such as glass or treated plastic, is important in the response of the cells to shear stress (97). In a more detailed study involving three different cell types, Spier and co-workers (102,103) have shown that three important cell lines (VERO, BHK, and MRC5) require different levels of shear stresses (varying from 30 to 55 dyne/cm$^2$) for detachment from different substrata (glass and polystyrene). They also showed that the required level of shear stress for cell detachment depends on the amount of serum in the medium and the prevailing pH. Some of their data are shown in Table 4.

It is now well documented that cell metabolism is also affected by fluid stresses. Stathopoulos and Hellums (99) found that the release of urokinase by HEK cells was not changed (compared to unstressed controls) at 2.6 dyne/cm$^2$ shear stress, but was over twice as high at 6.5 dyne/cm$^2$. At 13

**Table 4** Minimum Shear Stress Required for Cell Detachment

| | Critical shear stress (dyne/cm$^2$) | | |
|---|---|---|---|
| Substratum | BHK cells | Vero cells | MRC-5 cells |
| Glass | 43.9 | 55.0 | 47.9 |
| Polystyrene | 44.7 | 35.7 | 29.4 |

*Source:* From Ref. 102.

and 26 dyne/cm$^2$ the release rate decreased from the peak at 6.5 dyne/cm$^2$, but was still higher than the control. Human umbilical vein endothelial (HUVE) cells were shown to release five times greater amounts of prostacyclin than near-zero stress controls when exposed to steady shear stresses of 10 dyne/cm$^2$ (104). When the shear stress was varied as a 1-Hz square wave between 8 and 12 dyne/cm$^2$, prostacyclin production increased an additional 2.4 times. These results show that shear stress stimulates arachidonic acid metabolism by the endothelial cells (105). The mechanism by which prostacyclin release is increased has not yet been elucidated, but the importance of intracellular calcium has been established and some possible mechanisms have been suggested by a recent study (106). It has also been recently shown that physiological levels of arterial shear stress stimulate secretion of the protein tissue plasminogen activator (t-PA) by HUVE cells (107). The secretion of t-PA was selective since secretion of the t-PA inhibitor was not stimulated. This stimulation was found to be the result of increased t-PA mRNA levels (108). Cytoplasmic free calcium ions have been identified as mediators of the cell response to fluid shear (109), and stretch-activated calcium channels of the cell membrane have been shown to be mechanotransducers of hydrodynamic stress (89). The importance of cell attachment and the key role of the extracellular matrix regarding the ability of cells to withstand and respond to fluid forces have been further strengthened by the recent finding that laminar shear stresses inhibit the synthesis and release of fibronectin, and that sheared cells exhibit greater strength of attachment than unsheared cells (110). Shear stresses also affect cell proliferation as evidenced by changes in DNA synthesis rates (measured by tritiated thymidine incorporation). At laminar shear levels above 30 dyne/cm$^2$, proliferation decreases for subconfluent endothelial monolayers, but not for confluent monolayers (98). In contrast, even low levels of turbulent stresses affect cell proliferation in confluent monolayers (111). These results are consistent with the findings on how cell shape, cytoskeletal structure, cell attachment, and the extracellular matrix are important regulators of cell proliferation, metabolism, and function (112–115; see also Sec. 4).

## 5.2 Fluid-Mechanical Considerations in Nonporous Microcarrier Bioreactors

We will consider first the most widely used nonporous microcarriers. Cells grow only on the surface of these beads. In microcarrier bioreactors, theoretical considerations suggest that three possible mechanisms may damage cells on microcarriers (116): (1) direct interaction between the microcarrier and small turbulent eddies, (2) collisions between microcarriers, and (3) collisions between microcarriers and bioreactor internals (impeller and probes). The three mechanisms may not be important in all cases, but when they do cause cell damage or death, they act simultaneously. The first two mechanisms result from the structure of turbulence in an agitated reactor, specifically the eddies in the dissipation range of the turbulence spectrum. Concepts from isotropic turbulence can be used in agitated reactors to the extent that they pertain to small eddies (in the inertia and dissipation ranges), whereby the structure of turbulence is independent of the conditions of formation (116,117). The smallest eddies in the dissipation range of the spectrum are known as Kolmogorov-scale eddies and are characterized by the following size ($\eta$) and velocity ($v$):

$$\eta = (\nu^3/\epsilon)^{1/4} \qquad v = (\epsilon\nu)^{1/4} \tag{1),(2}$$

where $\nu$ is the kinematic viscosity of the medium and $\epsilon$ is the rate of viscous energy dissipation in the turbulence per unit mass of fluid. The power input ($P$) through the impeller can be obtained by direct measurement of either motor power consumption or torque times rotational speed of the agitator shaft. It may also be estimated from the impeller parameters. This is based on the formulae

$$P = N_p \rho_f n^3 d_i^5 \qquad \epsilon = P/\rho_f V \tag{3),(4}$$

in which $N_p$ is the dimensionless impeller power number; $\rho_f$ is the fluid density; $n$ is the impeller rotational rate (in rev $s^{-1}$); $d_i$ is the impeller diameter; and $V$ is the liquid volume in which the energy dissipation takes place. $N_p$ is a function of impeller geometry, impeller Reynolds number

$$Re = nd_i^2/\nu \tag{5}$$

and whether the vessel is baffled or not. $N_p$ is usually in the range 0.5–3.0 and is approximately constant in baffled vessels for $Re > 1000$; turbulent flow is taken to prevail in such an agitated reactor at $Re$ values above 1000 (116). The dissipation volume ($V$) in Eq. (4) may be taken as the total liquid volume of the bioreactor ($V_r$) to obtain an average value of $\epsilon$ in the system. However, it is well established that the dissipation rate is much higher around the impeller than elsewhere in a stirred vessel. As a result, in the vicinity of the impeller the turbulent power dissipation rate (a determinant of the intensity of turbulence and the size and velocity of the smallest eddies) becomes (117–119)

$$\epsilon_i = (P/\rho_f V_r)(V_r/d_i^3) = P/\rho_f d_i^3 = N_p n^3 d_i^2 \tag{6}$$

Our data suggest that $\epsilon$ should be best calculated from Eq. (6) (118,119). Experimental evidence suggests that the first two mechanisms are responsible for cell damage to different extents depending on the agitation intensity (118–120). The interaction of the microcarriers with the turbulent eddies can be described, in a first approximation, by $\eta/d$, the ratio of smallest eddy size to the bead diameter:

$$\frac{\eta}{d} = \frac{\nu^{3/4}}{\epsilon^{1/4} d} = \frac{\nu^{3/4}}{(N_\rho n^3 d_i^2)^{1/4} d} \tag{7}$$

Growth and death rate data show a good correlation with this parameter (119,121), but this does not prove that this is the only or predominant mechanism of cell damage. Death rates were measured after the cultures had reached high cell densities and were switched to media that support cell maintenance without further growth (119).

Modeling and predicting cell damage due to the interactions between beads poses several difficulties. First, it is not known whether the frequency or the severity of the interaction is the most important factor. We know that both are important, contributing to a different extent depending on the situation, but the quantitative relation remains unknown. Second, calculating the frequency and the energy of the interactions is a difficult fluid-mechanical problem that can be addressed only in an approximate fashion. We decided to correlate our data based on the turbulent collision severity (TCS) defined as

$$\begin{aligned} \text{TCS} &= (\text{kinetic energy})(\text{collision frequency/vol.})/N \\ &= \left(\frac{m v_{br}^2}{2}\right)\left(\frac{v_{br}\alpha^2}{d^4}\right) \Big/ \left(\frac{\alpha}{(\pi/6)d^3}\right) = v_{br}^3 \left(\frac{\pi^2 \rho_b d^2 \alpha}{72}\right) \end{aligned} \tag{8}$$

which represents collision energy per bead per time, where $v_{br}$ is the root mean square relative velocity of the beads, $\rho_{br}$ is the bead density (g/cm$^3$), $\alpha$ is the volume fraction of beads in the system, and $N$ is the number of beads per volume. The relative velocity ($v_{br}$) of the beads can be characterized in two ways. One is by the velocity $v = (\epsilon\nu)^{1/4}$ of the smallest eddies in turbulence since the beads have nearly the same size and density as those eddies. If the eddies are much larger than the beads, the relative velocity between neighboring beads is not well described using this assumption. Using the eddy velocity for $v_{br}$ gives the following result for an eddy-based TCS:

$$\text{TCS}_e = (\epsilon\nu)^{3/4} \left(\frac{\pi^2 \rho_b d^2 \alpha}{72}\right) \tag{9}$$

Alternatively, collisions between beads can be caused by a shear-based mechanism. Given two beads in a shear field, the relative velocity between the beads will equal the distance between the streamlines along which the beads are each moving times the local velocity gradient across those streamlines. Beads moving on streamlines less than one bead diameter apart can therefore collide, and the velocity of that collision is characterized by $(\gamma d)$, where $\gamma$ is the local velocity gradient or shear rate. In the smallest turbulent eddies $\gamma$ = eddy velocity/eddy size = $(\epsilon/\nu)^{1/2}$, so we can define a distinct shear-based TCS as:

$$\mathrm{TCS}_s = \left[\left(\frac{\epsilon}{\nu}\right)^{1/2} d\right]^3 \left(\frac{\pi^2 \rho_b d^2 \alpha}{72}\right) = \left(\frac{\epsilon}{\nu}\right)^{3/2} \left(\frac{\pi^2 \rho_b d^5 \alpha}{72}\right) \tag{10}$$

Experimental data of growth rates and death rates using bovine embryonic kidney cells can be correlated quite well using either of the two TCS expressions (118,119). These correlations do not prove, however, if any of these collision-model hypotheses are indeed valid, and what extent of cell damage can be attributed to bead-bead collisions. Since smaller beads increase the $\eta/d$ ratio and decrease the bead to bead collision severities [Eqs. (9) and (10)], the experimental finding (119) that smaller-diameter microcarriers give a higher growth rate of cells can be viewed as confirming that the models and hypotheses are indeed consistent with the experimental findings. Similarly, increased medium viscosity also caused higher growth rates (119,122), and this supports the eddy/bead size ratio and shear-based TCS mechanisms, but is opposite to the prediction of the eddy-velocity-based TCS. It has been well established (120) that both bead collisions and eddy-to-bead interactions are important and that the former interactions contribute more to cell damage at higher agitation intensities. Thus, the effects of bead size and medium viscosity do not have to be explained by one or the other mechanism alone and are also likely to vary considerably with agitation intensity. This has been confirmed experimentally in our laboratory, as we discuss next.

We have carefully investigated the quantitative dependency of cell growth and death on the viscosity and the other aforementioned parameters (123). Bovine embryonic kidney (BEK) cells growing in spinner flask microcarrier cultures were used in these studies. Growth rates were measured in exponential, simultaneous batch cultures. Death rates were measured after the cells were grown to high densities and had been switched to medium that allows cell maintenance without further growth. Results show that the effect of viscosity on specific growth and death rates depends on the level of agitation; increasing agitation amplifies the dependence of specific growth and death rates on viscosity. The specific death rates varied with viscosity to the $-0.65$ power at low agitation levels (120 rpm) and up to the $-2.0$ power at high agitation levels (160 rpm). The specific growth rates increased with viscosity to the 0.55 power at 120 rpm and to the 0.76 power at 140 rpm. At agitation levels of 100 rpm or lower there is no effect of viscosity on specific growth and death rates. These data show that there is a strong cross-parametric dependence of the death rate on the viscosity and the agitation intensity, which can be characterized by the agitation energy input per unit fluid volume, $\epsilon$. A model based on the turbulent energy content of eddies of length scales on the order of magnitude of the microcarrier diameter and lower has been developed which explains the functional dependence of specific growth and specific death rates on viscosity. The model has good predictive capabilities as well, but additional experimental data and model improvements are necessary.

To summarize, cells in microcarrier bioreactors are exposed to forces primarily due to interactions of beads with individual small eddies and bead-to-bead collisions. The first type of interactions are of higher frequency but lower severity compared to the second. In some cases, beads may also collide with the impeller and probes in the reactor. Such events are of lower frequency but potentially high severity. We know that the length of exposure to such events is important for cell injury to demonstrate itself, so frequency of exposure is very important. The severity of exposure is also critical. In light of the discussion in Sec. 5.1 on the types of cell responses to various levels and frequencies of fluid forces, we would expect a large spectrum of effects on cells in microcarrier bioreactors. So far, other than macroscopic cell death or reduction of cell growth, such effects have not been thoroughly investigated and none have been found. Preliminary results from our laboratory suggest that such effects are not only present, but also important for a number of systems. We still do not know with any degree of certainty how cells are damaged in microcarrier bioreactors due to any of the three aforementioned mechanisms. In all three mechanisms, local shear and normal forces will shear the cell with respect to its substratum or will push the cell against the substratum. The cell will resist the shearing due to its attachment, but in the process its membrane and/or cytoskeletal integrity may be partially affected and the protein bridges through which it attaches to the substratum may be damaged. Normal forces will be resisted similarly, but intuitively, the effect on the attachment bridges is likely to be less important. Because a fair amount of exposure time (say, 20–120 min, depending on the agitation intensity) is necessary for cell damage (i.e., cell death or growth retardation), it is apparent that a repeated exposure to a damaging event is necessary. Upon repeated exposure to the same or different damaging events, eventually either or both of two things can happen: the protein bridges through which the cell attaches to the substratum will be severely damaged and the cell will detach from the substratum, or the cell membrane and cell components will be severely damaged. Either of the two possibilities will affect the cell integrity and proliferation, as was discussed in Sec. 4. Death could be the result of either of the two possibilities. Furthermore, detachment may lead to cell death because the probability of reattachment will be minimal under severe agitation conditions and the cells cannot grow well or at all in the absence of attachment. Alternatively, cell injury can lead to metabolic retardation, which may result in cell detachment and death.

The way cell injury in microcarrier bioreactors has been assessed so far is relatively crude: by cell death in media that do not allow cell proliferation, and by reduction in growth in regular growth media. The reason for this is the fact that, in the beginning, one must address first the problem in a global fashion to resolve the issue of whether there are fluid effects, and what parameters affect cell injury. There have been only two systematic attempts to address this problem (Wang, Croughan, and co-workers; Cherry, Papoutsakis, and co-workers), and despite the progress made, our understanding is still incomplete, if not primitive. The difficulties derive from both the fluid-mechanical and the biological components of the problem. Given the fact that simpler transport problems (e.g., oxygen transport) in bioreactors have been resolved after 40+ years of research by hundreds of investigators, it is not surprising that the present problem is not yet satisfactorily understood.

## 5.3 Cell Attachment and Sensitivity to Agitation Damage in Microcarrier Bioreactors

How cells on microcarriers respond to fluid-mechanical forces depends both on the type of cell and on the quality of cell attachment to the beads. The dependence on the former, although not systematically examined, is well known. For example, fibroblasts and fibroblast-like cells [e.g., Chinese hamster ovary (CHO) cells] are considerably more resistant to damage than epithelial cells (e.g., VERO cells and the BEK cells mentioned above). In addition, cells from certain organs such as the liver are much more shear sensitive than other cells such as fibroblasts. Human cells are also more shear sensitive than other animal cells. We have discussed examples for the aforementioned cases earlier (Sec. 31 of Ref. 117). The effect of the quality of attachment has not been extensively studied but is well documented in at least one study (124). It was shown that chick embryo fibroblasts are much more resistant to agitation after a 24-h attachment to microcarriers under stationary culture conditions than cells exposed to agitation shortly after attachment. This is now a well-known fact for a variety of other cells and has become a part of standard protocols suggested by various microcarrier manufacturers. In our laboratory, for example, we found that with both CHO and VERO cells good cell attachment is very important for the cells in withstanding shear forces. So, especially for cells that sometimes do not look healthy in the inoculum, a longer attachment period with intermittent mild agitation is employed. The importance of cell attachment is also indirectly demonstrated by the fact that the same cells show different resistance to the same level of agitation depending on the type of bead they have been grown on. Since cells attach differently to various substrata, again the quality of attachment emerges as an important determinant of the shear sensitivity issue. The importance of cell-to-substratum attachment in the ability of cells to withstand fluid forces in well-defined fluid fields was discussed in Sec. 5.1.

### 5.4 Microcarrier and Cell Aggregation: Effects of Bioreactor Flow and Nutrition

Aggregation of cells and microcarrier beads in bioreactors is a well-known but poorly studied phenomenon. It has many practical implications and applications. Cells in certain parts of the cell or bead aggregates may be starved of oxygen and other nutrients, and this will affect their viability, growth, and productivity in terms of excreted products. On the other hand, cells that are part of an aggregate may create a potentially beneficial microenvironment due to the release of many growth factors and other glycoproteins, and also due to cell-to-cell and cell-to-ECM interactions. Cells and beads aggregate through the same mechanisms that cells attach to solid substrata or become part of a tissue. So, in principle, the conditions that enhance cell attachment ought to promote bead and/or cell aggregation. All the experimental evidence suggests that this is indeed the case, as we shall discuss shortly.

Bead bridging is commonly observed in microcarrier cultures (117,118, 125–127), and it has several interesting hydrodynamic and physiological implications. Bridging occurs when two beads collide and one or more cells at the point of impact stick to the other bead (118). All microscopic observations suggest that the formation of a bridge requires the impact of a bead with a fairly high cell coverage with one with low or zero coverage (118). Soon a ring of three or more cells forms between the beads, leaving frequently (but not always) a bare circle perhaps 20–50 $\mu$m in diameter where the beads are in actual contact. In other cases, the two beads are not in actual contact and are connected through a double layer of cells or a large cellular clump. The latter case is particularly prevalent in the case of transformed cells (e.g., CHO, BHK, and various malignant cells) which overgrow to form multiple layers and large cellular masses. Figure 4 (a and b) show typical bridging of this category and some of the first category. Normal and normal-like cells form bridges through the first mechanism only, because of strict contact inhibition and the inability to form anything more than a monolayer. Pictures of this type of bridging can be found in Ref. 118. Several beads may aggregate in this fashion, with the beads multi-

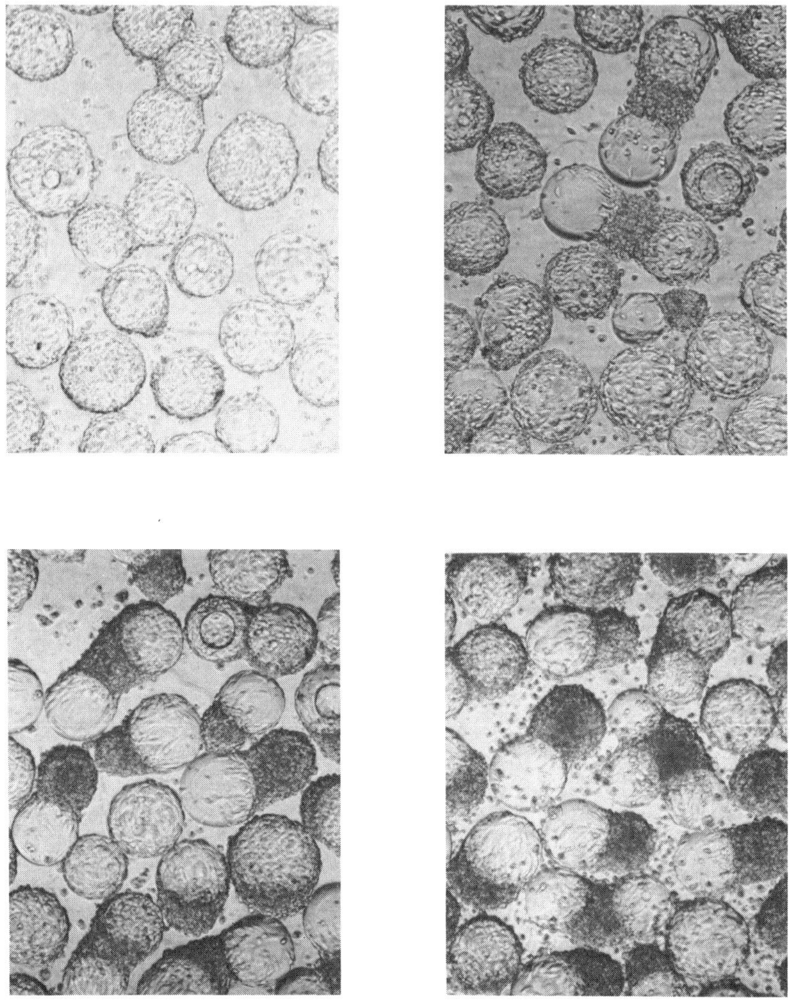


**Figure 4** Photomicrographs (100×) of recombinant CHO cells in protein proc tion medium in microcarrier culture at 115 rpm. Photographs taken at (a) 53 h, 112 h, (c) 188 h, and (d) 314 h. (Taken from Ref. 61.)

ply bridged to each other. Clump formation is more prevalent at lower levels of agitation, with clump size increasing as agitation speed decreases. Figure 5 shows this dependency on agitation for recCHO cells. This figure also shows that bridging increases linearly with time, but we do not know if this is the general kinetic form of bridging. Bridging decreases with increased agitation presumably because higher levels of agitation may reduce the probability of aggregate formation and/or increase the probability of breaking the formed bridge. The initial collision between two beads may be more energetic so there is a lesser probability that a cell will adhere to the colliding bead. Clumps tend to be compact rather than elongated or branched in structure as random attachment would make them. This suggests greater removal of singly bridged beads compared to multiply connected ones. The formation of large clumps, containing six or 10 or even dozens of beads, is fluid-mechanically equivalent to having microcarriers with effective diameters two or more times that of the individual beads. By the eddy/bead size ratio, cells growing on the outer surface of these large clumps would be expected to be particularly subject to hydrodynamic damage. We found no visual evidence of this with bovine embryonic kidney cells (which exhibit strict contact inhibition), although the measured net

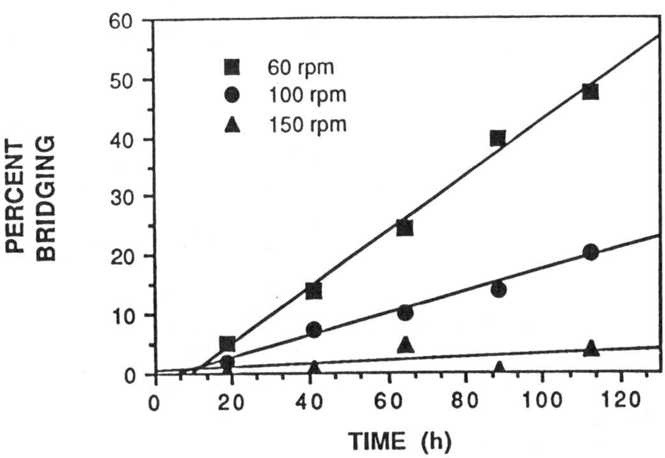


**Figure 5** Effect of agitation rate on bridge formation. Recombinant CHO cells growing on microcarriers in growth medium (5% calf serum) in spinner flasks at agitation rates of 60, 100, and 150 rpm. A bridge is a growth of cells between two microcarriers that holds the microcarriers together. Percent bridging is the number of bridges between microcarriers divided by the number of microcarriers. (Taken from Ref. 61.)

growth rate was lower at minimal agitation levels where clump formation was significant (118). Dissolved oxygen and pH levels were controlled, and the clumps were only up to about 10 beads in size, so mass transfer problems were not the probable cause of this decreased growth rate. The actual mechanism may be the death of cells on the clump exterior or the death of bridge cells when fluid forces or collisions manage to break a bridge apart. For the case of transformed CHO cells, we have visual evidence that when bead aggregates are formed, the cells tend to disappear from the externally exposed surface of the microcarriers and tend to grow in the bridging area between beads, thus forming large cellular masses as time progresses (Fig. 4, b and d). Photographic evidence suggests that the cellular masses grow more elongated and larger with time, until one of the beads is removed and the cellular mass becomes more spherical, remaining attached to only one bead (Fig. 4, c and d). Eventually, some cellular masses detach completely from all beads and cells grow in this aggregate form with no attachment to solid support. These cellular masses are approximately equal in size to the microcarriers. All available evidence from our laboratory (61) shows very high viability for the cells in these large cellular masses, which is somewhat surprising due to their large size. Visual evidence suggests that cells grow preferentially as part of these cellular masses rather on the microcarriers. We theorize that this is due to two reasons. First, cells are less susceptible to fluid (local shear) forces because these masses are elastic and the cell aggregate can deform under a stress without transmitting substantial stresses to individual cells. Second, cells in these cellular masses create a potentially beneficial microenvironment due to released autocrine growth factors as in the case of three-dimensional tissues. Cell-to-cell contact may also play an important role in this preferential aggregation. As long as there are no mass transfer problems either for the cell nutrients or for the protein products, these cellular masses are apparently beneficial for bioprocessing. This is because cells in these aggregates can condition their local microenvironment more effectively and thus require fewer specialized nutrients (growth factors) in the medium. Second, these aggregates can be retained in bioreactors for prolonged protein expression more easily than single cells and without the need to add more microcarriers. We understand that a number of companies operate their cell-culture reactors under conditions that enhance the formation of these cellular masses.

Other than agitation, in view of the parameters that affect cell adhesion (Sec. 2), one would expect that medium composition will play a significant role in promoting or controlling bead and cell aggregation. The presence of certain growth factors or proteins may enhance or inhibit cell adhesion, and this would be reflected in the degree of aggregation. Indeed, it was found that for recCHO cells, increasing (low) serum concentrations increases bead and cell aggregation, but at higher concentrations serum decreases aggregation (61).

In terms of modeling, the presence of bead aggregates complicates further the situation because of the larger variation of effective microcarrier size. This will alter the characteristics of the eddy/bead, bead/bead, and bead/internals interactions, and thus cells will experience an even larger variation of forces depending on whether they are part of a cell aggregate, bead aggregate, or attached on a single microcarrier.

## 6 IMPORTANCE OF CELL ADHESION IN TISSUE ENGINEERING: BONE MARROW CULTURES

Tissue engineering is an emerging field in which heterogeneous cell populations are maintained together in culture. When successful, the cells interact in a way that mimics the in vivo system, thereby generating both a useful in vitro model and a method of production for novel cells and cell products. These cultures are very complex due to the number of different cells and factors present, and therefore a generalized discussion is not easily generated. In order to effectively discuss the importance of cellular adhesion in tissue engineering, a specific example will be the focus of discussion. The problems encountered in the culture of hematopoietic cell populations are analogous to problems encountered with other tissue engineering applications as well (e.g., liver). Since the hematopoietic system is very extensive, complex, and probably not widely familiar to many readers, an introduction to the topic is in order.

Hematopoiesis, the regulated production of mature blood cells, is a complex scheme of multilineage differentiation that occurs mainly in the bone marrow of adult mammals. The many types of mature blood cells that exist in the circulation are all believed to be derived from a small common population of pluripotent stem cells, which form during one short interval in early embryonic life and maintain hematopoiesis thereafter through an extensive capacity for self-renewal. These pluripotent stem cells reside predominantly in the bone marrow of adult mammals. Many mature blood cells are short-lived and require continuous replacement by the hematopoietic system even under normal, unstressed conditions. In addition, the hematopoietic system is capable of rapid, but controlled fluctuations in response to a wide variety of emergency situations ranging from blood loss to infection. The study of this complex system has been greatly facilitated by the use of in vitro bone marrow cultures that retain some important physiological characteristics of bone marrow. Furthermore, there is much interest in the clinical application of such a system for patients with a wide range of medically important diseases from anemia to cancer and AIDS (128,129).

There are eight major types of mature blood cells in the hematopoietic system, which may be seen along the right-hand side of Figure 6. For traditional reasons, the blood cell population is broken down into the myeloid and lymphoid lineages. The myeloid lineage includes erythrocytes, monocytes, the granulocytes (neutrophils, eosinophils, and basophils), and platelets (derived from noncirculating megakaryocytes). Thymus-derived lymphocytes (T cells) and bone-marrow-derived lymphocytes (B cells) make up the lymphoid lineage. An incredible diversity of function is demonstrated in the hematopoietic system, and mature blood cells are highly specialized, carrying out a bewildering array of tasks throughout the body. All the cells are found in the circulation, and many of them are found in other tissues as well. For example, monocytes may leave the circulation and seed in various tissues where they serve different purposes. Kupffer cells (liver), alveolar macrophages (lung), osteoclasts (bone), and microglial cells (brain) are all cells derived from monocytes that have seeded in their respective tissues. The vast majority of mature blood cells are destined to remain functionally active for only a few hours or weeks before being destroyed and broken down (131). In fact, an average human must produce $3.7 \times 10^{11}$ mature blood cells per day to replace the hematopoietic cells lost due to natural wastage (132).

The staggering number of mature cells that must be continuously produced under normal unstressed conditions are derived from progenitor (or precursor) cells. These progenitor cells are unipotential or bipotential and are therefore capable of undergoing proliferation, differentiation, and development into only one or two of the mature cell types. These progenitor cells are designated by the term "colony-forming unit," CFU (or "colony-forming cell," CFC), because of their ability to form colonies of mature cells in semisolid agar culture. To specify the type of progenitor, a suffix is simply added to the CFU- designation. For example, granulocyte/macrophage colony-forming units (CFU-GM) proliferate and develop into mature neutrophils and macrophages. Erythroid colony-forming units (CFU-E) undergo growth and hemoglobinization to form mature erythrocytes. Similarly, other lineage-restricted progenitor cells have been described that give rise to eosinophils (CFU-Eos), basophils (CFU-Bas), and megakaryocytes (CFU-Meg) (133). In adult animals, these myeloid progenitor cells are located mainly in the bone marrow, with small populations in the spleen and the circulation. It therefore follows that the bone marrow is the major site of myeloid blood cell production in adults. B-cell progenitors are found in the bone marrow, spleen, and lymphoid tissues, while T-cell progenitors are formed only in the thymus and from there may migrate to other lymphoid

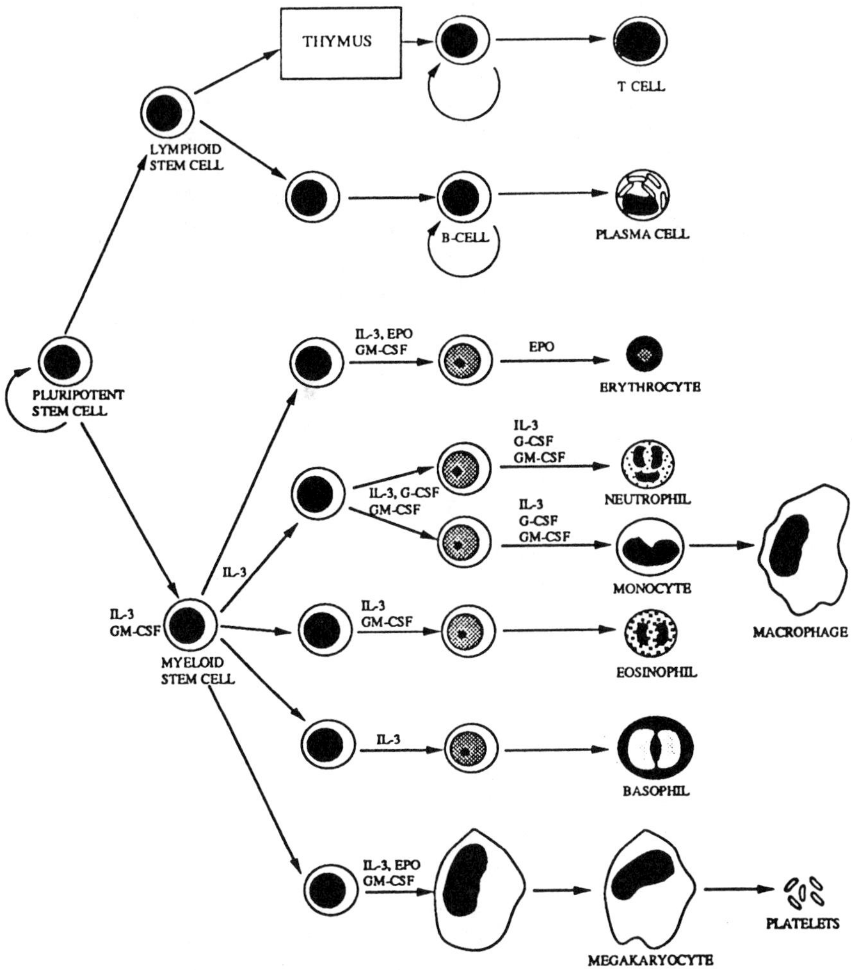


**Figure 6** Differentiation scheme of hematopoiesis. Pluripotent stem cells may undergo self-renewal or differentiation into immature progenitor cells. These progenitor cells proliferate and differentiate into the mature blood cells that are found in the circulation. The points of action of the major colony stimulating factors (CSFs) are also shown. (Adapted from Ref. 130.)

tissues. Like the mature cells they produce, most progenitor cells are short-lived because as they proliferate, they concomitantly undergo development and lose their proliferative potential.

To fulfill the continuous demand for mature cells, progenitor cells must in turn be continuously generated by more primitive cells that have the capacity to persist throughout the lifespan of the animal. These primitive cells capable of extensive self-renewal are called hematopoietic stem cells. The pyramidal structure of the hematopoietic system from the stem cell, to progenitor cells, to mature blood cells is shown in Figure 6. There are many steps in the pathway from stem to mature cells, which allows for a great amplification of cell numbers and many levels of control. Individual stem cells can generate several hundred progenitor cells within 7–14 days, and individual progenitor cells can generate up to 10,000 mature progeny within the same interval (134). Under normal conditions, most stem cells are not engaged in active proliferation and reside in the $G_0$ state outside of the proliferative cell cycle for several days during which they have time to repair DNA damage and maintain the genetic integrity of the stem cell population (135). Evidence for the existence of hematopoietic stem cells was first provided by Till and McCulloch (136). In their experiments, lethally irradiated mice were injected with $10^4$ to $10^5$ bone marrow cells from healthy syngeneic donor mice. Some of the injected cells seeded in the spleen and gave rise to macroscopically discernible hematopoietic colonies containing cells of many lineages and resulted in the restoration of the mouse hematopoietic system. The cells capable of forming colonies in the spleens of lethally irradiated recipient mice are called colony-forming unit-spleen (CFU-S) cells. CFU-S may be detected in the bone marrow (0.05% of cells), spleen (0.002% of cells), and circulation (0.0001% of cells) of adult mice (134, 137). The rarity of these cells is not surprising considering the great amplification that differentiation and proliferation provide in the production of mature cells. In fact, the stem cell reserve is probably sufficient for several lifespans even under conditions of extreme hematopoietic stress (138). The loss of cells from the stem cell compartment by differentiation to form progenitor cells is balanced by the production of new stem cells via self-renewal. Although the process that controls self-renewal and differentiation of stem cells is not understood, actions of hematopoietic growth factors, bone marrow stromal cells, and the extracellular matrix are known to be crucial in the regulation of hematopoiesis.

There is a family of glycosylated extracellular proteins that regulate the production and functional activity of hematopoietic cells. These proteins were initially discovered during the culture of hematopoietic cells in semi-solid agar, which required specific conditioned media supplementation for the formation of colonies. As a result, these substances are called colony stimulating factors (CSFs). The list of CSFs includes interleukin-1 through -10 (IL-1 through IL-10), erythropoietin (Epo), and granulocyte-macrophage colony stimulating factor (GM-CSF). These factors are capable of controlling proliferation, differentiation, and mature cell function of many target cells, as indicated in Figure 6. The continuous presence of CSFs in the concentration range of 1–100 pM is absolutely required for the in vitro proliferation of hematopoietic progenitor cells in the absence of bone marrow stromal cells.

The four types of bone marrow stromal cells are the endothelial cell, adventitial reticular cell, macrophage, and adipocyte. Due to the structure of the marrow, hematopoietic cells have a close structural and functional relationship with stromal cells. Gap junctions have been seen between stromal and hematopoietic cells (139), and there are many other reasons to believe that the stromal cells are a very vital part of the hematopoietic system. First, hematopoietic cells must be in direct contact with stromal cells to facilitate their survival, growth, and development both in vivo and in vitro (140–143). Second, discrete stromal elements show characteristic interactions with the different maturing myeloid cells, and with one another. Third, mice with the Steel mutation show a lack of hematopoiesis due to a defect in the stromal cell environment rather than in the hematopoietic cell population (144). Stromal cells are known to secrete CSFs and will usually exhibit increased secretion as a result of cell-cell interactions, or after in vitro manipulations such as irradiation, cell line generation, and transformation (135). The adhesive properties of stromal cells and hematopoietic cells are very important for the maintenance of hematopoiesis both in vivo and in vitro, and a discussion of these properties will give an introduction of the applications of cellular adhesion in the area of tissue engineering.

Long-term in vitro hematopoiesis was first achieved by Dexter and coworkers by using a two-phase culture system featuring the establishment of a bone-marrow-derived stromal layer during the first 3 weeks of culture (145). This adherent feeder layer is then capable of supporting hematopoiesis for several months after the flask is recharged with new, nonadherent bone marrow cells after the initial 3-week period. Many of these hematopoietic stem and progenitor cells become lodged on, within, and under the adherent layer and are slowly released into the culture medium as differentiation proceeds (141,146). The attachment (or homing) of the stem cells to the stromal cells is critical for the maintenance of viability and the stimulation of development of the stem cells (142). Although CSFs are known to be required for the growth of the hematopoietic cells, soluble CSFs are rarely detected in these long-term bone marrow (LTBM) cultures. It is believed that the stromal cells are able to synthesize and present membrane-bound CSFs to the hematopoietic cells in an active form, and that these growth factors are therefore only able to act on hematopoietic cells that are attached to the stromal cells (142). In fact, the recently described "juxtacrine" interactions between membrane-bound growth factors and their receptors resulting in cell-cell adhesion and intercellular stimulation of proliferation may be one of the mechanisms of stromal-hematopoietic cell adhesion (80). In addition, it is known that CSFs may bind in an active form to components of the extracellular matrix that are secreted by the stromal cells (147,148). The stromal-hematopoietic cell interaction must therefore be maintained in any culture system used for in vitro hematopoiesis. As yet, only the use of static small-scale cultures has been documented for the growth of these heterogeneous cell populations. An example of the attachment and intimate interactions of adjacent cells in these multilayer cultures may be seen in Figure 7. These LTBM cultures are believed to retain many of the features important for in vivo hematopoiesis, and have been used extensively to study the regulatory interactions between stromal cells and hematopoietic cells.

Although endothelial cells play a critical role in vivo by controlling the egress of mature blood cells into the circulation from the marrow cavities through transient migration pores, they are not believed to be a crucial component for establishing in vitro hematopoiesis (149). Adventitial reticular cells are fibroblasts that extend fine cytoplasmic processes forming an extensive network within which the hematopoietic cells proliferate and differentiate. These cells are preferentially associated with developing granulocytic cells. Hemonectin, a protein of the marrow ECM that specifically binds granulocytic precursors, has been implicated in the localization of granulopoiesis that occurs in localized compartments both within the bone marrow and in LTBM cultures (150). Mature granulocytes were found to adhere less strongly to hemonectin, and this loss of attachment may be the signal that allows mature granulocytes to enter the circulation. A similar mechanism has already been documented for the release of mature erythrocytes into the circulation (151). As these cells mature, expression of the fibronectin receptor is lost such that the maturing erythrocytes become nonadherent and are able to leave the marrow. In addition to its role as an anchorage protein, fibronectin also increases the sensitivity of immature erythrocytes to erythropoietin, a major erythroid growth factor (152). Erythropoiesis proceeds in isolated islets surrounding an associated macrophage nurse cell, which produces the necessary erythropoietin for the development of erythrocytes (153). The role of adipocytes that form in these cultures is unclear. In vivo, the large presence of fat cells in marrow (yellow marrow) corresponds to low hematopoietic activity as compared with mar-

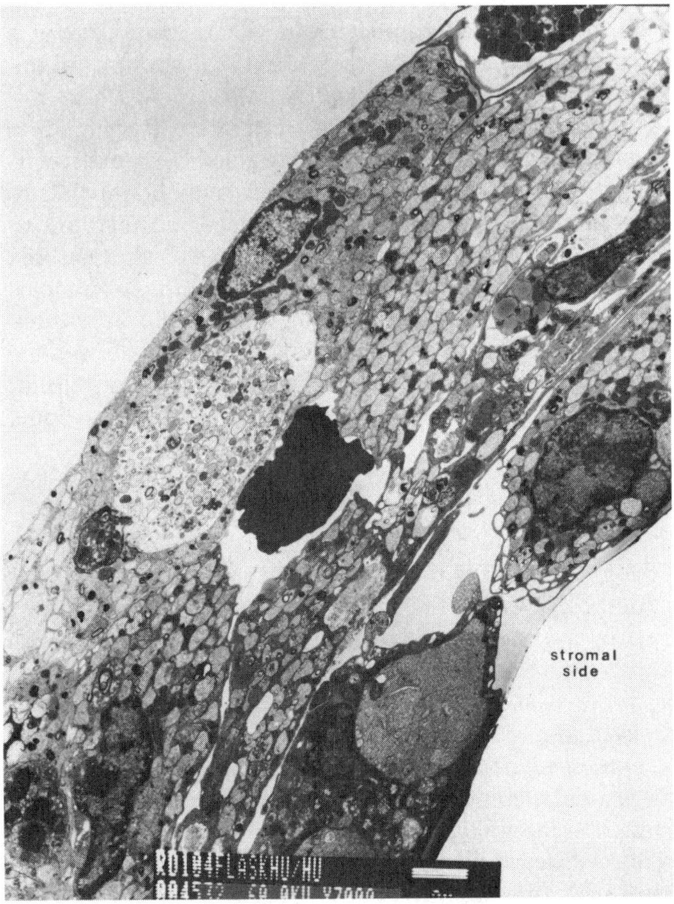


**Figure 7** Cross-section electron micrograph of the stromal layer of a murine long-term bone marrow culture. The stromal side, which had been attached to the substratum before processing, is indicated. The layer is several cells deep and adjacent cells exhibit very close membrane-membrane interactions. These direct cell-cell attachment interactions are thought to be important in the regulation of hematopoiesis. (Courtesy of M. R. Koller and M. E. Hunter.)

row containing fewer fat cells (red marrow) (154). In contrast, most investigators report that the formation of adipocytes in vitro is required for a stroma that will support hematopoiesis (140,145,155–157), while only a few report that adipocytes are not important for LTBM culture (158,159).

The role of the different stromal-hematopoietic cell interactions requires further study in order to understand what the hematopoietic cells are receiving from the stromal cells. Once this has been done, it may be possible to eliminate the stromal cells and simply grow homogeneous populations of hematopoietic stem cells by adding the required factors. Until then, any attempt to scale-up these bone marrow cultures will require some novel engineering allowing adequate feeding and mixing of the cultures without disrupting the stromal-hematopoietic cell adhesion interaction. The ideas presented here for the tissue engineering of in vitro bone marrow cultures also apply to the culture of other heterogeneous systems such as bone, cartilage, and liver.

## 7 SUMMARY

We have reviewed the general forces through which cells interact with substrata in their first nonspecific contact. The complex, fast-emerging biology of specific cell adhesion and the structure of the extracellular matrix were reviewed in substantial detail, and the most updated conceptual model of biological cell adhesion was assembled from past efforts and new literature data. The chemistries of the various possible substrata for cell adhesion have been reviewed extensively in the past, and here only a brief summary was presented, with particular emphasis on the materials for traditional and porous microcarriers. The fascinating molecular and cellular implications of cell adhesion were reviewed in detail to establish that cell adhesion and the extracellular matrix provide more than structural support for the cells and their assemblies, and that in fact they constitute fundamental regulators of cell function, metabolism, and differentiation. We reviewed the fluid-mechanical mechanisms of cell damage in microcarrier systems and provided experimental evidence for the importance of the cell-adhesion quality in the ability of cells to withstand fluid forces in bioreactors. We provided evidence that the interplay of cell adhesion and fluid forces is likely to produce cell responses more complex than that of simple life and death, and we suggested that such responses are awaiting investigation and exploration for new applications and culturing possibilities. We also reviewed the experimental evidence on the importance of cell adhesion in cell and microcarrier aggregation and discussed the implications of such aggregation on the culturing environment and the operation of bioreactors. Finally, we discussed the possible implications of cell adhesion as it relates to the developing field of tissue engineering, using the example of bone marrow culture, which involves a large variety of cells and constitutes one of the most complex cell culture systems.

## ACKNOWLEDGMENTS

This research was supported in part by the National Science Foundation (USA) under Grant EET-8896100 (Presidential Young Investigator Award to ETP) and by matching funds from the Monsanto Corporation and the Eli Lilly Company.

## REFERENCES

1. Edelman, G.M. (1989). Topobiology, *Sci. Am.*, *260*: 76.
2. Oster, G. (1989). Cell motility and tissue morphogenesis, in *Cell Shape: Determinants, Regulation, and Regulatory Role* (W.D. Stein and F. Bronner, eds.), Academic Press, New York, p. 33.
3. Dembo, M., and Bell, G.I. (1987). The thermodynamics of cell adhesion. *Curr. Top. Memb. Transp.*, *29*: 71.
4. Sherbet, G.V., Lakshmi, M.S., and Rao, K.V. (1972). Characterisation of ionogenic groups and estimation of the net negative electric charge on the surface of cells using natural pH gradients, *Exp. Cell Res.*, *70*: 113.
5. Barngrover, D. (1986). Substrata for anchorage-dependent cells, in *Mammalian Cell Technology* (W.G. Thilly, ed.), Butterworths, Boston, p. 133.
6. Bongrand, P., and Bell, G.I. (1984). Cell-cell adhesion: parameters and possible mechanisms, in *Cell Surface Dynamics: Concepts and Models* (A. Perelson, C. DeLisi, and F. Wiegel, eds.), Marcel Dekker, New York, Chapter 14.
7. Parsegian, V.A. (1973). Long-range physical forces in the biological milieu, *Annu. Rev. Biophys. Bioeng.*, *2*: 221.
8. Aplin, J.D., and Hughes, R.C. (1981). Cell adhesion on model substrata: threshold effects and receptor modulation, *J. Cell Sci.*, *50*: 89.
9. Napper, D.H. (1977). Steric stabilization, *J. Colloid Interface Sci.*, *58*: 390.
10. Grinnell, F. (1976). The serum dependence of baby hamster kidney cell attachment to a substratum, *Exp. Cell Res.*, *97*:265.
11. Hannan, G.N., and Reilly, W. (1988). Adsorption from fetal calf serum of collagen-like proteins which bind fibronectin and promote cell attachment, *Exp. Cell Res.*, *178*: 343.
12. Pena, S.G.J., and Hughes, R.C. (1978). Fibronectin-plasma membrane interactions in the adhesion and spreading of hamster fibroblasts, *Nature*, *276*: 80.
13. Hewitt, A.T., Kleinman, H.K., Pennypacker, J.P., and Martin, G.R. (1980). Identification of an adhesion factor for chondrocytes, *Proc. Natl. Acad. Sci. USA*, *77*: 385.
14. Couchman, J.R., Höök, M., Rees, D.A., and Timpl, R. (1983). Adhesion,

growth, and matrix production by fibroblasts on laminin substrates, *J. Cell Biol.*, *96*: 177.

15. Barnes, D.W., Silnutzer, J., See, C., and Shaffer, M. (1983). Characterization of human serum spreading factor with monoclonal antibody, *Proc. Natl. Acad. Sci. USA*, *80*: 1362.
16. Lazarides, E., and Burridge, K. (1975). α-Actinin: immunofluorescent localization of a muscle structural protein in nonmuscle cells, *Cell*, *6*: 289.
17. Geiger, B. (1979). A 130K protein from chicken gizzard: its localization at the terminal of microfilament bundles in cultured chicken cells, *Cell*, *18*: 193.
18. Burridge, K., and Connell, L. (1983). A new protein of adhesion plaques and ruffling membranes, *J. Cell Biol.*, *97*: 359.
19. Chen, W.-T., and Singer, S.J. (1982). Immunoelectron microscopic studies of the sites of cell-substratum and cell-cell contacts in cultured fibroblasts, *J. Cell Biol.*, *95*: 205.
20. Bershadsky, A.D., Tint, I.S., Neyfakh, A.A., and Vasiliev, J.M. (1985). Focal contacts of normal and RSV-transformed quail cells, *Exp. Cell. Res.*, *158*: 433.
21. Sanger, J.W., Sanger, J.M., and Jockusch, B.M. (1983). Differences in the stress fibers between fibroblasts and epithelial cells, *J. Cell Biol.*, *96*: 961.
22. Burridge, K., Fath, K., Kelly, T., Nuckolls, G., and Turner, C. (1988). Focal adhesions: transmembrane junctions between the extracellular matrix and the cytoskeleton, *Annu. Rev. Cell Biol.*, *4*: 487.
23. Heath, J.P., and Dunn, G.A. (1978). Cell to substratum contacts of chick fibroblasts and their relation to the microfilament system. A correlated interference-reflexion and high-voltage electron-microscope study. *J. Cell Sci.*, *29*: 197.
24. Small, J.V., Isenberg, G., and Celis, J.E. (1978). Polarity of actin at the leading edge of cultured cells, *Nature*, *272*: 638.
25. Molony, L., McCaslin, D., Abernethy, J., Paschal, B., and Burridge, K. (1987). Purification and characterization of talin from chicken gizzard smooth muscle, *J. Biol. Chem.*, *262*: 7790.
26. Wilkins, J.A., Chen, K.Y., and Lin, S. (1983). Detection of high molecular weight vinculin binding proteins in muscle and nonmuscle tissues with an electroblot-overlay technique, *Biochem. Biophys. Res. Commun.*, *116*: 1026.
27. Burridge, K., and Mangeat, P. (1984). An interaction between vinculin and talin, *Nature*, *308*: 744.
28. Bretscher, A. (1981). Fimbrin is a cytoskeletal protein that cross-links F-actin in vitro, *Proc. Natl. Acad. Sci. USA*, *78*: 6849.
29. Bennet, J.P., Zaner, K.S., and Stossel, T.P. (1984). Isolation and some properties of macrophage α-actinin: evidence that it is not an actin gelling protein, *Biochemistry*, *23*: 5081.
30. Hayman, E.G., Pierschbacher, M.D., Suzuki, S., and Ruoslahti, E. (1985). Vitronectin—a major cell attachment-promoting protein in fetal bovine serum, *Exp. Cell. Res.*, *160*: 245.
31. Avnur, Z., and Geiger, B. (1981). The removal of extracellular fibronectin from areas of cell-substrate contact, *Cell*, *25*: 121.

32. Burridge, K., and Fath, K. (1989). Focal contacts: transmembrane links between the extracellular matrix and the cytoskeleton, *BioEssays*, *10*: 104.
33. Ruoslahti, E. (1988). Fibronectin and its receptors, *Annu. Rev. Biochem.*, *57*: 375.
34. Ruoslahti, E. (1988). Structure and biology of proteoglycans, *Annu. Rev. Cell Biol.*, *4*: 229.
35. Ruoslahti, E., and Pierschbacher, M.D. (1987). New perspectives in cell adhesion, *Science*, *238*: 491.
36. McDonald, J.A. (1989). Receptors for extracellular matrix components, *Am. J. Physiol.*, *257*: L331.
37. Argraves, G.S., Dickerson, K., Burgess, W.H., and Ruoslahti, E. (1989). Fibulin, a novel protein that interacts with the fibronectin receptor beta subunit cytoplasmic domain, *Cell*, *58*: 623.
38. Horwitz, A., Duggan, K., Buck, C., Beckerle, M.C., and Burridge, K. (1986). Interactions of plasma membrane fibronectin receptor with talin—a transmembrane linkage, *Nature*, *320*: 531.
39. Barsky, S.H., Rao, C.N., Williams, J.E., and Liotta, L.A. (1984). Laminin molecular domains which alter metastasis in a murine model, *J. Clin. Invest.*, *74*: 843.
40. Clegg, D.O., Helder, J.C., Hann, B.C., Hall, D.E., and Reichardt, L.F. (1988). Amino acid sequence and distribution of mRNA encoding a major skeletal muscle laminin binding protein: an extracellular matrix associated protein with an unusual COOH-terminal polyaspartate domain, *J. Cell Biol.*, *107*: 699.
41. Pfaffle, M., Ruggiero, F., Hofmann, H., Fernandez, M.P., Selmin, O., Yamada, Y., Garrone, R., and von der Mark, K. (1988). Biosynthesis, secretion and extracellular localization of anchorin CII, a collagen-binding protein of the calpactin family, *EMBO J.*, *7*: 2335.
42. Edelman, G.M. (1987). CAM's and Ig's: cell adhesion and the evolutionary origins of immunity, *Immunol. Rev.*, *100*: 11.
43. Kleinman, H.K., Klebe, R.J., and Martin, G.R. (1981). Role of collagenous matrices in the adhesion and growth of cells, *J. Cell Biol.*, *88*: 473.
44. Rojkind, M., Gatmaitan, Z., Mackensen, S., Giambrone, M-A., Ponce, P., and Reid, L.M. (1980). Connective tissue biomatrix: its isolation and utilization for long-term cultures of normal rat hepatocytes, *J. Cell Biol.*, *87*: 255.
45. Rappaport, C., Poole, J.P., and Rappaport, H.P. (1960). Studies on properties of surfaces required for growth of mammalian cells in synthetic medium, *Exp. Cell Res.*, *20*: 465.
46. Panina, G.F. (1985). Monolayer growth systems: multiple processes, in *Animal Cell Biotechnology*, Vol. 1 (R.E. Spier and J.B. Griffiths, eds.), Academic Press, London, p. 211.
47. Spier, R.E. (1980). Recent developments in the large scale cultivation of animal cells in monolayers, *Adv. Biochem. Eng.*, *14*: 119.
48. Amstein, C.F., and Hartman, P.A. (1975). Adaptation of plastic surfaces for tissue culture on glow discharge, *J. Clin. Microbiol.*, *2*: 46.

49. Grinnell, F., Tobleman, M.Q., and Hackenbrock, C.R. (1972). Studies on cell adhesion. II. Adhesion of cells to surfaces of diverse chemical composition and inhibition of adhesion by sulphydryl binding reagents, *Arch. Biochem. Biophys.*, *153*: 193.
50. Jensen, M.D. (1981). Production of anchorage-dependent cells—problems and their possible solutions, *Biotechnol. Bioeng.*, *23*: 2703.
51. Harris, A. (1973). Behavior of cultured cells on substrata of variable adhesiveness, *Exp. Cell Res.*, *77*: 285.
52. Levine, D.W., Wang, D.I.C., and Thilly, W.G. (1979). Optimization of growth surface parameters in microcarrier cell culture, *Biotechnol. Bioeng.*, *21*: 821.
53. McKeehan, W.L., McKeehan, K.A., and Ham, R.G. (1981). The use of low-temperature subculturing and culture surfaces coated with basic polymers to reduce the requirement for serum macromolecules. In *The Growth Requirements of Vertebrate Cells In Vitro* (C. Waymouth, R.G. Itam, and P.G. Chapple, eds.), Cambridge University Press, Cambridge, p. 118.
54. Ingber, D.E., and Folkman, J. (1989). How does extracellular matrix control capillary morphogenesis? *Cell*, *58*: 803.
55. van Wezel, A.L. (1985). Monolayer growth systems: homogeneous unit processes, in *Animal Cell Biotechnology*, Vol. 1 (R.E. Spier and J.B. Griffiths, eds.), Academic Press, London, p. 265.
56. Sharp, L.H., and Schonhorn, H. (1964). In *Contact Angle, Wettability, and Adhesive Joints* (R.F. Gould, ed.), American Chemical Society, Washington, D.C., p. 189.
57. Maroudas, N.G. (1977). Sulphonated polystyrene as an optimal substratum for the adhesion and spreading of mesenchymal cells in monovalent and divalent saline solutions, *J. Cell. Physiol.*, *90*: 511.
58. Baier, R.E. (1972). The role of surface energy in thrombogenesis, *Bull. NY Acad. Med.*, *48*: 257.
59. Vroman, L. (1967). Surface activity in blood coagulation, in *Blood Clotting Enzymology* (W.H. Seegers, ed.), Academic Press, New York, p. 279.
60. Shahar, A., and Reuveny, S. (1987). Nerve and muscle cells in microcarriers in culture, *Adv. Biochem. Eng. Biotechnol.*, *34*: 33.
61. Borys, M.C. (1990). Effect of microcarrier culture conditions on recombinant protein expression by CHO cells, M.S. thesis, Northwestern University, Evanston, IL.
62. Koller, M.R., Mehrotra, Y., Unverzagt, K.L., Papoutsakis, E.T., and Maples, P.B. (1990). Production of recombinant murine IL-3 using microcarrier cultures of transfected COS 7 cells. Paper BIOT 171, National Meeting of the American Chemical Society, Washington, D.C.
63. Stoker, M.G.P., and Rubin, H. (1967). Density dependent inhibition of cell growth in culture, *Nature*, *215*: 171.
64. MacPherson, I., and Montagnier, L. (1964). Agar suspension culture for the selective assay of cells transformed by polyoma virus, *Virology*, *23*: 291.
65. Folkman, J., and Moscona, A. (1978). Role of cell shape in growth control, *Nature*, *273*: 345.

66. Folkman, J., and Greenspan, H.P. (1975). Influence of geometry on control of cell growth, *Biochim. Biophys. Acta, 417*: 217.
67. Dulbecco, R., and Stoker, M.G.P. (1970). Conditions determining initiation of DNA synthesis in 3T3 cells, *Proc. Natl. Acad. Sci. USA, 66*: 204.
68. Otsuka, H., and Moskowitz, M. (1975). Arrest of 3T3 cells in G1 phase in suspension culture, *J. Cell. Physiol., 87*: 213.
69. Benecke, B.-J., Ben-Ze'ev, A., and Penman, S. (1978). The control of mRNA production, translation and turnover in suspended and reattached anchorage-dependent fibroblasts, *Cell, 14*: 931.
70. Farmer, S.R., Ben-Ze'ev, A., Benecke, B-J., and Penman, S. (1978). Altered translatability of messenger RNA from suspended anchorage-dependent fibroblasts: reversal upon cell attachment to a surface, *Cell, 15*: 627.
71. Benecke, B-J., Ben-Ze'ev, A., and Penman, S. (1980). The regulation of RNA metabolism in suspended and reattached anchorage-dependent 3T6 fibroblasts, *J. Cell Physiol., 103*: 247.
72. Farmer, S.R., Wan, K.M., Ben-Ze'ev, A., and Penman, S. (1983). Regulation of actin mRNA levels and translation responds to changes in cell configuration, *Mol. Cell. Biol., 3*: 182.
73. Reiter, T., Penman, S., and Capco, D.G. (1985). Shape-dependent regulation of cytoskeletal protein synthesis in anchorage-dependent and anchorage-independent cells, *J. Cell Sci., 76*: 17.
74. Cervera, M., Dreyfuss, G., and Penman, S. (1981). Messenger RNA is translated when associated with the cytoskeletal framework in normal and VSV-infected HeLa cells, *Cell, 23*: 113.
75. Dike, L.E., and Farmer, S.R. (1988). Cell adhesion induces expression of growth-associated genes in suspension-arrested fibroblasts, *Proc. Natl. Acad. Sci. USA, 85*: 6792.
76. O'Neill, C.H., Riddle, P.N., and Jordan, P.W. (1979). The relation between surface area and anchorage dependence of growth in hamster and mouse fibroblasts, *Cell, 16*: 909.
77. Ingber, D.E., Madri, J.A., and Folkman, J. (1987). Endothelial growth factors and extracellular matrix regulates DNA synthesis through modulation of cell and nuclear expansion, *In Vitro Cell Dev. Biol., 23*: 387.
78. Hirst, R., Horwitz, A., Buck, C., and Rohrschneider, L. (1986). Phosphorylation of the fibronectin receptor complex in cells transformed by oncogenes that encode tyrosine kinases, *Proc. Natl. Acad. Sci. USA, 83*: 6470.
79. Lee, D.C., Rose, T.M., Webb, N.R., and Todaro, G.J. (1985). Cloning and sequence analysis of a cDNA for rat transforming growth factor-$\alpha$, *Nature, 313*: 489.
80. Anklesaria, P., Teixido, J., Laiho, M., Pierce, J.H., Greenberger, J.S., and Massague, J. (1990). Cell-cell adhesion mediated by binding of membrane-anchored transforming growth factor-$\alpha$ to epidermal growth factor receptors promotes cell proliferation, *Proc. Natl. Acad. Sci. USA, 87*: 3289.
81. Watt, F.M., Jordan, P.W., and O'Neill, C.H. (1988). Cell shape controls terminal differentiation of human epidermal keratinocytes, *Proc. Natl. Acad. Sci. USA, 85*: 5576.

82. Watt, F.M. (1986). The extracellular matrix and cell shape, *TIBS*, *11*: 482.
83. West, C.M., DeWeerd, H., Dowdy, K., and DeLaPaz, A. (1984). A specificity for cellular fibronectin in its effect on cultured chondroblasts, *Differentiation*, *27*: 67.
84. Gospodarowicz, D., Greenburg, G., and Birdwell, C.R. (1978). Determination of cellular shape by the extracellular matrix and its correlation with the control of cellular growth, *Cancer Res.*, *38*: 4155.
85. Grant, D.S., Tashiro, K-I., Segui-Real, B., Yamada, Y., Martin, G.R., and Kleinman, H.K. (1989). Two different laminin domains mediate the differentiation of human endothelial cells into capillary-like structures in vitro, *Cell*, *58*: 933.
86. Madri, J.A., Pratt, B.M., and Tucker, A.M. (1988). Phenotypic modulation of endothelial cells by transforming growth factor-β depends upon the composition and organization of the extracellular matrix, *J. Cell Biol.*, *106*: 1375.
87. Buck, C., and Horwitz, A.F. (1987). Cell surface receptors for extracellular matrix molecules, *Annu. Rev. Cell Biol.*, *3*: 179.
88. Breuer, D., and Wagener, C. (1989). Activation of the phosphatidylinositol cycle in spreading cells, *Exp. Cell Res.*, *182*: 659.
89. Lansman, J.B., Hallam, T.J., and Rink, R.J. (1987). Single stretch-activated ion channels in vascular endothelial cells as mechanotransducers, *Nature*, *325*: 811.
90. Dennerll, T.J., Joshi, H.C., Stell, V.L., Buxbaum, R.E., and Heidemann, S.R. (1988). Tension and compression in the cytoskeleton of PC-12 neurites II: quantitative measurements, *J. Cell Biol.*, *107*: 665.
91. Tomasek, J.J., and Hay, E.D. (1984). Analysis of the role of microfilaments and microtubules in acquisition of bipolarity and elongation of fibroblasts in hydrated collagen gels, *J. Cell Biol.*, *99*: 536.
92. Miller, A.O.A., Menozzi, F.D., and Dubois, D. (1989). Microbeads and anchorage-dependent eukaryotic cells: the beginning of a new era in biotechnology, *Adv. Biochem. Eng. Biotechnol.*, *39*: 73.
93. Griffiths, B. (1990). Advances in animal cell immobilization technology, in *Animal Cell Biotechnology*, Vol. 4 (R.E. Spier and J.B. Griffiths, eds.), Academic Press, New York, p. 149.
94. Eskin, S.G., Ives, C.L., McIntire, L., and Navarro, L.T. (1984). Response of cultured endothelial cells to steady flow, *Microvasc. Res.*, *28*: 87.
95. Ives, C.L., Eskin, S.G., and McIntire, L. (1986). Mechanical effects on endothelial cell morphology: in vitro assessment, *In Vitro Cell Dev. Biol.*, *22*: 500.
96. Ives, C.L., Eskin, S.G., McIntire, L., and DeBakey, M.E. (1983). The importance of cell origin and substrate in the kinetics of endothelial cell alignment in response to steady flow, *Trans. Am. Soc. Artif. Org.*, *29*: 269.
97. Levesque, M.J., and Nerem, R.M. (1985). The elongation and orientation of cultured endothelial cells in response to shear stress, *J. Biomech. Eng.*, *107*: 341.
98. Levesque, M.J., Sprague, E.A., Schwartz, C.J., and Nerem, R.M. (1989).

The influence of shear stress on cultured vascular endothelial cells: the stress response of an anchorage-dependent mammalian cell, *Biotechnol. Prog.*, *5*: 1.

99. Stathopoulos, N.A., and Hellums, J.D. (1985). Shear stress effects on human embryonic kidney cells in vitro, *Biotechnol. Bioeng.*, *27*: 1021.
100. Franke, R.P., Graefe, M., Schnittler, H., Seiffge, D., and Dreckhahn, D. (1984). Introduction of human vascular endothelial stress fibers by fluid shear stress, *Nature*, *307*: 648.
101. Wechezak, A.R., Viggers, R.F., and Sauvage, L.R. (1985). Fibronectin and F-actin redistribution in cultured endothelial cells exposed to shear stress, *Lab. Invest.*, *53*: 639.
102. Spier, R.E., Crouch, C.E., and Fowler, H. (1987). A study of the magnitude of the physical forces involved in the interaction of animal cells to their supporting substrata, *Dev. Biol. Stand.*, *66*: 255.
103. Crouch, C.E., Fowler, H.W., and Spier, R.E. (1985). The adhesion of animal cells to surfaces: the measurement of critical surface shear stress permitting attachment or causing detachment, *J. Chem. Tech. Biotechnol.*, *35B*: 273.
104. Frangos, J.A., Eskin, S.G., McIntire, L.V., and Ives, C.L. (1985). Flow effects of prostacyclin production by cultured human endothelial cells, *Science*, *227*: 1477.
105. McIntire, L.V., Frangos, J.A., Rhee, B.G., Eskin, S.G., and Hall, E.R. (1987). The effect of fluid mechanical stress on cellular arachidonic acid metabolism, *Ann. NY Acad. Sci.*, *516*: 513.
106. Bhagyalakshmi, A., and Frangos, J.A. (1989). Mechanism of shear-induced prostacyclin production in endothelial cells, *Biochem. Biophys. Res. Commun.*, *158*: 31.
107. Diamond, S.L., Eskin, S.G., and McIntire, L.V. (1989). Fluid flow stimulates tissue plasminogen activator secretion by cultured human endothelial cells, *Science*, *243*: 1483.
108. Diamond, S.L., Sharefkin, J.B., Dieffenbach, C., Frasier-Scott, K., McIntire, L.V., and Eskin, S.G. (1990). Tissue plasminogen activator messenger RNA levels increase in cultured human endothelial cells exposed to laminar shear stress, *J. Cell Physiol.*, *143*: 364.
109. Ando, J., Komatsuda, T., and Kamiya, A. (1988). Cytoplasmic calcium response to fluid shear stress in cultured vascular endothelial cells, *In Vitro Cell Dev. Biol.*, *24*: 871.
110. Gupte, A., and Frangos, J.A. (1990). Effects of flow on the synthesis and release of fibronectin by endothelial cells, *In Vitro Cell. Dev. Biol.*, *26*: 57.
111. Davies, P.F., Remuzzi, A., Gordon, E.J., Dewey, Jr., C.F., and Gimbrone, M.A., Jr. (1986). Turbulent fluid shear stress induces vascular endothelial cell turnover in vitro, *Proc. Natl. Acad. Sci. USA*, *83*: 2114.
112. Ben-Ze'ev, A. (1989). Cell shape and cell contacts: molecular approaches to cytoskeleton expression, in *Cell Shape: Determinants, Regulation, and Regulatory Role* (W.D. Stein and F. Bronner, eds.), Academic Press, New York, p. 95.
113. Pesciotta-Peters, D.M., and Mosher, D.F. (1989). Fibronectin as a trans-

ducer of tension, in *Cell Shape: Determinants, Regulation, and Regulatory Role* (W.D. Stein and F. Bronner, eds.), Academic Press, New York, p. 121.

114. Hay, E.D., and Svoboda, K.K. (1989). Extracellular matrix interaction with the cytoskeleton, in *Cell Shape: Determinants, Regulation, and Regulatory Role* (W.D. Stein and F. Bronner, eds.), Academic Press, New York, p. 147.
115. Farmer, S.R., and Dike, L.E. (1989). Cell shape and growth control: role of cytoskeleton-extracellular matrix interactions, in *Cell Shape: Determinants, Regulation, and Regulatory Role* (W.D. Stein and F. Bronner, eds.), Academic Press, New York, p. 173.
116. Cherry, R.S., and Papoutsakis, E.T. (1986). Hydrodynamic effects on cells in agitated tissue culture reactors, *Bioproc. Eng.*, *1*: 29.
117. Cherry, R.S., and Papoutsakis, E.T. (1990). Understanding and controlling injury of animal cells in bioreactors, in *Animal Cell Biotechnology*, Vol. 4 (R.E. Spier and J.B. Griffiths, eds.), Academic Press, New York, p. 72.
118. Cherry, R.S., and Papoutsakis, E.T. (1988). Physical mechanisms of cell damage in microcarrier cell culture bioreactors, *Biotechnol. Bioeng.*, *32*: 1001.
119. Cherry, R.S., and Papoutsakis, E.T. (1989). Growth and death rates of bovine embryonic kidney cells in turbulent microcarrier bioreactors, *Bioproc. Eng.*, *4*: 81.
120. Croughan, M.S., Hamel, J-F.P., and Wang, D.I.C. (1988). Effects of microcarrier concentration in animal cell culture, *Biotechnol. Bioeng.*, *32*: 975.
121. Croughan, M.S., Hamel, J-F.P., and Wang, D.I.C. (1987). Hydrodynamic effects on animal cells grown in microcarrier cultures, *Biotechnol. Bioeng.*, *29*: 130.
122. Croughan, M.S., Sayre, E.S., and Wang, D.I.C. (1989). Viscous reduction of turbulent damage in animal cell culture, *Biotechnol. Bioeng.*, *33*: 862.
123. Lakhotia, S., and Papoutsakis, E.T. (1992). Agitation induced cell injury in microcarrier cultures. The protective effect of viscosity is agitation-intensity dependent: experiments and theory, *Biotechnol. Bioeng., 39*: 95.
124. Sinskey, A.J., Fleischaker, R.J., Tyo, M.A., Giard, D.J., and Wang, D.I.C. (1981). Production of cell-derived products: virus and interferon, *Ann. NY Acad. Sci.*, *369*: 47.
125. Mered, B., Albrecht, P., and Hopps, H.E. (1980). Cell growth optimization in microcarrier culture, *In Vitro*, *16*: 859.
126. Varani, J., Dame, M., Veals, T.F., and Wass, J.A. (1983). Growth of three established cell lines on glass microcarriers, *Biotechnol. Bioeng.*, *25*: 1359.
127. Scattergood, E.M., Schlaback, A.J., McAleer, W.J., and Hilleman, M.R. (1980). Scale-up of chick cell growth on microcarriers in fermentors for vaccine production, *Ann. NY Acad. Sci.*, *413*: 332.
128. Clark, S.C., and Kamen, R. (1987). The human hematopoietic colony-stimulating factors, *Science*, *236*: 1229.
129. Freundlich, N. (1989). Patient, heal thyself—with lab-grown cells, *Business Week*, March 20: 148.
130. Golde, D.W., and Gasson, J.C. (1988). Hormones that stimulate the growth of blood cells, *Sci. Am.*, *259*: 62.

131. Cronkite, E.P., and Feinendegen, L.E. (1976). Notions about human stem cells, *Blood Cells, 2*: 263.
132. Dexter, T.M., and Spooncer, E. (1987). Growth and differentiation in the hemopoietic system, *Annu. Rev. Cell Biol., 3*: 423.
133. Dexter, T.M., Ponting, I.L.O., Roberts, R.A., Spooncer, E., Heyworth, C., and Gallagher, J.T. (1988). Growth and differentiation of hematopoietic stem cells, in *Cell Physiology of Blood* (R.B. Gunn and J.C. Parker, eds.), Rockefeller University Press, New York, p. 19.
134. Metcalf, D. (1984). *The Hemopoietic Colony Stimulating Factors*, Elsevier, New York.
135. Lajtha, L.G. (1979). Stem cell concepts, *Differentiation, 14*: 23.
136. Till, J.E., and McCulloch, E.A. (1961). A direct measurement of the radiation sensitivity of normal mouse bone marrow cells, *Radiat. Res., 14*: 213.
137. Spangrude, G.J., Heimfeld, S., and Weissman, I.L. (1988). Purification and characterization of mouse hematopoietic stem cells, *Science, 241*: 58.
138. Boggs, D.R., Boggs, S.S., Saxe, D.F., Gress, L.A., and Carfield, D.R. (1982). Hematopoietic stem cells with high proliferation potential assay of their concentration in marrow by the frequency and duration of cure of $W/W^v$ mice, *J. Clin. Invest., 70*: 243.
139. Allen, T.D., and Dexter, T.M. (1984). The essential cells of the hemopoietic microenvironment, *Exp. Hematol., 12*: 517.
140. Kodama, H., Amagai, Y., Koyama, H., and Kasai, S. (1982). A new preadipose cell line derived from newborn mouse calvaria can promote the proliferation of pluripotent hemopoietic stem cells in vitro, *J. Cell. Physiol., 112*: 89.
141. Kodama, H., Sudo, H., Koyama, H., Kasai, S., and Yamamoto, S. (1984). In vitro hemopoiesis within a microenvironment created by MC3T3-G2/PA6 preadipocytes, *J. Cell. Physiol., 118*: 233.
142. Roberts, R.A., Spooncer, E., Parkinson, E.M., Lord, B.I., Allen, T.D., and Dexter, T.M. (1987). Metabolically inactive 3T3 cells can substitute for marrow stromal cells to promote the proliferation and development of multipotent haemopoietic stem cells, *J. Cell. Physiol., 132*: 203.
143. Spooncer, E., Heyworth, C.M., Dunn, A., and Dexter, T.M. (1986). Self-renewal and differentiation of interleukin-3-dependent multipotent stem cells are modulated by stromal cells and serum factors, *Differentiation, 31*: 111.
144. Williams, D.A., Ropsenblatt, M.F., Beier, D.R., and Cone, R.D. (1988). Generation of murine stromal cell lines supporting hematopoietic stem cell proliferation by use of recombinant retrovirus vectors encoding simian virus 40 large T antigen, *Mol. Cell. Biol., 8*: 3864.
145. Dexter, T.M., Allen, T.D., and Lajtha, L.G. (1977). Conditions controlling the proliferation of haemopoietic stem cells in vitro, *J. Cell. Physiol., 9*: 335.
146. Yamazaki, K., Roberts, R.A., Spooncer, E., Dexter, T.M., and Allen, T.D. (1989). Cellular interactions between 3T3 cells and interleukin-3-dependent multipotent haemopoietic cells: a model system for stromal-cell-mediated haemopoiesis, *J. Cell. Physiol., 139*: 301.
147. Roberts, R., Gallagher, J., Spooncer, E., Allen, T.D., Bloomfield, F., and

Dexter, T.M. (1988). Heparan sulfate bound growth factors: a mechanism for stromal cell mediated haemopoiesis, *Nature*, *332*: 376.

148. Gordon, M.Y., Riley, G.P., Watt, S.M., and Greaves, M.F. (1987). Compartmentalization of a haematopoietic growth factor (GM-CSF) by glycosaminoglycans in the bone marrow microenvironment, *Nature*, *326*: 403.
149. Berneman, Z.N., Chen, Z.-Z., Ramael, M., Van Poucke, K., Korthout, M., van Bockstaele, D.R., and Peetermans, M.E. (1989). A quantitative and dynamic study of endothelial cells and megakaryocytes in human long-term bone marrow cultures, *Leukemia*, *3*: 61.
150. Campbell, A.D., Long, M.W., and Wicha, M.S. (1987). Haemonectin, a bone marrow adhesion protein specific for cells of granulocyte lineage, *Nature*, *329*: 744.
151. Patel, V.P., Ciechanover, A., Platt, O., and Lodish, H.F. (1985). Mammalian reticulocytes lose adhesion to fibronectin during maturation to erythrocytres, *Proc. Natl. Acad. Sci. USA*, *82*: 440.
152. Weinstein, R., Riordan, M.A., Wenc, K., Zhou, M., and Dainiak, N. (1989). Dual role of fibronectin in hematopoietic differentiation, *Blood*, *73*: 111.
153. Crocker, P.R., and Gordon, S. (1987). Studies on the interaction between murine resident bone marrow macrophages and haematopoietic cells, *NATO ASI Series*, *H8*: 259.
154. Tavassoli, M., Maniatis, A., and Crosby, W.H. (1974). Induction of sustained hemopoiesis in fatty marrow, *Blood*, *43*: 33.
155. Fliedner, T.M., Calvo, W., Klinnert, V., Nothdurft, W., Prümmer, O., and Raghavachar, A. (1985). Bone marrow structure and its possible significance for hematopoietic cell renewal, *Ann. NY Acad. Sci.*, *459*: 73.
156. Itoh, K., Tezuka, H., Sakoda, H., Konno, M., Nagata, K., Uchiyama, T., Uchino, H., and Mori, K.J. (1989). Reproducible establishment of hemopoietic supportive stromal cell lines from murine bone marrow, *Exp. Hematol.*, *17*: 145.
157. Riley, G.P., and Gordon, M.Y. (1987). Characterization of cultured stromal layers derived from fetal and adult hemopoietic tissues, *Exp. Hematol.*, *15*: 78.
158. Greenberger, J.S., Fitzgerald, T.J., and Anklesaria, P. (1989). Recent studies of the hematopoietic microenvironment in long-term bone marrow cultures, *Immunol. Res.*, *8*: 226.
159. Whitlock, C.A., Tidmarsh, G.F., Müller-Sieburg, C., and Weissman, I.L. (1987). Bone marrow stromal cell lines with lymphopoietic activity express high levels of pre-B neoplasia-associated molecule, *Cell*, *48*: 1009.

# 4
# Surface Immobilization of Plant Cells

**Jean Archambault**
*University of Quebec at Trois-Rivières, Trois-Rivières, Quebec, Canada*

The immobilization of plant cells represents an interesting process alternative for the production of valuable phytochemicals, which has been barely explored because of the limited availability of efficient culture systems that can be scaled up. Most immobilization techniques developed for plant cells, involving entrapment in polysaccharide gels, preformed membranes, or foam particles, have serious drawbacks. Recently, the adhesive properties of plant cells were exploited in a surface immobilization technique that was successfully scaled up in laboratory-size bioreactors. The state of development of this last technology is reviewed with emphasis on immobilization processes, bioreactor development and experimentation, and physiological behavior of cultured plant cell biofilms.

## 1 INTRODUCTION

### 1.1 From Plant and Plant Products to Plant Cell Culture

Plants are constantly involved in our daily lives for nutritive, medicinal, or decorative purposes. Plant-derived complex chemicals of commercial interest are used principally by the pharmaceutical, cosmetic and food industries. In 1986, these natural products comprised 25% of the pharmaceutical presciptions in the United States, which accounted for 3–8 billion dollars in sales (1).

These unique phytochemicals belong to a vast class of compounds known as secondary metabolites. Primary metabolism involves the biochemical reactions that produce components essential for survival. Secondary metabolism procures compounds that serve no clear function to the plant, yet they derive from common precursors found in the primary metabolic pathways. The plant's propensity to produce secondary metabolites probably evolved from environmental pressures during the course of natural section. They emerged in plants generally at low yields ($< 1$–5% W/W) to serve as defense molecules against predators, microbial or fungal invasions, insect repellants/attractants, and so forth.

The conventional supply of these fine chemicals remains limited and uncertain. Whole plants require long cultivation periods (months to years), and they are susceptible to capricious environmental conditions. In addition, geographical and political restraints may limit their accessibility.

The complex structure of these compounds prevents their economical chemical synthesis, and the feasibility of the multienzymatic cloning ($> \sim 5$–15 enzymes) required for their biosynthesis by microorganisms remains to be demonstrated. The culture of plant tissues and cells in vitro has been suggested for the last 30–40 years as the most probable alternate and stable source of these products. The technique of plant cell culture is well known and described in a variety of books (2,3, and other refs.). It involves initially the dedifferentiation and subsequently the growth of plant cells obtained from differentiated viable excised tissues that are surface sterilized and placed on a solidified medium containing mainly growth regulators, carbohydrates, vitamins, inorganic salts, and a gelling agent. The resulting callus culture of plant cells is maintained by periodical ($\sim$3–6 weeks) subculturing onto similar solidified media. Suspension culture is achieved by transferring and growing callus cells into liquid media of similar formulations as used for callus cultures except for the gelling agent. Periodic ($\sim$1–3 weeks) subculturing ($\sim$10–30% V/V) ensures continued growth and maintenance of these suspension cultures. Under proper care, the resulting plant cell lines can generally be maintained for years, depending on the plant species under study. Information on the physiology of plant and plant cells can be obtained from the literature.

The complex phytochemicals of interest are often not produced in significant amounts, if at all, under normal growth conditions in these cultures. Various product induction schemes have been developed with some success to stimulate secondary metabolism, including mainly variation in growth regulators, nutritional starvation, osmotic and other stresses (elicitation), and organ culture.

## 1.2 Commercial Exploitation of Plant Cell Culture

No real commercial success has yet been reported for the production of valuable phytochemicals through plant cell culture (4,5). Means to induce the biosynthesis of desired products (secondary metabolites) have still not been devised (for example: vincristine, vinblastine, codeine, and morphine), or product yields and productivities are considered too low for economical production or cannot be scaled up satisfactorily (6). Still, plant cell culture may represent the only technological and economical route to the industrial production of some of these high-value compounds, such as ginkgolides (7), taxol (8), and so forth.

The well-publicized case of shikonin production is particularly revealing. High-product yields [up to 23% of the biomass dry weight (dw)] have been obtained in a large-scale (750 L) batch suspension process. The production rate was reported to be 5 kg shikonin per 3 weeks of culture for a total revenue of ~$20,000 per batch (4). Delays in approval of this product, limited market size [150 kg/year (4)], and the high operating cost of this culture system (≥$500–1000/day) may all have contributed to Mitsui's decision to stop this production (9).

Obviously, the market size (volume × price) must be sufficient to justify either low-volume expensive production runs or the economies of scale of larger processes and/or a multiple (plant) products business strategy. In addition, there is an urgent need for more integrated approaches to develop commercially viable plant-cell based bioprocesses. And the most critical components of these bioprocesses remain the basic culture technique and bioreactor and their interaction with the living plant cells.

## 1.3 The Scale-up of Plant Cell Culture

The metabolism and production capabilities of plant cells cultured in vitro are significantly affected by the type, scale, and physical environment of the culture system used (6,10–13). In fact, this is an essential part of the development of plant cell lines and product induction schemes, which is largely underestimated by many research groups involved in this field (14–16). A productive cell line that cannot grow and/or produce satisfactorily in a large-scale culture system is of no use. The proper culture technique, system, and conditions must be selected and tuned to the cell system and production behavior so that productivity can be maximized.

Culture conditions in shake flasks are highly variable, limiting [nutrient consumption, mixing, dissolved oxygen (DO), and carbon dioxide ($DCO_2$), pH etc.], and often far from optimized. Bioreactors, on the other hand, offer tremendous opportunities for monitoring and controlling the culture's physical and chemical environment and the cell's metabolism even at small scale ($\geq 2$ L). Typically, however, the performance of large-scale plant cell suspensions barely reaches that of shake flask cultures (17). Obviously, the versatility and potential of bioreactors have not yet been exploited for plant cells, mainly because of the limited availability of suitable culture, monitoring, and control systems and related knowledge. For example, the effect of mixing shear on the culture of high-density ($\geq$10–20 g dw/L) shear-sensitive and pseudoplastic plant cell suspensions (18–20) has not been properly exploited for bioreactor design and/or selection (5,11,21) until recently (22). This puts in doubt the usefulness of airlift, the preferred configuration, and conventional stirred tank bioreactors for (high density) plant cell suspension culture-based bioprocesses (13). Forced gassing is another characteristic of large-scale systems ($\geq 2$ L) whose effects on plant cell culture are yet to be fully understood (23–28).

### 1.4 Plant Cell Immobilization

Immobilization is a particularly attractive alternative for the large-scale culture of plant cells, especially when considering their natural tissue-forming tendency and the problems of suspension culturing. Obviously, growth and production kinetics as well as product release need to be matched to this type of culture system. The superior performance of an immobilization-based process over traditional batch (or continuous) suspension culture systems is expected to result from easier hydrodynamic handling and control of the culture and, mainly, in maintaining viable, productive, and product-releasing cells over a long production time. This allows amortizing the high cost of raising the inoculum of the main process bioreactor over a larger quantity of product and simplifying downstream processing. Previous studies have shown that immobilized plant cells can be more productive than suspended cells and can be maintained viable, productive, and induced to release their precious secondary metabolites over long periods of time (up to 6 months) under proper culture conditions limiting growth and stimulating production (29–33). Most techniques developed to immobilize plant cells involve their entrapment within membranous structures and have various limitations (31,34).

The culture of plant cells as a biofilm attached to a suitable surface may represent a more promising system than other culture techniques. This contribution addresses the state of development of this technology. In Sec. 2, the field of plant cell immobilization is reviewed with major emphasis on the surface immobilization of plant cells (SIPC). In Sec. 3, SIPC bioreactors development and experimentation are discussed. Finally, the use of this technology is considered within the context of plant-cell-based bioprocesses.

## 2 PLANT CELL IMMOBILIZATION TECHNOLOGY

### 2.1 Entrapment Techniques

In their review, Hulst and Tramper (34) identified entrapment in gels, in preformed membranes, and in polyurethane foam particles as the main techniques developed to immobilize plant cells. The first and most widely used approach involves mixing the cell suspension with a solution of a natural polysaccharide, adding a gelling agent, and forming hardened beads or sheets (35) of uniform dimensions ($\leq$3–5 mm) to prevent mass [oxygen (36)] transfer limitations and culture heterogeneity. In the other cases, plant cells are introduced into membrane devices [hollow fiber cartridge (37), sandwiched between flat membranes (38)] or mixed with polyurethane foam particles ($\sim 1\ cm^3$), into which they penetrate and grow.

These techniques have serious practical drawbacks. The large-scale ($\geq$10–1000 L) sterile and consistent production of a uniform population of small gel beads equally loaded with cells and hardened under mild conditions represents an interesting challenge, not to mention the design of a suitable culture system (31). Other limitations of the gel entrapment technique include biomass leakage and breakage upon growth and limited biomass concentration per unit volume of bioreactor (39). Furthermore, many of the polysaccharides used to immobilize plant cells have chemical structures similar to biotic elicitors (40), which may affect the metabolism of entrapped cells and, consequently, the short- and long-term performance of the immobilized culture. Immobilization in polyurethane foam particles is a simpler, milder technique. However, the basic entrapment of cells from a suspension is slow (7–24 days) (31,41), and proper scale-up, hydrodynamic handling, and mass transfer control of the resulting culture system are also interesting challenges (33). Finally, the usefulness of membrane devices for the large-scale culture of plant cells is doubtful in view of the biomass loading and mass transfer problems associated with this technique and the high cost and limited availability of equipment.

### 2.2 Surface Immobilization

The successful immobilization of plant cells onto solid surfaces of various configurations and their growth as a continuous biofilm fully bathed in the culture medium have been reported mainly by this author (42), Radvanyi et al. (43), and Kargi et al. (44). The technique described by the last group consists of immobilizing plant cells on gelatin or alginate beads in a packed column and on agar layers. The use of eliciting polysaccharides as the immobilization surface may affect the culture physiology as noted above. The scale-up potential of this technique, including the uniform inoculation and growth of the biofilm over the immobilizing surface under flow conditions, remains to be demonstrated and consequently will not be discussed further.

The main contribution of Dicosmo's group was the development of a thermodynamic model based on surface chemistry to explain the initial static interaction of plant cells with immobilizing surfaces (45). We concentrated our efforts on developing a technique, from the selection of suitable immobilizing materials (39) to the design and testing of efficient bioreactors that can be scaled up (46), to culture plant cell biofilms. In the following sections, the main aspects of this technology are reviewed.

### 2.2.1 Surface Immobilizing Materials

Obviously, suitable materials must be easily available and handled, sterilized, inert, nontoxic, mechanically stable, and biocompatible with plant cells. It may also be desirable that they not affect the cell's metabolism (e.g., polysaccharides). Results from the work of Dicosmo et al. suggested the use of low-energy hydrophobic surfaces ($\sigma_s$) that favored efficient early static attachment of plant cells (47) in view of their lower surface energy ($\sigma_c$ $\sim$40–70 $N.m^{-2}$) as compared to mammalian and microbial cells (47,48). However, the free energy of adhesion is significantly affected by various culture parameters more or less controllable, including the cell species and line under study, the growth stage of the inoculum, and the medium composition, which are difficult to quantify and model. In addition, this concept does not take into account the polysaccharide-generating capacity of cultured cells, the size distribution of inoculated cell aggregates, or the culture system geometry and hydrodynamics.

We found the micro- and macroconfiguration of the immobilizing material to be also of prime importance for the immobilization efficiency (ease, efficacy, and uniformity of cell attachment) of the culture system. Fibrous polymeric mats with large pores (80–130 $\mu$m) retained plant cells more efficiently than various porous plastic sheets and membranes [polypropylene, polyethylene, and polyvinylidene fluoride (39)]. A similar approach was reported by Facchini and Dicosmo (49). In addition, the ease of formability of the selected fibrous mat allowed its shaping into a volumetrically efficient [high-gross immobilizing area (A) per volume of culture system (V)] vertical spiral configuration (A/V $\sim$1.0–1.2 $cm^{-1}$). This matrix structure was placed into suitably designed bioreactors (see Fig. 1 and Sec. 3.1) whose immobilization efficiency was better than observed for SIPC flask cultures (see Sec. 3.2).

### 2.2.2 Surface Immobilization Processes

The attachment of plant cells and aggregates onto suitable surfaces in an actual (dynamic) culture system is a continuous process occurring mainly in three stages. At inoculation, the first interaction of the culture broth with

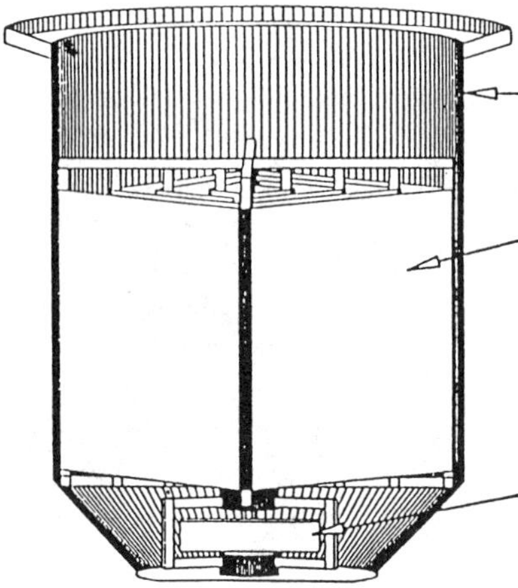


**Figure 1** Vertical square spiral immobilizing structure.

the configured immobilizing matrix is governed by the hydrodynamics and operating conditions of the bioreactor, which must ensure uniform circulation and distribution of cells and aggregates over the immobilizing surface to allow subsequent attachment phases to take place efficiently.

The second stage involves the initial attachment of the cells and aggregates to the surface as described by Facchini et al. (45) but under flow conditions. The cells do not seem to participate actively in this phase of immobilization. At least four factors were found to play key roles in promoting cell deposition and formation of first links with the immobilizing material within the matrix structure. Some characteristics of the inoculated suspension, including the cell species and line cultured and its polysaccharide-generating capacity, its growth stage, and aggregate size distribution, affect this early attachment. For example, *Catharanthus roseus* (fine suspensions) (46), *Ginkgo biloba* (mixed suspensions) (50), and *Tripterygium wilfordii* (fine suspensions) (51) cells immobilized faster (1–24 h) in bioreactors than *Vitis vinifera* (very fine suspensions) (unpublished results), *Glycine max* (small aggregate suspensions) (46), and various *Papaver* species (mixed suspensions) cultures (unpublished results). In this last case, however, high variability in immobilization performance (speed and strength of attachment) was observed between the different species studied.

Results from our first studies using flask cultures indicated that early stationary-phase (10-day-old) *C. roseus* cell suspensions immobilized more efficiently (60%) than younger (6–8-day-old: 40–55%) or older (14 days: 40–55%) inocula (39). This requirement was challenged in bioreactors and was found less critical, at least for *C. roseus* cells, underlining the importance of suitable culture system design (46). Best inoculum growth stage, biomass concentration, and volumetric ratio may need to be determined for each plant cell species and line to be immobilized. Obviously, large aggregate suspensions immobilized faster but less uniformly than fine suspension inocula. Subsequent growth, especially under controlled dissolved oxygen conditions (see Sec. 3.3.2), yielded relatively uniform biofilms.

Other factors inducing early immobilization of plant cells included low-flow and -shear mixing conditions and the macroconfiguration of the immobilizing surface, which helped trap and retain aggregates. Also, short-range, favorable free energy of adhesion conditions [use of low $\sigma_s$ (polymeric) material when $\sigma_c < \sigma_L$ (medium surface tension) or high $\sigma_s$ materials when $\sigma_c > \sigma_L$] was found by Facchini et al. (45) to promote early attachment of plant cells to polymeric surfaces. This energy balance is affected by factors similar to those noted above. For example, high $\sigma_c$ cell species (*Papaver somniferum*: $\sigma_c \sim 70$ N.m$^{-2}$ attached less efficiently to the polymeric surfaces tested ($\sigma_s \leq 67$ N.m$^{-2}$) while *C. roseus* cells ($\sigma_c \sim 53$ N.m$^{-2}$) displayed good adhesive properties (48). The effect of the inoculated cells' age could not be distinguished so clearly in these studies (45,48) since the experimental protocol prevented assessing together the effect of $\sigma_c$, the cell's polysaccharide-generating capacity, and the chemistry of the medium ($\sigma_L$). However, a peak of adhesion was observed for late exponential/early stationary phase cultures (45).

Once in contact with the immobilizing material, it appears that plant cells of studied species play an active role in promoting the attachment process (stage 3). They seem to adapt their shape, plasticizing at growth and rigidifying in place afterward, to the microconfiguration of the surface (39). In addition, there is conclusive evidence that cells from various species [*C. roseus, N. tabacum, G. max* (39), *H. lupulus, B. vulgaris, C. pubescens,* and *D. carota* (52,53)], generate, release, and/or secrete some mucilaginous film around themselves and between their cell wall and the immobilizing surface. This mucilage is suspected to be made of polysaccharides and/or proteins similar to those released by suspension-cultured plant cells and/or to the middle lamella intracellular pectic substances occurring in plants (54). These compounds may serve as a gap filler and/or gluing agent, strengthening the bonds between the first layer of cells and the immobilizing material and, subsequently, ensuring the cohesiveness of the growing biofilm.

The adhesion strength of surface immobilized *C. roseus, N. tabacum,* and *G. max* cell biofilms was assayed by submitting strips of biomass-loaded materials to two-phase flow in a vertical column (ID ~2.5 cm). The biofilms resisted stripping from the immobilizing material under liquid flow conditions of up to 22 L/min while air was sparged as a rate of 0.5 L/min.

This was confirmed in bioreactors whose operating conditions were much milder but sufficient for efficient mass transfer (46). However, the bond strength and biofilm cohesiveness appear to be plant cell species and/or biofilm growth stage dependent, as some *V. vinifera* and *Papaver* species immobilized cultures, although displaying satisfactory culture performance, resulted in less cohesive biofilms at harvesting than other species studied.

## 3 SIPC CULTURE ENGINEERING

### 3.1 Bioreactor Development

The attachment of cells of various plant species, including the highly sensitive *Papaver* cells, onto stationary fibrous mats occurred naturally and proved to be mild, efficient, and simple in flasks (39). The selected immobilizing material did not require surface treatment. In addition, it did not seem to affect the medium composition and chemistry or the cell's metabolism.

The scale-up of this culture technique into a suitable bioreactor required taking into account various factors. Results from flask cultures showed that cell attachment and biofilm growth were more efficient on fixed fibrous surfaces than on floating particles, which resulted in biomass attrition and abrasion. The matrix arrangement into a vertical spiral configuration provided for an easily and uniformly accessible high-immobilization area per unit volume of bioreactor.

The hydrodynamics of the bioreactor had to result in uniform low shear pumping, mixing, and distribution of the sensitive inoculated cells throughout the immobilizing structure with minimal foaming. Subsequently, the flow pattern of the medium had to permeate readily, rapidly, and uniformly the whole biomass-loaded matrix configuration without creating dead zones while ensuring sufficient mass transfer to and from the immobilized plant cell biofilm. Obviously, the bioreactor needed to provide for aeration and reliable sterile containment of the culture, as well as industrial scale-up potential.

The fixed vertical spiral-wound structure of the immobilizing material placed in a modified airlift bioreactor, with the riser tube located at the center of the spiral, offered a best-overall configuration meeting these requirements. A related small-scale (2 L) bioreactor (Fig. 1) was developed, with a similar immobilizing structure placed in a low-H/D (1.5) mechanically-stirred and air-sparged vessel (46).

The sizing of the bioreactor internals was governed primarily by four parameters. The theoretical maximum wet biofilm thickness for retained high viability depended on the oxygen requirement of the cells for best performance of the culture, on the oxygen diffusivity within the biofilm, and on the bulk dissolved oxygen (DO) concentration or the oxygen transfer capacity of the bioreactor to the biofilm. This proved not to be as simple as expected since the oxygen uptake rate (OUR) of cultured plant cells, including plant cell biofilms, was found to vary with culture conditions (see Sec. 3.3.2 and 3.3.3), with growth stage, and, most probably, within the biofilm. In addition, the metabolism of plant cell biofilms seemed to adapt to low, but nondetrimental, bulk DO concentration (see Sec. 3.3.2). An average thickness of 3 mm yielded highly viable immobilized cultures of high oxygen-demanding *C. roseus* cells (OUR ~0.4–0.9 mM/gdw/h) (28).

The W/D ratio of the plant cell biomass, the second important design parameter, is affected by many factors, including the cell species and line cultivated, the growth stage of the culture, and the medium osmotic pressure. Its value ranges generally from 10 to 50. This ratio, in combination with the immobilizing material and biofilm thicknesses and the spacing between adjacent biofilm layers (~0.4 cm) selected to allow for sufficient liquid circulation for mixing and mass transfer, determines the A/V and, consequently, the maximum biomass concentration that permits efficient operation of the bioreactor. Obviously, this calls for strict control over biofilm growth without hindering secondary metabolism and productivity, which remains a difficult physiological problem to be solved for each process case (30). Actual laboratory-scale (2–20 L) SIPC bioreactors were designed with total spacings between adjacent immobilizing surfaces of 1.0–1.2 cm and maximum A/V of 1.2 $cm^{-1}$ to yield a theoretical maximum immobilized biomass (W/D ~10) concentration of 36 g dw/L. This compares to a theoretical concentration of 60 g dw/L for a similar suspension culture assuming a 40% porosity of the wet biomass packed bed. As stated previously (39), the overall biomass concentration, calculated over the entire effective volume of the bioreactor, of plant cell immobilization culture systems will always be inferior to or barely reach that of suspension culture systems. This results from the volume of the immobilizing matrix. Concentrations of up to 23.5 g dw/L have already been achieved for *V. vinifera* immobilized cultures (W/D ~12.4) without overgrowth or bridging between adjacent biofilm layers and with the immobilizing structure loaded with wet biomass occupying 40% of the total culture volume.

In view of the low maximum volumetric oxygen consumption of typical surface-immobilized plant cells (*C. roseus* cells: ~2–3 mM/L/h), the modified airlift bioreactor with a low riser-to-downcomer area ratio of 0.03, equipped with an horizontal flow diverting plate [outside diameter of 1.2 × (riser tube ID)] located at a distance of 1 riser-tube diameter above the riser tube, was found to have sufficient oxygen transfer capacity ($k_La$ ~7–10$h^{-1}$) at reasonable aeration rates of 0.3–0.5 VVM for this type of culture. This range of aeration rates resulted in mixing times within the immobilizing structure of the 6-L bioreactor of ~70 s, which are faster than plant cell metabolic rates (~days), including OUR (≥300 s) (46). Higher or lower oxygen transfer rates and operation at constant DO concentration were achieved by manipulating the composition of the sparged gases without affecting the mixing efficiency of the culture system.

The high immobilization efficiency of the SIPC bioreactors resulted from the combination of the parameters discussed previously. This process was further improved by the low slenderness ratio (H/D ~0.9–1.8) of the spiral structure and by operating at low aeration rates (~0.2–0.4 VVM) and liquid level (to the top of the spiral structure) during the immobilization phase (≤24–48 h) of the culture.

## 3.2 SIPC Culture Systems Experimentation

The testing of the SIPC bioreactors revealed other interesting features of this immobilization technique. Larger-scale (2–20 L) cultures proved to be more representative of biofilm cultivation of plant cells and performed better, in terms of immobilization process efficiency and growth behavior, than SIPC flask cultures. Most cells (≥90%) from inoculum suspensions of all plant cell species cultivated in these bioreactors attached easily, readily (≤24–48 h), and with relative uniformity onto the immobilizing material. At the end of all cultures, 95–99% of the biomass present in the bioreactor was attached to the support, forming a continuous biofilm and leaving the medium free of cells. The unattached biomass remained in the limited foam layer of the cultures (46).

In all cases, suspension inocula of proper quality (high viability, good growth characteristics, of late exponential to early stationary growth phase, etc.), sufficient volume (10–30% V/V) and biomass content (to yield initial immobilized culture concentrations ≥1–2 g dw/L), and suitable bioreactor operating conditions (temperature, aeration rate, etc.) resulted in immobilized cultures consuming main extracellular nutrients (carbohydrates, ammonium, phosphate, and nitrate) and growing, on a dry biomass basis, at rates 50–100% of those of similar shake flask suspension cultures (28,39,46, 50,51). These growth rates may be improved upon further optimization of inoculating and operating conditions. However, all SIPC flask and bioreactor cultures resulted in significantly lower efficiencies of carbohydrate utilization for plant cell dry biomass formation (≤0.2–0.4 g dw/g CHO) than flask and bioreactor grown suspension cultures of the same plant cell species (0.5–0.6 g dw/g CHO). This suggests higher respiration rates and/or different carbon storage or utilization patterns, biomass elemental composition, growth process, and/or metabolism of immobilized as compared to suspension-cultured plant cells (see Sec. 3.3.3). This may limit further growth improvement of plant cell biofilms.

The viability of SIPC cultures was evaluated by monitoring main nutrients consumption and release ($NH_4$ and $PO_4$), indicative of possible cell lysis (55), and on-line respiration during the whole culture duration, by dry biomass increase, and by cellular staining and microscopic observation at the end of the cultures. In most cases, SIPC bioreactor cultures were highly viable ($\geq 90\%$) and compared well to similar suspension cultures grown in shake flasks as long as inoculating and operating (temperature, oxygen and carbon dioxide partial pressures, etc.) conditions were satisfactory and essential nutrients were available extra- or intracellularly (28,39,46,50,51).

## 3.3 SIPC Biofilm Physiology

From flask to large-scale bioreactor cultures, there is evidence that the technique, system, and operating conditions used to culture plant cells affect their performance (7,10–13). This is particularly important for (surface-) immobilized plant cells and upon scale-up. Certain groups may have concluded too soon that the metabolisms of immobilized and free suspension cultured plant cells and aggregates are similar (32,33). Proper comparison of the performance of both culture systems may require the development and optimization of different operational schemes, depending on the particular growth and secondary metabolism behavior of plant cells cultured according to each technique.

In this context, our approach was to gain a better understanding of the growth of plant cell biofilms (biomass formation pattern, main nutrients uptake and release, effect of gaseous phase, respiration, etc.) as compared to the simple benchmark system of suspension cultures grown in shake flasks. This represents a starting point toward the development of SIPC based bioprocesses. Off- and on-line monitoring of significant features of the behavior of immobilized cultures (28) will allow characterizing the important process parameters of the culture's physiological state before, during, and after stimulation of secondary metabolism. This will provide for improved reproducibility and permit process modeling, control, and optimization.

### 3.3.1 *Growth Behavior*

The slower biomass growth of SIPC cultures seems to display a linear behavior (46). The slower consumption of main extracellular nutrients of immobilized cultures shows a linear pattern after lags of various durations linked to adaptation or other uptake processes such as phosphate accumulation, extracellular sucrose hydrolysis prior to monosaccharide uptake [except for *T. wildordii* cells (51)], and ammonium uptake prior to nitrate utilization, depending on cell species and culture conditions. This slower and apparently linear growth pattern of SIPC cultures may result from limited external mass transfer or restricted diffusion of nutrients, including oxygen, through the biofilm. Increased mixing, however, did not improve growth. Diffusion within plant cell biofilms compares to that of microbial films and calcium alginate, which are of the order of that observed for water ($\sim 2.4 \times 10^{-9}$ $m^2/s$) (56). This yields a characteristic diffusion time [(biofilm thickness)$^2$/diffusion coefficient] for a 3-mm wet biofilm of 3.8 h, which is faster than the highest nutrient consumption rates of SIPC cultures ($\leq$1–3 g/L/day) (30). Oxygen availability could be limiting. However, this may not be as restrictive as expected and previously reported (36), as discussed in the next section.

A different explanation for this behavior involves the particular growth mechanism of plant cells combined with the almost unidirectional and space-limited expansion of a SIPC biofilm. The growth of plant cells occurs through successive phases of cell expansion, mitosis, and finally formation of a common cell plate between the two nearly formed cells. The two daughter cells remain attached to each other within the biofilm and, theoretically, in communication through openings in the cell plate called plasmodesmata (54). This growth is, most probably, restricted to the biofilm external cell layers as subsurface cells experience volumetric or steric hindrance pressures, nutrient gradients, and/or intercellular signals inhibiting expansion or division and inducing adaptive metabolic changes limiting further growth. According to this physical model, the biofilm would be made of two layers. The external layer would be meristematic, growing at a high rate depending on culture conditions (nutrients, including oxygen, availability) because of efficient mixing and mass transfer. Cells of the internal layer would gradually enter stationary phase under volumetric pressures and restricted nutrient availability. This pattern would result in the observed biofilm behavior of linear growth and nutrient consumption, lower internal oxygen requirements (see next section), and declining and lower final biomass yield on carbohydrates since, as growth proceeds, more carbohydrates are consumed for maintenance purposes, as compared to a suspension-cultured biomass growing freely in three dimensions (see Sec. 3.3.3).

### *3.3.2 Gassing Requirements*

The supply of sufficient oxygen to any biofilm growing and/or surviving under aerobic conditions remains one of the most difficult and critical bioengineering concern in the development of an immobilized bioprocess. Obviously, this originates from the low solubility of this essential nutrient, its required continuous supply, the numerous transfer steps involved between the three physical phases of the culture, and the difficulty of measuring the dissolved oxygen concentration in the biofilm under dynamic bioreactor operating conditions. In the case of plant cell cultures, this operation may be easier, on the one hand, because of their lower (< ~10%) oxygen requirement as compared to microbial cultures. On the other hand, however, the limited gas dispersion energy that can be applied to these cultures, their possible hyperventilation (see below), and their high sensitivity to environmental upsets complicate further this mass transfer operation.

In addition, plant cells cultured in bioreactors under forced gassing appear to display some requirement for a suitable carbon dioxide partial pressure in, and/or transfer balance [$(k_La)_{O_2} \sim (k_La)_{CO_2}$ (22,28)] between, the gas and liquid (and possibly biofilm) phases for best growth. This effect has been clearly demonstrated for suspension (23–27) and surface-immobilized (28) *C. roseus* cultured cells but remains to be generalized for other plant cell species.

The viability and growth of *C. roseus* SIPC bioreactor cell cultures were significantly affected by their gas phase $CO_2$ partial pressure (28). Injection of 2% (V/V) $CO_2$ in the sparged air of 20-L SIPC bioreactor cultures (constant aeration rate of 0.51 VVM, $k_La$ ~$8.0h^{-1}$) resulted in efficient biomass formation (growth rate and biomass yield), but still lower than for shake flask cultures, and high viability retention as long as extracellular nutrients were available and medium DO concentration was above 20–30% of air saturation. Air (0.03% $CO_2$) and 5% $CO_2$ cultures grew less efficiently and showed declining viability after maximum respiration even though nutrients and oxygen were not limiting. A similar growth behavior was observed for *C. roseus* cell suspension cultures carried out in bioreactors, which was ascribed to culture hyperventilation (loss of essential gasses including $CO_2$) when sparged with air (26) and stimulation of primary metabolic enzymes when aerated with 2% $CO_2$-supplemented air (27). The viability and DO concentration of these bioreactor suspension cultures were generally not reported. The poor performance of cultures sparged with high $CO_2$ content ($\geq$3%) can be attributed to toxic levels of this respiratory by-product for *C. roseus* cells (28).

Few, if any, studies have been reported on the effect of DO concentration on the growth of suspension or immobilized plant cell cultures performed in bioreactors under constant mixing conditions. The author's group carried out this study for *C. roseus* SIPC 6-L airlift bioreactor batch cultures performed at a constant aeration rate of 0.36 VVM ($k_La$ ~$6.4h^{-1}$) and constant levels of DO by varying the inlet gas composition in air, nitrogen, oxygen, and $CO_2$ (57). Cultures grew well at DO concentrations above 10% and up to 150% of air saturation, the highest concentration tested. Growth, main nutrients (carbohydrates and nitrate) uptake, and biomass yield appeared to follow saturation kinetics with respect to DO concentration. The apparent half-maximum and the maximum growth rates were at DO concentrations of 60% and 90–100%, respectively. Injection of 2% $CO_2$ in 90% DO controlled cultures improved growth marginally (~10%).

Two other results of this study revealed important features of the biofilm physiology. All cultures performed at DO concentrations of 30% and above were highly viable (≥90%). Robins et al. observed, for 8 $mm^3$ polyurethane foam particles loaded with *C. pubescens* cells, a rapid decline in DO concentration from 100% to 5% of air saturation within 1 mm of the biofilm thickness using microelectrodes, and retained high viability (80%) (53). Culture conditions, external mixing, and mass transfer efficiencies and rates were not reported. These results, although those of Robins et al. were for a different system, support the concept of adaptive capabilities and behavior of biofilm-cultured plant cells. The second interesting result observed was more uniform growth of the biofilm under controlled DO concentration conditions than in uncontrolled cultures.

Few reports have been published on the modeling of oxygen transport in an immobilized plant cell biomass. Hulst et al. (36) estimated that unlimited respiration of *D. carota* cells entrapped in Ca alginate required bead diameters not to exceed 1.4–3.8 mm assuming a constant bulk DO concentration of 95% of air saturation and an effective diffusion coefficient ($D_e$) in the immobilizing matrix-biomass of 80% of that measured in water ($D_w$). Mavituna et al. (33) reported $D_e$ values of 200–2000% of $D_w$ for glucose diffusing in *C. frutescens* cells immobilized in polyurethane foam particles. Kargi et al. (58) estimated similar high $D_e$ values for $O_2$ diffusing in a 2-5-mm horizontal biofilm of *C. roseus* cells maintained under slow metabolism and/or of questionable viability. These high apparent, rather than effective, measured or estimated diffusion coefficients suggest that other, more significant mass transfer mechanisms that only passive diffusion are involved in the transport of oxygen (and other nutrients) through a plant cell biofilm, such as extra- and/or intracellular (cytoplasmic streaming combined to plasmodesmata communication) convection.

Our group is completing a modeling study on the transport oxygen in a SIPC biofilm within the context of the bioreactor work presented above (57). Our approach differs for that of these authors and is based on the bilayer model discussed in Sec. 3.3.1. Growth is only limited by DO availability and depends on DO concentration in the biofilm. Dynamic external and internal (assuming passive diffusion at $D_e = D_w$) oxygen transport is coupled to consumption to estimate the biofilm average DO concentration and growth.

Simulated results from this simple mass transfer model correlate well with experimental findings and support the adaptive bilayer behavior proposed for a growing SIPC biofilm in Sec. 3.3.1. Dissolved oxygen diffusing in a plant cell biofilm appears not to be limiting to its survival as long as it is not too thick (≤3 mm), $O_2$ maintenance consumption rate is of the order of external/internal supply rate, and the bulk DO concentration is maintained above a minimal level (~10–20% of air saturation). Growth, on the other hand, depends on the actual biofilm DO concentration above a minimal value, which results from the balance between external/internal supply and consumption rates.

### 3.3.3 Respiration Rates

All SIPC flask and bioreactor cultures of the various plant cell species tested, including those supplemented with 2% $CO_2$ and performed at high DO concentration (≥90%), resulted in decreasing yields of biomass on carbohydrates during growth and lower (~10–25%) final biomass yields as compared to shake flask grown suspension cultures (28,39,46,50,51,57). This supports and may be explained, at least partly, by the biofilm bilayer model described above. As growth proceeds, the proportion of the biomass in stationary phase (internal layer) increases relative to the growing external layer, and consequently, more carbohydrates are consumed for maintenance.

Recent findings involving SIPC bioreactor *C. roseus* cell cultures shed new light on these results. It appeared that the specific average respiration rates, measured on-line and calculated relative to the estimated increasing biomass concentration, of 20-L cultures sparged with air without DO concentration control were significantly higher (maximum $q_{CO_2}$ ~1.1 mM/g dw/h) (28) than those reported for similar suspension cultures carried out in bioreactors (maximum $q_{CO_2}$ ~0.2–0.5 mM/g dw/h) (59–61). Immobilized cultures performed at controlled high DO concentrations (≥90%) (57) or supplemented with 2% or 5% $CO_2$ without DO concentration control respired at rates (maximum $q_{CO_2}$ ~0.5–0.6 mM/g dw/h) lower than air-sparged immobilized cultures but higher than air-sparged suspension cultures (28). Mass balances on carbon consumed for respiration and biomass production have confirmed that the low biomass yields on carbohydrates of immobilized cultures resulted from their higher specific respiration rates (28).

Hyperventilation combined with a rapid decline in DO concentration, from high (DO ≥90%) to low (DO ~20–30%) growth-promoting conditions, of 20-L immobilized cultures sparged with air without DO control resulted, most probably, in physiological stresses yielding high transient respiration rates. These three effects may have caused severe carbon dioxide and/or oxygen limitations within the biofilm, resulting in poor growth and rapid loss of viability of these cultures. Carbon dioxide supplemented (2%) immobilized cultures displayed improved growth, as observed for suspension cultures (26), and lower respiration rates under the same oxygen supply rates and concentration conditions. This seems to have prevented the detrimental conditions suggested above and/or induced a more robust metabolism (27) of immobilized cells at these culture conditions.

On the other hand, immobilized cultures carried out at constant DO concentration showed increasing growth rates and biomass yields on carbohydrates, and consequently decreasing respiration rates, with increasing DO concentration up to 90–100% of air saturation (at $\mu_{max}$) (57). Obviously, the availability of this essential nutrient limited growth and affected respiratory activities. The immobilized cells appeared to adapt to the culture conditions by maintaining some physiological equilibrium between growth and respiration, governed by dissolved oxygen (and other nutrients) availability above critical levels ($\geq$10–20% of air saturation) which allowed maintenance of culture viability. This behavior correlates well with the mass transfer model discussed in the previous section. Similar studies on the viability, growth, and respiratory behavior of plant cell suspension cultures performed under various gassing regimes including at controlled DO concentration have not been reported.

This respiratory behavior of SIPC biofilms may depend on plant cell species as well as culture system and conditions. The biological model system used [*C. roseus* cells of line MCR17 (39)] may not be fully representative of all plant cell species, because of its ease of culture, or may be considered as a worst-case situation in view of its high specific respiration rate (0.4–0.9 mM $O_2$/g dw/h) (28) as compared to other plant cell species (0.03–0.7 mM $O_2$/g dw/h) (53). This behavior, however, does differ significantly from most of the few results reported on the respiration of immobilized plant cells, which show rates similar to those of suspension cultured cells using off-line measurements (32,33,53). On-line respiration and carbohydrate mass balance studies may reveal novel features of these immobilized cultures as well.

Furthermore, the performance of plant cells immobilized by entrapment in various polysaccharide gels was related to their respiration rate and to the diffusion of oxygen through the resulting matrix (36). Culture survival required gel bead diameters not to exceed 3 mm, with little reference to continuous oxygen supply or bulk liquid phase DO concentration and mixing/transfer conditions. The difficulties and limits of this particular immobilization technique can be better understood in the context of the results discussed above and when considering the probable eliciting properties and, consequently, stressing and respiration-increasing activities of many of the polysaccharide gels used to entrap plant cells.

### 3.3.4 Secondary Metabolites Production

Few studies have yet been reported on the production of secondary metabolites by surface-immobilized plant cells. First results indicated at least similar production of indole alkaloids from SIPC bioreactor and shake-flask-suspension *C. roseus* cell cultures. In addition, some alkaloids were detected in the medium of immobilized cultures, contrary to suspension cultures, under apparently nonlytic conditions (62). Others have reported similar metabolite release for other immobilization systems (63,64).

Further studies revealed some interesting points about the experimental methodology and this production (65–67). First, truly valid comparison with suspension culture requires proper design and optimization, including product release and continuous on-line harvesting, of a bench-scale immobilization process, which has not yet been worked out, to fully evaluate the advantages, potential, and productivity of this culture technique. Similar work may be necessary for suspension cultures if they are to used as benchmark systems. Nonetheless, the kinetics of indole alkaloids production in unoptimized 0.2-L shake-flask-suspension and 2-L SIPC bioreactor, *C. roseus* cell cultures using Zenk's Alkaloid Production Medium (68) were found to be similar except for two points. The total production of known alkaloids was the same for both culture systems. However, the total production of extracted and ultraviolet-absorbing compounds obtained from two-stage suspension cultures was higher than from two-stage immobilized cultures. This was ascribed to the low and unoptimized gassing regime (low DO concentration and $CO_2$ deprivation) of the larger immobilized cultures. More interesting, however, was the release of up to 40% of the produced alkaloids into the medium of immobilized cultures under nonlytic conditions. No alkaloid was detected in the medium of suspension cultures. Subsequently, first studies have shown that total alkaloid production can be released from immobilized *C. roseus* cultures under apparently nondamaging stimulation conditions (67).

Obviously, the release of phytochemicals of interest into the medium is an essential prerequisite to the development of bioprocesses based on immobilized plant cells, which now appears not to be as limited as previously suggested.

Much work remains to be undertaken to fully evaluate the industrial potential of the surface immobilized plant cell culture technology for the production of valuable secondary metabolites. In fact, there is a wide range of physiological states between callus, suspension, and biofilm controlled culture of plant cells and the original plant, which has not yet been explored but is surely full of promises for the production of complex phytochemicals of industrial interest.

## 4 TOWARD SIPC-BASED BIOPROCESSES

The advantages of the surface immobilized plant cell culture technique have been reviewed and discussed. Its three drawbacks—biofilm diffusion, growth control, and product release—may not be as limiting as originally suggested. Product release was observed and may be further stimulated by mild means. However, this must be coupled to on-line harvesting and preliminary purification/concentration to be process efficient. Obviously, growth control will require careful nutritional balance studies for active, but limited biomass formation and efficient secondary metabolites production for each process case.

The real challenge remains the development of an integrated cost-effective bioprocess for the production of valuable secondary metabolites. The main cost factors of this production appear to be related to the slowness of the growth and production phases as well as to downstream processing steps. The selection of cell lines of good growth and production potential, even though these characteristics are often in inverse relationship, in both shake flask and (immobilization) bioreactors represents a first phase in this development. Subsequently, suitable suspension culture systems of increasing volume are required to efficiently raise the volume of the inoculum in an optimized physiological state for the main process bioreactor.

At this point, the use of an immobilization bioreactor as the main process unit may offer significant operational and cost advantages over batch, semi-, or continuous suspension culture systems, upon proper process development, control, and optimization and if the problems of growth control, product release, and production cycling can be resolved. There is now evidence that these difficulties can be overcome. The main operational advantages of an immobilization-based process include the ease of control over the culture's hydrodynamics, mixing shear and mass transfer, and the efficient separation of the gas, liquid, and biomass phases for manipulation. Lower production costs are expected from the increased productivity of the process upon reusing the same biomass in successive production cycles and from simpler product harvesting and associated first concentration and purification steps. Other factors whose effects remain to be ascertained include the physiological and productive behavior of an optimized plant cell biofilm and the continuous removal of released products.

## ACKNOWLEDGMENTS

The author thanks all the participants and associates of the research team who have contributed to the development of these technologies at the National Research Council of Canada and École Polytechnique de Montréal, especially Prof. C. Chavarie. He is also grateful for useful editorial comments from Dr. Ian Reid and to Mrs. T. Goneau for typing the manuscript.

## REFERENCES

1. Walker, A. (1987). *New Biotech.*, September: 9–12.
2. Wetter, L.R., and Constabel, F. (1982). *Plant Tissue Culture Methods*, National Research Council of Canada, Plant Biotechnology Institute, Saskatoon, Saskatchewan, Canada.
3. Dixon, R.A. (1985). *Plant Cell Culture. A Practical Approach*, IRL Press, Oxford, WA.
4. Curtin, M.E. (1983). Harvesting profitable products from plant tissue culture, *Bio/Technology*, *1*: 649.
5. Kreis, W., and Reinhard, E. (1989). The production of secondary metabolites by plant cells cultivated in bioreactors, *Planta Med.*, *55*: 409.
6. Scragg, A.H., Morris, P., Allan, E.J., Bond, P., Hegarty, P., Smart, N.J., and Fowler, M.W. (1987). The effect of scale up on plant cell culture performance, in *Plant and Animal Cells: Process Possibilities* (C. Webb and F. Mavituna, eds.), Ellis Horwood, Chichester, UK, p. 77.
7. Braquet, P. (1988). The ginkgolides. From Chinese pharmacopeia to a new class of pharmacological agents: the antagonists of platelet-activating factor, in *Ginkgolides: Chemistry, Biology, Pharmacology and Chemical Perspective* (P. Braquet, ed.), J.R. Praus Science, Barcelona, p. XV.
8. Anon. (1990). Taxol: hope for cancer patients; fear for trees, *Biotechnol. Newswatch*, *10*(13): 2.
9. Rajnchapel-Messai, J. (1988). Cellules végétales en quête de métabolites, *Biofutur*, July–August: 23.
10. Smart, N.J., and Fowler, M.W. (1981). Effect of aeration on large scale culture of plant cells, *Biotechnol. Lett.*, *3*: 171.
11. Fowler, M.W. (1982). The large scale cultivation of plant cells, *Prog. Ind. Microbiol.*, *16*: 207.
12. Allan, E.J., and Scragg, A.H. (1986). Comparison of the growth of *Cinchona ledgeriana* Moens suspension culture in shake flasks and 7 litre airlift bioreactors, *Biotechnol. Lett.*, *8*: 635.
13. Thomas, N.H., and Jones, D.A. (1987). Fluid dynamic considerations in airlift and annular vortex bioreactors for plant cell culture, *Ann. NY Acad. Sci.*, *506*: 171.
14. Ellis, B.E. (1986). Production of plant secondary metabolites without plants: a perspective, *Biotech. Adv.*, *4*: 279.
15. Payne, G.F., Shuler, M.L., and Brodelius, P. (1987). Large scale plant cell culture, in *Large Scale Cell Culture Technology* (B.K. Lydersen, ed.), Hanser Pub., Munich, p. 194.
16. Constabel, F. (1988). Principles underlying the use of plant cell fermentation for secondary metabolite formation, *Biochem. Cell Biol.*, *66*: 658.
17. Scragg, A.H., Morris, P., Allan, E.J., Bond, P., and Fowler, M.W. (1987). Effect of scale-up on serpentine formation by *Catharanthus roseus* suspension cultures, *Enzyme Microb. Technol.*, *9*: 619.
18. Tanaka, H. (1982). Oxygen transfer in broths of plant cells at high density, *Biotechnol. Bioeng.*, *24*: 425.

19. Tanaka, H. (1987). Large-scale cultivation of plant cells at high density: a review, *Process Biochem.*, August: 106.
20. Scragg, A.H., Allan, E.J., Bond, P.A., and Smart, N.J. (1986). Rheological properties of plant cell suspension cultures, in *Secondary Metabolism in Plant Cell Cultures* (P. Morris, A. Scragg, A. Stafford, and M.W. Fowler, eds.), Cambridge University Press, Cambridge, UK., p. 178.
21. Panda, A.K., Mishra, S., Bisaria, V.S., and Bhojwani, S.S. (1989). Plant cell reactors—a perspective, *Enzym. Microb. Technol., 11*: 386.
22. Jolicoeur, M., Chavarie, C., Carreau, P.J., and Archambault, J. (1992). Development of a helical-ribbon impeller bioreactor for high-density plant cell suspension culture, *Biotechnol. Bioeng.*, *39*: 511.
23. Smart, N.J., and Fowler, M.W. (1984). Mass cultivation of *Catharanthus roseus* cells using a nonmechanically agitated bioreactor, *Appl. Biochem. Biotechnol.*, *9*: 209.
24. Maurel, B., and Pareilleux, A. (1986). Carbon dioxide fixation and growth of heterotrophic cell suspension of *Catharanthus roseus, J. Plant Physiol.*, *122*: 347.
25. Ducos, J.P., and Pareilleux, A. (1986). Effect of aeration rate and influence of $pCO_2$ in large-scale cultures of *Catharanthus roseus* cells, *Appl. Microbiol. Biotechnol.*, *25*: 101.
26. Hegarty, P.K., Smart, N.J., Scragg, A.H., and Fowler, M.W. (1986). The aeration of *Catharanthus roseus* L.G. Don suspension cultures in airlift bioreactors: the inhibitory effect at high aeration rates on culture growth, *J. Exp. Bot.*, *37*: 1911.
27. Ducos, J.P., Ferron, G., and Pareilleux, A. (1988). Growth and activities of enzymes of primary metabolism in batch cultures of *Catharanthus roseus* cell suspension under different $pCO_2$ conditions, *Plant Cell, Tiss. Organ Cult.*, *13*: 167.
28. Archambault, J. (1991). Large scale (20 L) culture of surface-immobilized *Catharanthus roseus* cells, *Enzym. Microb. Technol.*, *13*: 882.
29. Brodelius, P. (1983). Production of biochemicals with immobilized plant cells: possibilities and problems, *Ann. NY Acad. Sci.*, *413*: 383.
30. Brodelius, P. (1983). Immobilized plant cells, in *Immobilized Cells and Organelles*, Vol. I (B. Mattiasson, ed.), CRC Press, Boca Raton, FL, p. 27.
31. Rhodes, M.J.C. (1985). Immobilized plant cell cultures, in *Topics in Enzyme and Fermentation Biotechnology*, Vol. 10 (A. Wiseman, ed.), Ellis Horwood, West Sussex, UK, p. 51.
32. Lindsey, K., and Yeoman, M.M. (1985). Immobilized plant cell culture systems, in *Primary and Secondary Metabolism of Plant Cell Cultures* (K.H. Neumann, W. Barz, and E. Reinhard, eds.), Springer Verlag, Berlin, p. 304.
33. Mavituna, F., Park, J.M., Wilkinson, A.K., and Williams, P.D. (1987). In *Plant and Animal Cells: Process Possibilities*. (C. Webb and F. Mavituna, eds.), Ellis Horwood, Chichester, UK, p. 92.
34. Hulst, A.C., and Tramper, J. (1989). Immobilized plant cells: a literature survey, *Enzyme Microb. Technol.*, *11*: 546.
35. Rosevear, A., and Lambe, C.A. (1985). Immobilized plant cells, *Adv. Biochem. Eng./Biotechnol.*, *31*: 37.

36. Hulst, A.C., Tramper, J., Brodelius, P., Eijkenboom, L.J.C., and Luyben, K.Ch.A.M. (1985). Immobilized plant cells: respiration and oxygen transfer, *J. Chem. Technol. Biotechnol.*, *35B*: 198.
37. Jose, W., Pedersen, H., and Chin, C. (1983). Immobilization of plant cells in a hollow-fiber reactor, *Ann. NY Acad. Sci.*, *413*: 409.
38. Hallsby, G.A., and Shuler, M.L. (1986). Altering fluid flow patterns changes pattern of cellular associations in immobilized tissue cultures, *Biotechnol. Bioeng. Symp.*, *17*: 741.
39. Archambault, J., Volesky, B., and Kurz, W.G.W. (1989). Surface immobilization of plant cells, *Biotechnol. Bioeng.*, *33*: 293.
40. Albersheim, P., and Darvill, A.G. (1985). Oligosaccharins, *Sci. Am.*, September: 58.
41. Rhodes, M.J.C., Smith, J.I., and Robins, R.I. (1987). Factors affecting the immobilization of plant cells on reticulated polyurethane foam particles, *Appl. Microbiol. Biotechnol.*, *26*: 28.
42. Archambault, J., Volesky, B., and Kurz, W.G.W. (1986). Surface immobilization of plant cells, in *VI International Congress on Plant Tissue and Cell Culture*, University of Minnesota, August 1986, p. 450.
43. Radvanyi, L., Dicosmo, F., and Absolam, D.R. (1986). Adhesion of *Catharanthus roseus* cells to polymer surfaces: surface thermodynamic aspects, in *VI International Congress on Plant Tissue and Cell Culture*, University of Minnesota, August 1986, p. 450.
44. Kargi, F., Friedel, I., and Maricic, K. (1988). Plant cell biofilm cultures, in *American Chemical Society 4th Congress of North America*, Toronto, June 1988.
45. Facchini, P.J., Neumann, A.W., and Dicosmo, F. (1988). Thermodynamic aspects of plant cell adhesion to polymer surfaces, *Appl. Microbiol. Biotechnol.*, *29*: 346.
46. Archambault, J., Volesky, B., and Kurz, W.G.W. (1990). Development of bioreactors for the culture of surface immobilized plant cells, *Biotechnol. Bioeng.*, *35*: 702.
47. Facchini, P.J., Neumann, A.W., and Dicosmo, F. (1989). The effect of inoculum concentration on the kinetics and extent of cultured plant cell adhesion to polymer surfaces under static conditions, *Biotechnol. Appl. Biochem.*, *11*: 74.
48. Facchini, P.J. (1990). Adhesion of various species of suspension-cultured plant cells to inert substrates: initial interactions, *FEMS Microbiol. Lett.*, *67*: 313.
49. Facchini, P.J., and Dicosmo, F. (1990). Immobilization of plant cells using glass fibres for the production of secondary metabolites, in *VII International Congress on Plant Tissue and Cell Cultures*, Amsterdam, June 1990, p. 339.
50. Carrier, D.J., Coulombe, P., Mancini, M., Neufeld, R., Weber, M., and Archambault, J. (1990). Immobilized *Ginkgo biloda* cell cultures, in *Progress in Plant Cellular and Molecular Biology* (H.J.J. Nijkamk, L.H.W. Van der Plas, and J. Van Aartrijk, eds.), Kluwer A.P., Dordrecht, p. 614.
51. Pépin, M.F., Chavarie, C., and Archambault, J. (1991). Growth and immobilization of *Tripterygium wilfordii* cultured cells, *Biotechnol. Bioeng.*, *38*: 1285.

52. Rhodes, M.J.C., Robins, R.J., Turner, R.J., and Smith, J.I. (1985). Mucilaginous film production by plant cells immobilized in a polyurethane or nylon matrix, *Can. J. Bot.*, *61*: 2357.
53. Robins, R.J., Parr, A.J., Richards, S.R., and Rhodes, M.J.C. (1986). Studies of environmental features of immobilized plant cells, in *Secondary Metabolism in Plant Cell Culture* (P. Morris, A.H. Scragg, A. Stafford, and M.W. Fowler, eds.), Cambridge University Press, Cambridge, UK, p. 162.
54. Alberts, B., Bray, D., Lewis, J., Raff, M., Roberts, K., and Watson, J.D. (1983). Special features of plant cells, in *Molecular Biology of the Cell* (Alberts et al., eds.), Garland, New York, p. 1099.
55. MacCarthy, J.J., Ratcliffe, D., and Street, M.E. (1980). The effect of nutrient medium composition on the growth cycle of *Catharanthus roseus* G. Don cells grown in batch cultures, *J. Exp. Bot.*, *31*: 1315.
56. Mavituna, F., and Park, J.M. (1987). Determination of the effective diffusion coefficient of glucose in callus tissue, *Chem. Eng. J.*, *34*: B1.
57. Archambault, J., Pouyez, C.M., Perrier, M., Pépin, M.F., and Chavarie, C. (1990). Effect of dissolved oxygen concentration on the growth of *Catharanthus roseus* surface immobilized cell cultures: a modelling and experimental study (in preparation).
58. Kargi, F., Ganapathi, B., and Moricic, K. (1990). Indole alkaloid formation by *Catharanthus roseus* cells in a biofilm reactor, *Biotechnol. Prog.*, *6*: 243.
59. Pareilleux, A., and Vinas, R. (1983). Influence of the aeration rate on the growth yield in suspension cultures of *Catharanthus roseus* (L) G. Don, *J. Ferment. Technol.*, *61*: 429.
60. Bond, P.A., Fowler, M.W., and Scragg, A.H. (1988). Growth of *Catheranthus roseus* cell suspensions in bioreactors: on-line analysis of oxygen and carbon dioxide levels in inlet and outlet gas streams, *Biotechnol. Lett.*, *10*: 713.
61. van Gulik, W.M., Meijer, J.L., Ten Hoopen, H.J.G., Luyben, K.Ch.A.M., and Libbenga, K.R. (1989). Growth of *Catharanthus roseus* cell suspension culture in a modified chemostat under glucose-limiting conditions, *Appl. Microbiol. Biotechnol.*, *30*: 270.
62. Archambault, J., Volesky, B., and Kurz, W.G.W. (1990). Production of indole alkaloids by surface immobilized *C. roseus* cells, *Biotechnol. Bioeng.*, *35*: 660.
63. Parr, A.J., Robins, R., and Rhodes, M.J.C. (1986). Product release from plant cells grown in cultures, in *Secondary Metabolism in Plant Cell Cultures* (P. Morris, M.W. Fowler, and A. Scragg, eds.), Cambridge University Press, Cambridge, UK, p. 174.
64. Payne, G.F., and Shuler, M.L. (1988). Selective adsorption of plant products, *Biotechnol. Bioeng.*, *31*: 922.
65. Tom, R., Jardin, B., Chavarie, C., and Archambault, J. (1991). Effect of culture process on alkaloid production by *C. roseus* cells. I. Suspension cultures, *J. Biotechnol.*, *21*: 1.
66. Tom, R., Jardin, B., Chavarie, C., Rho, D., and Archambault, J. (1991). Effect of culture process on alkaloid production by *C. roseus* cells. II. Immobilized cultures, *J. Biotechnol.*, *21*: 21.

67. Jardin, B., Tom, R., Chavarie, C., Rho, D., and Archambault, J. (1991). Stimulated indole alkaloid release from *C. roseus* immobilized cultures. First studies, *J. Biotechnol.*, *21*: 43.
68. Zenk, M.H., El-Shagi, H., Arens, H., Stockigt, J., Weiler, W., and Deus, B. (1977). Formation of the indole alkaloids serpentine and ajmalicine in cell suspension cultures of *Catharanthus roseus*, in *Plant Tissue Culture and Its Biotechnological Applications* (W. Barz, E. Reinhard, and M.H. Zenks, eds.), Springer-Verlag, New York, p. 27.

# 5
# Cell Aggregation and Sedimentation

**Robert H. Davis**
*University of Colorado, Boulder, Colorado*

The aggregation of cells into clumps or flocs has been exploited for decades in such applications as biological wastewater treatment, beer brewing, antibiotic fermentation, and enhanced sedimentation to aid in cell recovery or retention. More recent research has included the use of cell aggregation and sedimentation to selectively separate subpopulations of cells. Potential biotechnological applications include overcoming contamination, maintaining plasmid-bearing cells in continuous fermentors, and selectively removing nonviable hybridoma cells from perfusion cultures.

## 1 INTRODUCTION

Cell aggregation is the clumping together of otherwise discrete microbial or cultured cells from a dispersed state into aggregates containing many cells that are linked together. Aggregation is also commonly referred to as flocculation, coagulation, agglutination, agglomeration, and self-adhesion, although sometimes these terms are used for slightly different meanings (for example, agglutination often refers to the sexual association of cells). Because of their larger size, cell aggregates sediment due to gravity or centripedal forces much more rapidly than do single cells.

For decades, even centuries, microbial aggregation and sedimentation have played important roles in a variety of industrial processes such as beer and antibiotic fermentations, activated-sludge treatment of wastewater, and biomass recovery. As a result, a number of symposia and sessions have been devoted to the subject, as well as books written to review current research (1–3).

With the advent and growth of modern biotechnology during the past two decades, cell aggregation and sedimentation have taken on new and expanded importance (4). They are often exploited in the first step of product recovery, that of separating the biomass from the culture medium, either with or without the aid of centrifugation. Aggregation and sedimentation may also be used to maintain high cell densities and productivities in continuous fermentations, either by serving as a means for cell retention in fluidized-bed or tower fermentors or by aiding in cell recycle processes. In recent research, not as yet tested outside of the laboratory, aggregation and/or sedimentation has been demonstrated to be capable of separating cell subpopulations so that unproductive cells may be selectively removed from a bioreactor. This selective separation allows for high levels of productivity to be maintained for prolonged periods, despite the competitive disadvantage of the productive cells in the presence of unproductive contaminant or segregant cells.

It is not feasible to review all aspects of cell aggregation and sedimentation in this chapter. Instead, a brief overview of the fundamentals of these processes is given with liberal reference to other sources. This overview is followed by brief descriptions of traditional applications of cell aggregation and sedimentation, including beer fermentation, activated sludge processes, and cell harvesting and recycle. The chapter concludes with a review of recent research on modern biotechnological applications of cell aggregation and sedimentation exploiting selective cell separation technology.

## 2 FUNDAMENTALS OF CELL AGGREGATION AND SEDIMENTATION

### 2.1 Factors Affecting Cell Aggregation

Microbial aggregation is observed in virtually all microbial taxa (1): bacteria, yeasts, cellular slime molds, filamentous fungi, algae, and protozoa. In addition, many types of plant and animal cells aggregate or flocculate under favorable conditions in suspension cultures. Although bioflocculation is ubiquitous among microorganisms and cell cultures and has been widely studied, it remains poorly understood (5). Some progress has been made by drawing on the knowledge base of colloidal aggregation, but cell aggregation involves less well-defined systems and is considerably more complex.

Genetic, physiological, nutritional, environmental, and physical factors all come into play in determining whether a cell suspension will aggregate and to what degree.

### 2.1.1 Genetic Factors

Perhaps the most important factors affecting cell aggregation are genetic. Not all species aggregate, and different strains within a species differ in their capacity to aggregate. For example, *Saccharomyces cerevisiae* (brewer's or baker's yeast) may be purchased at the grocery store as a dry, powdered yeast, which is quite "powdery" or nonflocculent (Fig. 1a). In contrast, most yeasts of the same species used in beer fermentation are flocculent (Fig. 1b). Some progress has been made in identifying a gene, *FL01*, that is responsible for nonsexual flocculation of defined strains of *S. cerevisiae* (6,7). The exact role of this gene in promoting flocculation is not certain, but it is thought to code for proteinaceous structures associated with the cell wall that bind to $\alpha$-mannan cites on adjoining cells (8,9).

Whereas yeast cells form flocs due to interactions between walls of adjoining cells, some bacterial cells form flocs due to interactions between pili of adjoining cells. Type 1 pili of *Escherichia coli* are proteinaceous appendages about 7 nm in diameter and 2 $\mu$m long. The hydrophobic and adhesive nature of these pili gives rise to pili-pili bonds and consequently to the formation of small aggregates. Cells within normal piliation have approximately 200 pili on their surface and form small flocs containing less than 10 cells. However, there are mutant cells which lack the genes required for piliation and which do not flocculate, and there are hyperpiliated strains with pilus expression under plasmid control which form flocs on the order of 1 mm in size (10). The structural gene, *pilA*, which codes for the repeating polypeptide subunit of pili, pilin, has been identified (10,11), sequenced (12,13), and placed under the control of a *tac* operator and promoter (14). Another mechanism of bacterial flocculation, known as fluffing, is thought to involve a cell surface protein, a regulatory protein, or an enzyme involved in the synthesis of surface polysaccharides. Warne and Bowden (15) have isolated two strains of bacteria, one of which is Flu$^-$ and forms aggregates, and the other is Flu$^+$ and does not. Although the action of the *flu* operon is not well understood, it clearly involves a different mechanism than that of the *pil* operon, as Flu$^-$ bacterial cells do not have pili on their surfaces.

Researchers have isolated several genes involved in the synthesis of an exopolysaccharide from *Zoogloea ramigera* (16). Cellular production of this polysaccharide results in flocs as large as several millimeters in diameter. Thus, one practical means of controlling microbial aggregation which may be exploited in the future is the use of recombinant DNA technology

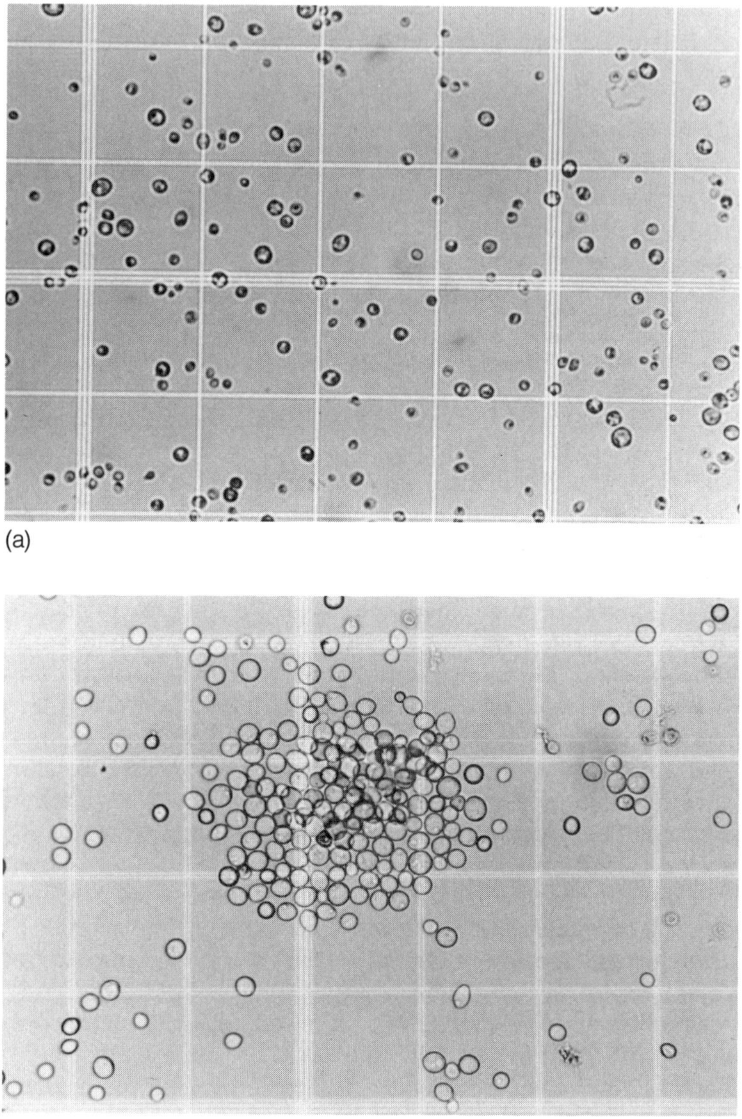

(a)

(b)

**Figure 1** Photomicrographs of (a) nonflocculent and (b) flocculent strains of *S. cerevisiae* yeast. The background is a hemacytometer slide with a 50-$\mu$m grid.

to manipulate genes that code for the production of substances which, in turn, promote or inhibit aggregation.

### 2.1.2 Chemical Factors

Chemical factors, such as pH, ion concentrations, and polymers, have also been shown to affect flocculation (1,5,17). In particular, bivalent cations such as calcium (and, to a lesser extent, magnesium and manganese) have been shown to play a key role in flocculation of brewer's yeast. One hypothesis is that $Ca^{2+}$ forms salt bridges by linking anionic components such as protein carboxyl groups and phosphomannan phosphate groups of adjacent cell walls (18). A second is that $Ca^{2+}$ ions may act as cofactors in the binding of proteinaceous, lectin-like components of the cell wall with $\alpha$-mannans or other carbohydrates (8). In contrast, monovalent cations have been shown to have an inhibitory effect on flocculation (17,19).

Two types of anionic groups have been proposed for the bridge formation between yeast cell walls involving polyvalent cations: the carboxyl groups found in acidic wall proteins, and the phosphodiester linkages found in the cell-wall phosphomannan. The weight of evidence suggests that the carboxyl groups are responsible for floc formation. Mill (20,21) found that flocculent cells treated with 1,2-epoxypropane lost their ability to flocculate. He assumed this is due to the esterfication of the surface carboxyl groups. He also showed that ferric salts and a variety of other multivalent metal cations promote flocculation. Perhaps the strongest evidence implicating surface carboxyl groups as being the prime players in flocculation is the work of Jayatissa and Rose (22). They excised phosphodiester groups from the surface of yeast cells using hydrogen fluoride, expecting to see a decrease in flocculation. Instead, removal of the phosphodiester groups led to an increase in flocculation! Beavan et al. (23) used electrophoretic mobility studies to assess the density of amino, phosphodiester, and carboxyl groups on the surface of the cell wall. They reported an increase in the density of surface carboxyl groups with increasing floc-forming ability and proposed that, during exponential growth, yeast cells have both types of anionic groups on their surface, giving rise to repulsive forces that keep the cells dispersed. The removal of the phosphodiester groups and the subsequent repulsive forces causes an increase in flocculation. Once cultures enter the stationary phase, the density of carboxyl groups increases in those yeasts capable of flocculation, and calcium-mediated flocculation occurs.

Not all researchers agree that surface carboxyl groups are responsible for yeast flocculation, however. Lyons and Hough (24–26) found that the density of phosphodiesters increased with the increase in floc-forming ability. This is in direct conflict with the results of the study by Beavan et al. (23). The majority of evidence, however, suggests that the carboxyl groups are the surface binding groups during flocculation (27). It is likely that acidic media proteins along with calcium combine to form the bonds between flocculent yeast cells. It has been observed that certain brewery strains are flocculent in protein-rich media but not in chemically defined media deficient in proteins (18,28). If a protein-rich medium is treated with protease, flocculent strains will no longer aggregate. These studies indicate that peptide, in addition to calcium, is necessary to form bridges between flocculent cells, at least in some strains.

Numerous researchers have shown that sugars, particularly mannose, inhibit yeast flocculation (22,29). The salt bridge theory cannot account for this behavior. They believe it is likely that secondary hydrogen bonding between mannose residues on the cell surfaces is also a significant factor in flocculation. Increased glucose concentrations have also been found to reduce the degree of flocculation in activated sludge (30,31).

In an effort to encompass all the experimental evidence for yeast flocculation, Miki et al. (32) proposed a flocculation mechanism by which lectin-like binding takes place between cells. According to their study, flocculation is mediated by $\alpha$-branched mannans and a protein present only on the surface of flocculent yeasts. The mannans are not strain specific and, in fact, are present on all yeasts and act as cognons. The protein behaves as a lectin-like cognor and is sensitive to proteases and the presence of calcium. Flocculation is inhibited until the fermentable sugars are consumed and the lectin is able to bind to the mannose residues on the yeast cell surface. An interesting implication of this theory is that flocculent cells should bind nonflocculent cells in a mixed culture since both have the necessary mannose residues.

Since the surface charge (usually negative) of microbial cells is affected by pH, it is not surprising that their ability to flocculate also depends on pH. Most researchers agree that optimal flocculation occurs at a slightly acidic pH, between 4.5 and 5.5 (33). There is also strong evidence of a synergistic effect of pH and divalent ion concentration, presumably because of the interplay between the anionic cell-wall groups and the divalent-ion bridge or cofactor participation (33–35). Unfortunately, the interaction is much more complex than the rules of colloid science for which destabilization (aggregation) occurs when the surface potential is reduced to zero or when the ionic strength is increased to compress the surrounding electrical double layer, thereby permitting attractive London–van der Waals forces to prevail (5).

A common method for flocculating charged colloidal particles involves the addition of polyelectrolytes of opposite charge from that of the cell surfaces (36). Addition of a cationic polyelectrolyte, chitosan, was found to flocculate *E. coli* bacteria, both by changing the charge of the cells and by modifying their hydrogen bonding characteristics (37). Polysaccharides and other polymers are also widely used to promote bioflocculation through a cell-cell bridging mechanism (5). The use of extracellular polymers is widely practiced for flocculating bacteria in activated sludge processes for wastewater treatment (5,38), and it has been shown that $Ca^{2+}$ ions also play a role in polymer-induced flocculation of activated sludge (39). Polymers, with or without additional ions, have also been used to increase flocculation of microbial cells (40), animal cells (41), and cell debris (42) to improve the efficiency of sedimentation or centrifugation for separating cellular material from its aqueous suspending medium. Bentonite clay, when added to a water treatment process, has been shown to aid in the sedimentation of various types of bacteria by promoting heterocoagulation (43).

### 2.1.3 Physiological Effects

In addition to the molecular architecture of the cell wall, flocculation also depends on the physiological state of the cells. In wastewater treatment processes, flocculation starts at the beginning of the stationary phase (17). In general, conditions that favor cell proliferation also favor cell dispersion or lack of flocculation (1). An advantage of aggregation in substrate-deficient environments is that cell death and lysis release nutrients that may be consumed by live cells in close proximity. Slime molds, such as *Dictyolstelium discoideum*, are free-living individual cells in the growth state and then aggregate and form multicellular organisms upon exhaustion of food supplies such as yeast and bacteria (44). Aggregation of microorganisms that feed on other microorganisms is advantageous because nutrients are less likely to be lost due to diffusion once extracellular enzymes from the micropredators have lysed the other organisms.

*S. cerevisiae* yeast cells have also been observed to flocculate as they are about to enter the stationary phase in batch cultures. In light of the previous discussion, it is thought that the onset of aggregation correlates with the appearance of an increased density of carboxyl groups on the cell surfaces. This may be the result of the insertion or rearrangement of acidic cell-wall proteins as the growth rate declines, but more likely is the result of an increase in the activity of certain enzymes that generate carboxyl groups in existing cell-wall proteins (27).

### 2.1.4 Physical Factors and Deflocculation

Mild agitation helps to promote cell-cell collisions and, hence, aggregation. On the other hand, strong agitation serves to disperse fragile flocs by shearing them apart (45,46). Deflocculation may also be induced by heating. Reversible deflocculation of various yeasts is significant at 60°C and complete at 80°C (20,46). Also, adding a chelating agent like EDTA which binds to $Ca^{2+}$ ions has been shown to inhibit aggregation of yeast, as have urea, mannose, and the addition of fermentable sugars (17,20,21,47). Many of these factors provide strong evidence that hydrogen binding is involved in flocculation (27).

## 2.2 The Fractal Nature of Cell Aggregates

Most naturally formed aggregates of cells have a loosely branched, porous structure, rather than the tightly packed structure typical of centrifuge pellets. Moreover, the porosity of the aggregates usually increases as the aggregate size increases. A quantitative understanding of the structure of cellular aggregates is important because the structure influences the sedimentation rates of aggregates and also the rate at which nutrients diffuse into the interior of an aggregate. The branched structure and increasing porosity with increasing size indicate that the aggregates may have a fractal nature.

A fractal is an object that contains substructures that are similar to the whole. As reviewed by Mandelbrot (48), fractal geometry has provided a basis for understanding a variety of irregular structures in nature (mountains, clouds, rivers, etc.), and it has been used to study phenomena such as turbulence, chaos, and structure formation in gels, macromolecules, and aggregates.

A fractal aggregate has the property that the number of primary particles ($N$) composing the aggregate increases in proportion to the linear dimension of the aggregate raised to a fractional power:

$$N = \alpha D^f, \tag{1}$$

where the fractional power $f$ is known as the fractal dimension, $D$ is a defined diameter or linear dimension of the aggregate, and $\alpha$ is a preexponential factor. For three-dimensional aggregates of colloidal particles, theory and experiment have shown that fractal dimension is in the range $1.75 \geq f \geq 2.05$, with the actual value depending on the processes by which the aggregate was formed (49). The observed fractal dimension less than the Euclidean value of 3.0 is indicative of the branched, open structure of the aggregates (Fig. 2).

Davis and Hunt (51) examined numerous flocs of *S. cerevisiae* yeast under a microscope and found them to have the loosely packed and branched structure typical of fractals. They measured the geometric mean diameter of each floc and then counted the number of cells in each floc either directly or after dispersion by heating. A log-log plot of aggregation number versus diameter is given in Figure 3 for two strains that they examined. Linear regression of these data yields a slope of $f = 1.79 \pm 0.10$ at the 90% confidence level. Similar measurements for a variety of strains and

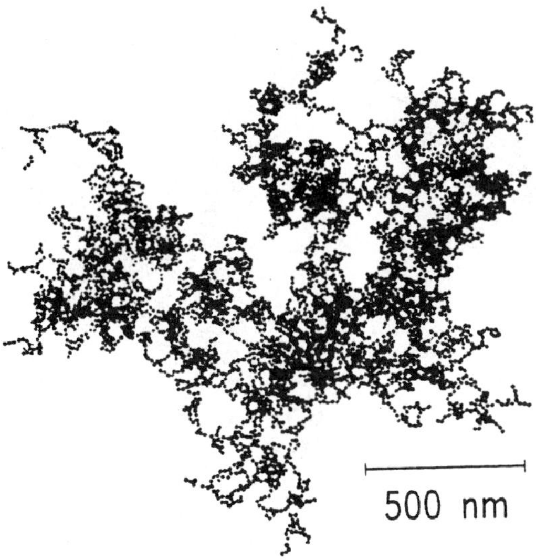


**Figure 2** Fractal aggregate of colloidal gold particles. (Reproduced from Ref. 50 by permission of the American Institute of Physics, copyright 1984.)

media yielded fractal dimensions in the range $1.75 \leq f \leq 2.25$ (51), in close agreement with the fractal dimensions reported for colloidal aggregates (49). Using a different strain that formed larger, but more tightly packed, flocs, Wilkinson and Logan (52,53) determined fractal dimensions of $f = 2.66 \pm 0.34$ for *S. cerevisiae* aggregates formed in test tubes. These flocs were roughly spherical, and the overlap of their range of fractal dimensions with the Euclidean value of 3.0 indicates that they may not be fractal. In contrast, they found that bacterial flocs of *Zoogloea ramigera* cultured in test tubes are more loosely packed, with fractal dimensions of $f = 1.79 \pm 0.28$ (53,54). Further data supporting the highly porous and fractal nature of bacterial flocs in activated sludge processes are presented by Li and Ganczarczyk (55–57). Marine snow (amorphous aggregates of marine microorganisms and debris) has also been found to have a porous, fractal structure (54,58). Additionally, the hydrodynamic environment can strongly influence aggregate structure. When *Z. ramigera* was cultured in a mixed Virtis reactor under conditions of high shear, its fractal dimension was found to increase to $2.87 \pm 0.29$ (52).

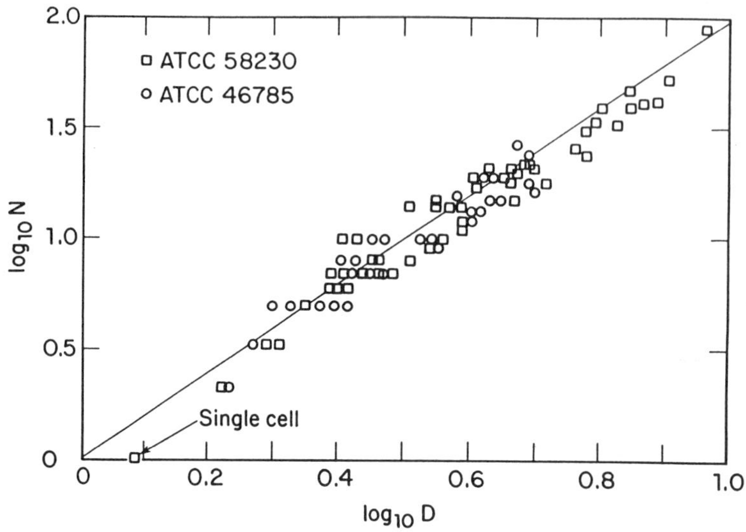


**Figure 3** Fractal relationship between yeast floc size and aggregation number. The unit measure of the geometric mean diameter, $D$, is 5 $\mu$m, and the solid line is the best-fit linear regression of the data. (Reproduced from Ref. 51 by permission of the American Institute of Chemical Engineers, copyright 1986 AIChE.)

## 2.3 Sedimentation Velocities of Cells and Aggregates

Stokes' law for the sedimentation velocity ($v$) of an isolated rigid sphere is (59):

$$v = (\rho_s - \rho)D^2 g/(18\mu), \tag{2}$$

where $\rho_s$ is the density of the sphere, $D$ is its diameter, $\rho$ is the fluid density, $\mu$ is the fluid viscosity, and $g$ is the gravitational acceleration constant. This equation applies provided that the Reynolds number, $Re = \rho Dv/\mu$, is less than about 0.1, a condition that is met for most cells and flocs. Also, to avoid hindered settling effects, the solids concentration must be less than about 1% by volume. For nondilute suspensions, the right-hand side of Eq. (2) should be multiplied by a hindred settling function, $f(c)$, where $c$ is the suspended solids volume fraction. A simple and widely used empirical correlation is that attributed to Richardson and Zaki (60):

$$f(c) = (1 - c)^n, \tag{3}$$

for which the exponent has been measured to be $n = 5.0$ for monodisperse suspensions at small Reynolds numbers (61). Although originally developed for monodisperse suspensions of spheres, Eq. (3) is often used to estimate hindered settling effects in polydisperse suspensions having particles with a range of sizes and shapes.

If we consider a single yeast cell with a cell density of $\rho_s = 1.1$ g/cm$^3$ and an effective diameter of $D = 5 \times 10^{-4}$ cm, settling in water ($\rho = 1.0$ g/cm$^3$, $\mu = 0.01$ g/cm/s) under normal gravity ($g = 981$ cm/s$^2$), then Eq. (1) indicates that its Stokes settling velocity is only $1.4 \times 10^{-4}$ cm/s, or 0.5 cm/h. Bacterial cells sediment considerably slower owing to their smaller size ($D = 1 \times 10^{-4}$ cm, typically). As a result, sedimentation-based removal of cells from suspension is usually accomplished by centrifugation rather than gravitational settling. In this case, the gravitational acceleration constant, $g$, is replaced by the centrifugal acceleration, $\omega^2 r$, in Eq. (2), where $\omega$ is the rotational velocity and $r$ is the distance from the axis of rotation. Typical centrifuges and ultracentrifuges operate in the range $\omega^2 r/g = 10^3 - 10^5$.

Another possibility for increasing the sedimentation rate of cells is to induce aggregation so that the effective size increases. Unfortunately, the settling velocity does not increase in proportion to the floc diameter squared, as might be inferred from Eq. (2). This is because the fractal nature of the flocs causes them to become more porous, or less dense, as the floc size increases. Activated sludge flocs have been found to settle with velocity proportional to $D^{0.55}$ (62), whereas marine snow settles with velocity proportional to $D^{0.26}$ (63). The small, fractional exponents are indicative of the fractal nature of cellular aggregates.

The effective relative density of a fractal aggregate is proportional to $D^{f-3}$ (51):

$$\rho_s - \rho = (\rho_p - \rho)6\alpha V_p D^{f-3}/\pi, \tag{4}$$

where $\rho_s$ is the effective density of the aggregate, and $\rho_p$ and $V_p$ are the density and volume, respectively, of the primary particles or cells comprising the aggregate. Substituting this relationship into Stoke's law yields (51):

$$v = \alpha V_p(\rho_p - \rho)g\, D^{f-1}/3\pi\mu, \tag{5}$$

indicating that the settling velocity of a fractal aggregate is proportional to $D^{f-1}$. This relationship assumes that the effective Stokes diameter and the aggregate diameter defined by Eq. (1) are equivalent, and that the floc settles as a rigid sphere. It does not take into account the reduced drag due to the bulk flow of liquid through the porous floc, but this effect is thought to be small (64).

## 2.4 Methods for Measuring Cell Aggregation

Measurements of the degree of aggregation in cell suspensions are important in the design of separation and fermentation processes that are based on aggregation and sedimentation. Measurements are also needed to assess the influence of various factors on promoting or inhibiting aggregation.

The simplest methods for measuring cell aggregation involve unaided visual observation, which provides a subjective or qualitative estimate of the relative degree of flocculation of a culture (28,65,66). Quantitative information may be obtained in a more tedious manner by microscopic enumeration with the aid of a hemacytometer slide (28). Simple sedimentation methods, such as the Burns test, in which the rate of clarification or sediment formation is estimated by allowing a suspension to settle for a specified time, provide an indirect measure of the degree of flocculation (67–69). They are based on the general principle that a higher degree of flocculation results in a faster sedimentation rate.

The visual and sedimentation methods cited above are useful in providing qualitative or semiquantitative information on the relative degree of flocculation of various cultures. For the proper design of cell separation and fermentation processes, however, more complete quantitative measurements are needed. For example, floc size and settling velocity distributions are needed to calculate the maximum throughput possible in a centrifuge used to harvest cells. In 1986 Davis and Hunt (51) developed a sedimentation/light extinction method for obtaining this information. As in earlier light extinction techniques (9,28,70), the cells in their growth medium or in a special salt or buffer solution are placed in an optical cuvette that is then gently shaken and inserted into a colorimeter. The light absorbance reading is related to the cell density by Beer's law or a calibration curve. A measure of the degree of flocculation is made by observing the change in turbidity over time as the flocs sediment out of suspension. Davis and Hunt have shown how the absorbance versus time curve may be converted into the floc size and velocity distributions using modified versions of Stokes' law and Beer's law (51). They modified Stokes' sedimentation law to account for the fractal structure of the flocs [see Eq. (5)] and Beer's light absorbance law to account for the fact that aggregated cells absorb less light than do the same number of free cells, since many of the cells in an aggregate are in the shadow of other cells.

Figures 4–6 show typical results of the sedimentation/light extinction method applied to a moderately flocculent strain of *S. cerevisiae*, ATCC 46785. In Figure 4 is plotted the light absorbance versus the time elapsed from the start of sedimentation. The flat section of the curve for small times represents the time required for the largest flocs to settle out of the region in and above the light beam. The absorbance then decreases as first larger and then smaller flocs settle past the beam. The absorbance approaches zero for long times since the fluid becomes clarified of cells at and above the level of the beam. Figure 5 shows the corresponding floc-settling velocity distribution. The solid line is that determined using the unmodified Beer's law with a constant extinction coefficient, whereas the dashed curve

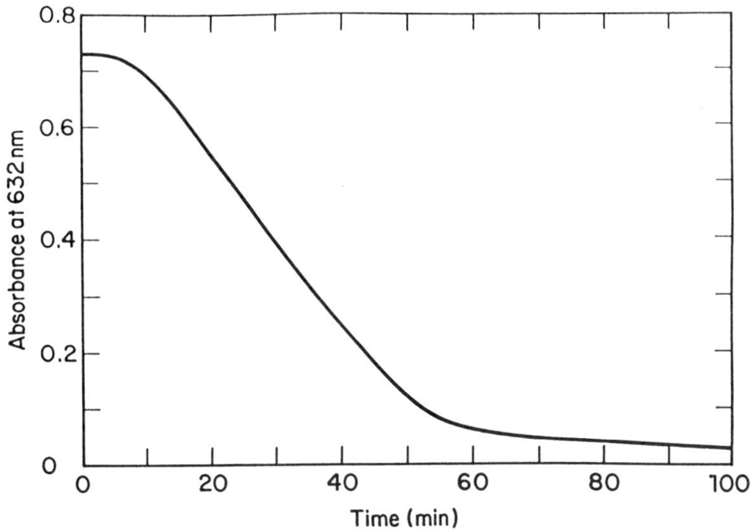


**Figure 4** Light absorbance as a function of time for sedimentation of ATCC 46785 yeast flocs past a beam located 1.0 cm below the meniscus. (Reproduced from Ref. 51 by permission of the American Institute of Chemical Engineers, copyright 1986 AIChE.)

represents the converged iteration using the modified Beer's law with a variable extinction coefficient which accounts for the shadowing effect. The settling velocity distribution was then converted into a floc size distribution function in terms of the aggregation number, $N$, representing the number of cells per floc. This was accomplished using Eqs. (1) and (5) to relate the floc settling velocity to its size. Tedious hemacytometer measurements of the floc sizes were also made, and they are shown in Figure 6 to be in good agreement with the results of the sedimentation/light extinction method. This method has also been used to determine the size distributions of bacterial flocs (30).

## 3 TRADITIONAL APPLICATIONS OF CELL AGGREGATION AND SEDIMENTATION

### 3.1 Beer Fermentation

It is generally agreed that beer originated in the early civilizations of Mesopotamia and Egypt, in about 4,000 B.C. (71). Beer is classically made by malting barley to generate enzymes that digest carbohydrates into ferment-

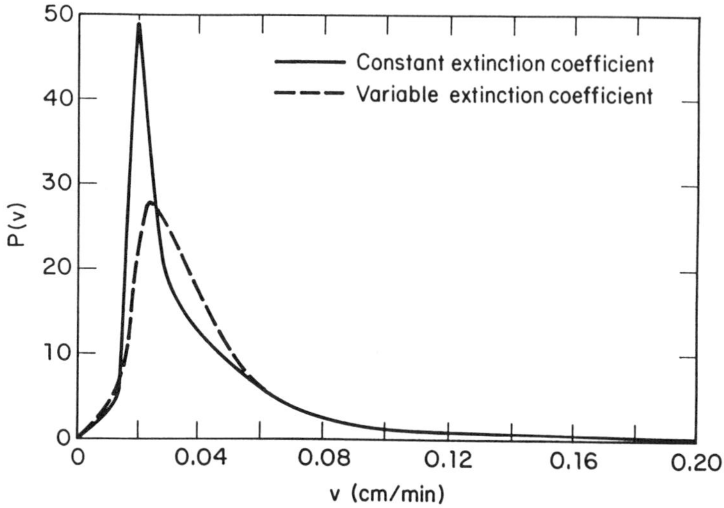


**Figure 5** Normalized probability density function of floc-settling velocities for ATCC 46785 yeast. (Reproduced from Ref. 51 by permission of the American Institute of Chemical Engineers, copyright 1986 AIChE.)

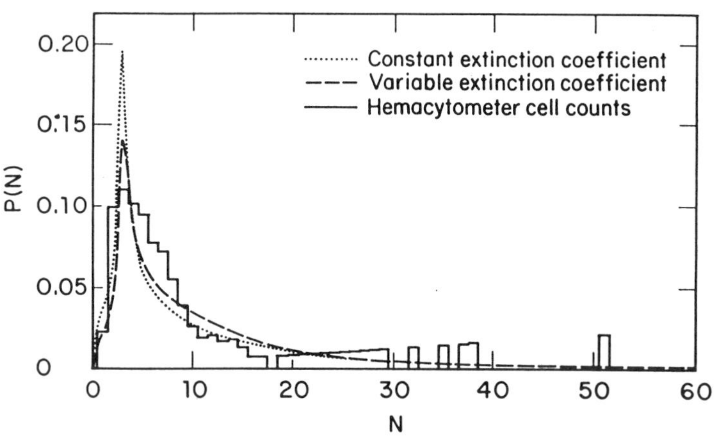


**Figure 6** Normalized probability density function of the floc aggregation numbers for ATCC 46785 yeast. (Reproduced from Ref. 51 by permission of the American Institute of Chemical Engineers, copyright 1986 AIChE.)

able sugars during mashing, brewing the malted barley extract together with hops and perhaps other grains such as corn or rice, and then fermenting the cooled brew (called "wort") with yeast.

Two species of the genus *Saccharomyces* are of primary importance in beer making. Strains of *S. cerevisiae* are the top fermenters that are used to make ale. Toward the end of fermentation, they aggregate together with the sparkle (small $CO_2$ bubbles) and rise to the top of the beer. Strains of *S. uvarum* (formerly classified as *S. carlsbergensis*) are the bottom fermenters that are used to make the lager beers that dominate the American market. They flocculate into loose clumps of 100 or more cells and sink to the bottom of the vessel when the fermentation is complete. The taxonomic propriety of separating these two strains as distinct is a matter of contention (1).

The importance of flocculation and sedimentation during beer fermentation was noted as early as 1876 by Pasteur (72). Beer fermentors generally do not include mechanical stirring. The yeast cells and small flocs remain in suspension owing to the natural stirring that results from carbon dioxide evolution and bubble formation as sugars are converted into alcohol. Near the end of the fermentation, the aggregation increases, perhaps due in part to the secretion of proteins by the cells, and the flocs settle to the bottom or rise to the top due to gravity as the natural stirring ceases (71). If a strain is too flocculent, it will settle prematurely and the fermentation will not be complete. This reduced attenuation results in a beer that has a higher specific gravity than desired, contains less alcohol, and is sweeter. On the other hand, if a strain is too powdery (nonflocculent) it will settle only slowly after the fermentation has slowed to the point where natural stirring by carbon dioxide evolution has stopped. The remaining yeast undergoes a secondary fermentation that produces off-flavors; their high concentration in suspension also reduces the efficiency of the subsequent filtration process to clarify the beer.

The effects of flocculation on beer fermentation are illustrated in Figure 7, which shows a plot of cell concentration versus fermentation time in a tubular fermentor for two different yeast strains (33). The Boulder "B" yeast is a powdery yeast for which the average aggregation number was determined by the method of Davis and Hunt (51) to be approximately $N = 5$ at the point of maximum flocculence ($t = 75$ h). This strain led to a vigorous fermentation in which 80% attenuation and a cell concentration of $1.9 \times 10^8$ cells/ml were reached in only 80 h. However, the individual yeast cells and small flocs settled only slowly, so a high concentration of $6 \times 10^7$ cells/ml remained in suspension after 360 h, even though 100% attenuation of the wort had been reached. In contrast, ATCC 48869 is a flocculent yeast for which the average aggregation number was approxi-

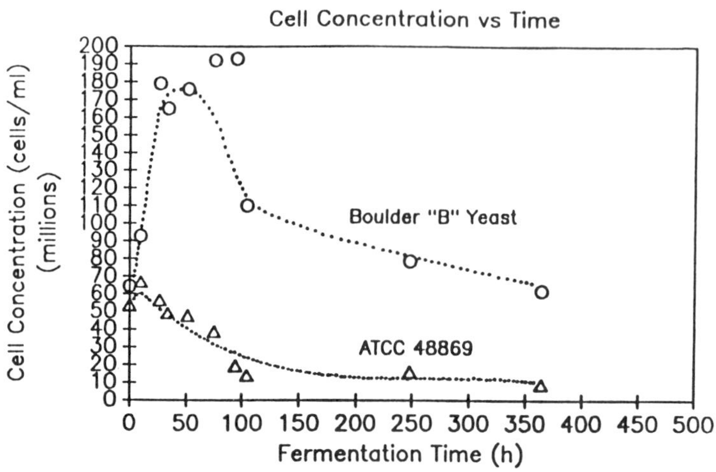


**Figure 7** Cell concentration versus time in a tubular fermentor for a flocculent yeast (Boulder "B") and a nonflocculent yeast (ATCC 48869). (From Ref. 33.)

mately $N = 200$ at the point of maximum flocculence ($t = 15$ h). This yeast flocculated prematurely, and the sedimentation of the flocs precluded a vigorous fermentation or the achievement of high cell density in the fermentor. The attenuation was only 60% after 80 h and 75% after 250 h. Neither of these strains represents the ideal behavior of remaining relatively unflocculated during the active fermentation and then flocculating and rapidly settling when the desired degree of fermentation is reached. Clearly, the ability of the brewer to assess and control yeast flocculation is an important factor in the production of beer with consistent quality.

## 3.2 Activated Sludge Processes

A primary natural function of microbes is their decomposition or organic materials. This function has been exploited by mankind for centuries to treat domestic and industrial wastes. In particular, activated sludge processes are commonly used in secondary wastewater treatment to decompose a variety of soluble and insoluble compounds that are not easily removed by sedimentation during primary treatment. In recent years, the use of activated sludge processes for decomposing a wide variety of toxic chemicals has also been widely studied (73–75).

Activated sludge typically consists of a mixed microbial community, together with colloidal particles and other suspended solids. There are two main types of bacterial subpopulations in activated sludge: filamentous bacteria and floc-forming bacteria (38). Bioflocculation is an essential feature of activated sludge processes. Whether it is carried out in a continuous plug flow reactor or in a sequencing batch reactor (76), a subsequent clarification zone or step is necessary to protect the receiving water stream from discharges of highly active biological solids and also to enable high concentrations of biomass to be recycled to the biologically active aeration zone or step. To achieve rapid and complete settling of solids in the clarifier, it is important that the sludge retain a high degree of flocculation. Moreover, flocculation aids in the adsorption of pollutants during the aeration/contact period (38).

Unfortunately, the aggregation and sedimentation properties of biomass in the activated sludge process are quite variable. Failure of an activated sludge to adequately flocculate and settle is known as "bulking." In general, a proliferation of the filamentous organisms is observed in bulking sludges (38,76). Thus, the stability of the flocculent character of activated sludge is expected to be improved by operating under conditions that control the growth of filamentous bacteria (73–75). Stable flocculation also depends on the synthesis of extracellular polymers by the floc-forming bacteria, and this may be manipulated by controlling the nutrient environment (30,76). Additionally, the floc size and sedimentation properties are strongly influenced by the hydrodynamic environment to which they are subjected (4,77). A better understanding of all these factors is currently helping to reduce the occurrence of bulking and to design clarifiers that consistently achieve the desired separation of solids from water.

## 3.3 Sedimentation-Based Cell Harvesting and Recycle

The first step in product recovery subsequent to fermentation is the separation of the cells from the bulk of the spent fermentation medium. Cell separations are also required if the biomass is to be concentrated and recycled to achieve high productivities in continuous fermentations. An alternative to cell recycle for continuous fermentations is to retain or immobilize cells within the fermentor. One method of cell retention involves using flocculent cells in a fluidized bed or tower fermentor in which the continuous upflow of fluid exerts a drag on the flocs that is balanced by the gravitational settling force acting on them (31).

A common method of cell harvesting, vacuum filtration, employs a filter aid such as diatomaceous earth which becomes associated with the biomass in the resulting filter cake. This is not desirable if the biomass itself is the product or if the intention is to recycle the biomass to the fermentor. Possible alternatives include membrane-based filtration, which does not employ a filter aid, and gravitational and centrifugal sedimentation.

Owing to the small size and relative density of single cells, gravity-based sedimentation is slow. Consequently, centrifuges are commonly used in the fermentation and biotechnology industries. Unfortunately, centrifugation is expensive when employed on the large scale. The cost associated with the centrifugal harvesting step of recombinant cells used to overexpress a pharmaceutical protein may be only a small fraction of the total cost of producing and purifying the final product, but this is not the case for low-value products such as microbial biomass.

Microbial biomass includes such diverse products as baker's yeast, bacterial insecticides, nitrogen-fixing bacteria, and single-cell protein (78). The term "single-cell protein" (SCP) refers to microorganisms such as algae, actinomycetes, bacteria, yeasts, molds, and higher fungi grown in large-scale fermentation systems for use as protein sources in human foods or animal feeds (40). Although this term was coined by C. L. Wilson at the Massachusetts Institute of Technology in 1966, people have eaten certain microorganisms as a portion of their diet since ancient times (79). Presently, the economies of SCP are not competitive with other low-cost protein sources, such as fish meal and soy beans, so it is primarily used to provide supplemental proteins and vitamins rather than as a primary food supply (79). Research continues on developing improved processes and exploiting cheap raw materials to reduce the cost of production of SCP.

One of the major problems with SCP production is the difficulty in harvesting cells due to slow sedimentation (79–82). In general, centrifuges are used for harvesting bacterial SCP, whereas algae and yeasts are often harvested by gravitational settling because of their larger cell sizes. In either case, the rate of cell harvesting may be increased by the addition of chemical polymeric flocculents, but this often leads to the unsatisfactory result of the artificial flocculent staying with the food or feed product (79,82). This problem has been circumvented by Imperial Chemical Industries (ICI) in the United Kingdom. The ICI process uses a pressure-cycle airlift fermentor to grow *Methylophilus methylotrophus* bacteria on methanol (78,82,83). Although operated only intermittently owing to market considerations, this process is capable of producing 6000 metric tons of dry product, known commercially as Pruteen, per month (79). An unusual feature of the process is the initial separation of the cells from their growth medium by the induction of flocculation and sedimentation so that a higher biomass slurry can be fed to the harvesting centrifuges than is typical of most SCP processes (79). A schematic of this process is shown in Figure 8. Moreover, an artificial flocculent is not required to induce agglomeration. Instead, partial cell lysis is achieved by heat and acid treatment (83). This presumably releases proteins and nucleic acids that serve to induce flocculation.

In addition to inducing flocculation, another method for effectively increasing the rate of sedimentation is the use of inclined settlers. The devices,

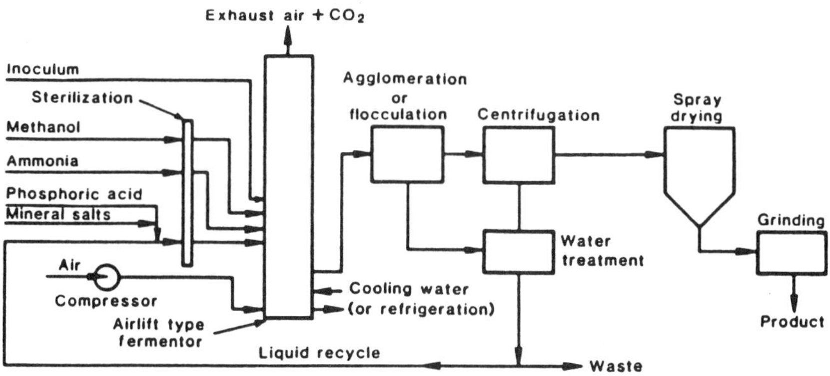


**Figure 8** Schematic of the ICI process for making single-cell protein from methanol. (From Ref. 40, copyright 1984 by the American Association for the Advancement of Science, reprinted with permission.)

known commercially as "supersettlers" or "lamella settlers," consist of sedimentation vessels having inclined walls. These settlers are composed of either a single narrow tube or channel inclined from the vertical or a large tank containing several closely spaced, tilted plates. Clarification rates in inclined settlers are often one or two orders of magnitude larger than those observed in vertical settlers. Clarifiers with inclined surfaces have been used for decades for the removal of solids from wastewater.

Enhanced sedimentation in vessels with inclined walls was first documented in 1920 by the physician Boycott (84) and recently reviewed by Davis and Acrivos (59). The enhancement in the sedimentation rate (the "Boycott effect") results from an increase in the surface area available for settling. When a vessel contains inclined surfaces, the particles settle not only onto the vessel bottom, but also onto the upward-facing walls. These particles then form thin sediment layers that slide rapidly toward the bottom of the vessel due to gravity, as shown in Figure 9. Simultaneously, a clarified fluid layer is produced beneath each downward-facing wall. These particle-free layers are more buoyant than the bulk suspension, and the fluid flows rapidly to the top. Inclined settlers may easily be adapted to continuous operation. The feed suspension may be introduced into the settler in a variety of ways, and the concentrated sediment is removed as a slurry from the bottom of the settler while the clarified fluid is removed from the top of the settler.

The clarification rate in inclined settlers is given by the PNK theory (85,86). This theory states that the rate of production of clarified fluid is equal to the vertical settling velocity of the particles multiplied by the hori-

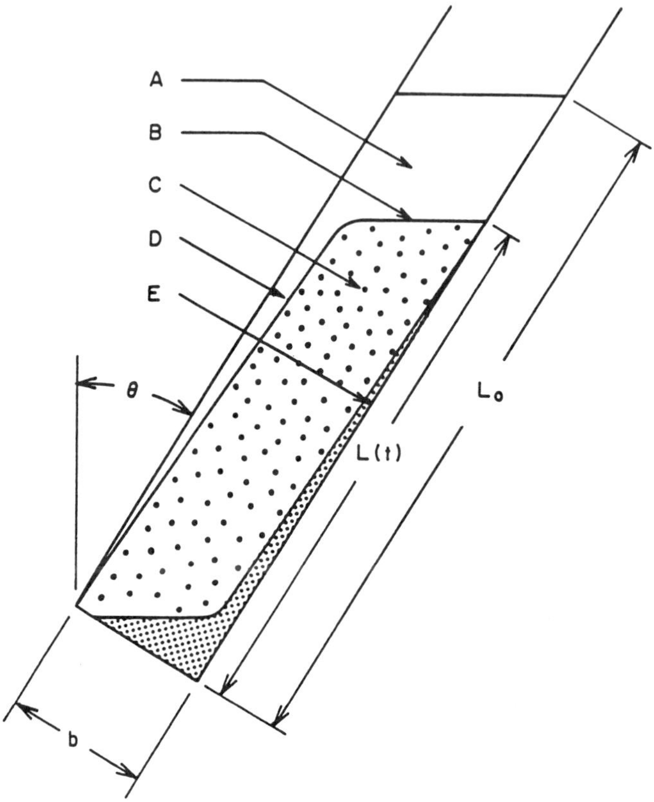


**Figure 9** Regions that develop during sedimentation in an inclined channel: (A) region of particle-free fluid above the suspension, (B) interface between the particle-free fluid and the suspension, (C) suspension, (D) thin, particle-free fluid layer beneath the downward-facing surface, (E) concentrated sediment. $L_o$ is the length of the portion of the vessel filled with suspension at time $t = 0$, $L(t)$ is this length at a later time, $\theta$ is the angle of inclination of the vessel walls from the vertical, and $b$ is the spacing between these walls. (Reproduced from Ref. 91 by permission of Society for Industrial Microbiology, copyright 1985.)

zontal projection of the channel area available for settling. Thus, for a single channel with the parallel-plate geometry of Figure 9,

$$S(v) = vw(L \sin \theta + b \cos \theta), \qquad (6)$$

where $S$ is the volumetric rate of production of fluid clarified of particles or cells with vertical settling velocity $v$, $\theta$ is the inclination angle of the plates from the vertical, $b$ is the spacing between the plates, $w$ is the width of the channel, and $L$ is the length of the portion of the vessel filled with suspension. The corresponding clarification rate for vertical settlers ($\theta = 0$) is $S = vwb$, and the ratio of the inclined to vertical settling rate is thus equal to $(L/b) \sin \theta + \cos \theta$. This ratio can be made very large by using high aspect ratio channels ($L/b \gg 1$) at moderate inclinations.

Equation (6) was originally proposed on purely geometric arguments; more recently, Acrivos and Herbolzheimer (87) have used the continuity equation of fluid mechanics to verify its validity. A summary of further mathematical theories for the flow of particles and fluid within inclined settlers can be found in the review by Davis and Acrivos (59). Equation (6) is strictly valid only if all the particles fall with the same velocity, $v$. If there is a distribution of shapes or sizes (as is typical in a suspension of microorganisms), then there will be an accompanying distribution of settling velocities. The sedimentation of such polydisperse suspensions in inclined channels has been described by Davis et al. (88).

The majority of fundamental experimental studies of sedimentation in inclined tubes and channels has focused on suspensions of inert solid particles, even though the enhancement effect was first documented for the sedimentation of red blood cells (84). A few recent experiments have demonstrated the utility of the Boycott effect for microbial suspensions (89–91). Davis and Birdsell (91) examined batch sedimentation of unflocculated suspensions of *S. cerevisiae* yeast in vertical and inclined settlers. They found that the time required for 75% of the cells in a culture to settle was reduced from over 33 h for a vertical settler to just over 0.5 h when the same settler was inclined at an angle of 60° from the vertical. Their settler had an aspect ratio of $L/b = 90$ initially, which decreased with time as the cells settled out of suspension. Very good agreement between theory and experiment was obtained.

Recently, Tabera and Iznaola (92) have designed an inclined settler for recycle of a flocculated yeast in a continuous ethanol fermentation process. Their settler has a compact lamella design employing several parallel inclined plates. This design gave an enhanced separation efficiency over that of a traditional vertical settler design. As a result of the improved biomass recycling achieved, the new fermentor employing the inclined settler achieved a 37% increase in ethanol productivity over that of the equivalent column fermentor with a vertical side branch for cell recycle (92). This improvement is primarily the result of the higher dilution rates that were made possible by the more efficient recycle system. In earlier work, Stephanopoulos et al. (93) attached an inclined settler sidearm directly to a laboratory fermentor (see Fig. 10). This allowed for an essentially cell-free fermentation broth to be withdrawn through the sidearm while maintaining a high cell density in the fermentor. Continuous flow fermentation experi-

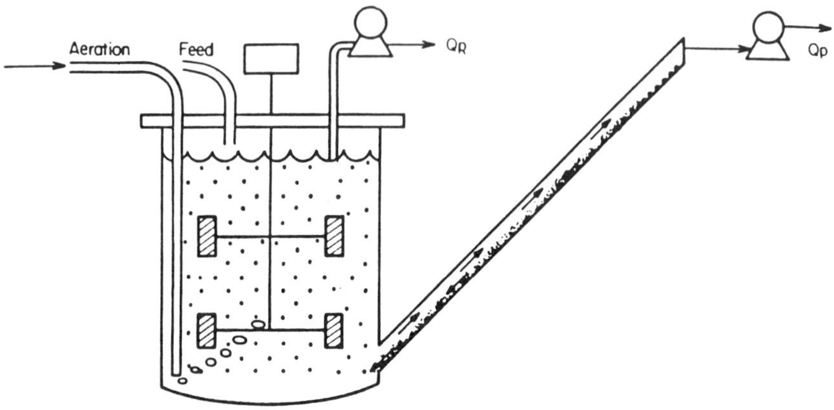


**Figure 10** Schematic of a continuous fermentor design employing an inclined settler sidearm for cell sedimentation and recycle. (Reproduced from Ref. 93 by permission of the American Institute of Chemical Engineers, copyright 1985 AIChE.)

ments with *S. cerevisiae* yeast showed that many-fold increases in the steady-state cell concentration could be achieved using high dilution rates that would lead to cell washout in an ordinary chemostat. Of course, the concept of cell recycle for increasing cell concentrations and productivities in continuous cultures is not new. It has been used for at least 70 years in activated sludge processes for wastewater treatment (94), and more recently in processes such as ethanol production by yeast (94,95) and L-sorbose, Z-ketogluconic acid, and cholera toxin production by bacteria (96,97). However, these processes require either very large conventional settling tanks or membrane-based crossflow filtration units. Inclined settlers offer a simple and compact alternative to these technologies.

## 4 FLOCCULATION AND SEDIMENTATION FOR SELECTIVE CELL SEPARATIONS AND RECYCLE

### 4.1 Fundamentals of Sedimentation-Based Selective Cell Separations

In the previous section, the use of inclined settlers for enhancing cell harvesting and recycle was described. These applications involve the nonselective separation of cells from their suspending fluid. *Selective* cell separations, in which different subpopulations of cells are separated from each other, may be of even greater importance in modern biotechnology because they can be used to selectively remove unproductive or parasitic cells from fermentation or cell-culture bioreactors.

Because of the importance to immunology, pathology, hematology, and other related fields, a host of selective cell separation methods have been developed, and a number of symposia and books have been devoted to the subject (98,99). These methods include differential sedimentation and centrifugation, partitioning in two-phase aqueous systems, affinity adhesion, magnetic cell sorting, optical cell sorting, field flow fractionation, and electrophoresis. In general, they have been developed for analytical purposes and have been limited to small-scale, high-resolution applications (98). Recently, Davis and co-workers (100–103) have shown that inclined settlers are capable of high-capacity separations of cell subpopulations with different settling velocities due to differences in cell size, density, shape, or flocculation character, although with a lower resolution than previous methods. Inclined settlers are particularly suitable for high-capacity separations of cell subpopulations because of their simple construction, ease of scale-up, and ability to be operated on a continuous basis.

The theory of an inclined settler operated under steady-state and transient conditions for inert particle classification by size has been developed previously (104). Davis et al. (100) extended the theory to steady-state operation of an inclined settler for selective cells separations, as shown in Figure 11. A cell suspension is fed into the settler at the volumetric flowrate $Q_f$. This feed suspension contains cells with a distribution of settling velocities owing to their distribution of sizes, densities, shapes, and/or flocculation character. This distribution is then separated by the inclined settler into an overflow stream containing a fine fraction of slower-settling cells, which do not settle out of suspension in the inclined settler, and an underflow stream containing a coarse fraction of faster-settling cells, which do settle out of suspension. The overflow rate and underflow rate are denoted as $Q_o$ and $Q_u$. The goal is to predict the composition of the fine and coarse fractions, given the composition of the feed suspension. The equations derived below are for a single subpopulation of cells. They may then be applied to each subpopulation present in the feed suspension, or to the entire cell population.

Steady-state mass balances on total suspension, total cells in the subpopulation, and those cells in the subpopulations that have settling velocity $v$, respectively, are given as follows:

$$Q_f = Q_o + Q_u, \tag{7}$$

$$Q_f X_f = Q_o X_o + Q_u X_u, \tag{8}$$

$$Q_f X_f P_f(v) = Q_o X_o P_o(v) + Q_u X_u P_u(v), \tag{9}$$

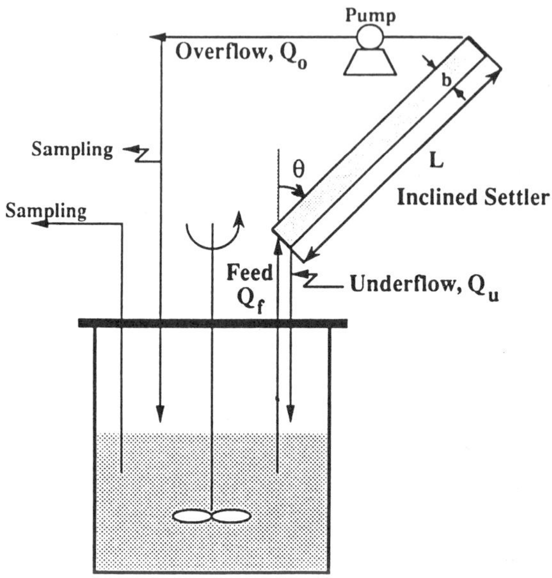


**Figure 11** Schematic of an inclined settler used for selective cell separations. (Reprinted from Ref. 100 with permission, copyright 1991 American Chemical Society.)

where $X$ is the total cell mass concentration of the subpopulation, and the subscripts $f$, $o$, and $u$ refer to the feed, overflow, and underflow streams, respectively. Also, $P(v)$ is the normalized probability density function, defined such that $P(v)dv$ is the fraction of cells by mass in a given stream that has settling velocities between $v$ and $v + dv$. $P(v)$ is normalized so that

$$\int_0^\infty P(v)dv = 1. \tag{10}$$

The probability density function in the feed stream, $P_f(v)$, can be determined either directly by allowing a sample to sediment in the presence of an optical density recording device (32) or indirectly by measuring the particle size distribution microscopically or with a particle size analyzer and then using an expression such as Eq. (2) for nonflocculent cells or Eq. (5) for flocculated cells, to relate the sedimentation velocity of each particle to its size.

The final mass balance required is that on cells entering the overflow stream. Material entering the overflow is comprised of a mixture of clarified fluid, which enters the overflow at the volumetric rate $S(v)$ given by

Eq. (6), and unsettled suspension, which enters the overflow at the volumetric rate $Q_o - S(v)$. A cell mass balance on the mixing of these two streams yields

$$Q_o X_o = (Q_o - S(v))X_f \qquad Q_o > S(v). \tag{11}$$

If $Q_o \leq S(v)$, all of the cells with vertical settling velocity $v$ settle out of suspension before reaching the overflow, and $X_o = 0$.

For dilute suspensions in which it is assumed that the cells settle without interfering with one another, Eq. (11) may be applied to any given fraction of the distribution, with the modification that the total cell concentration, $X$, is replaced by the mass fraction of cells having settling velocities between $v$ and $v + dv$, which is $XP(v)dv$. Therefore,

$$Q_o X_o P_o(v) = (Q_o - S(v))\, X_f P_f(v) \qquad S(v) < Q_o, \tag{12}$$

$$P_o(v) = 0 \qquad S(v) \geq Q_o. \tag{13}$$

Integrating Eq. (12) over all cell-settling velocities in the subpopulation, with the normalization constraint given by Eq. (4) applied, an expression for the total cell concentration in the overflow stream is obtained as

$$X_o = X_f \int_0^{v_o} \frac{(Q_o - S(v))}{Q_o} P_f(v)\, dv, \tag{14}$$

where $v_o$ is the cutoff sedimentation velocity and is defined by $S(v_o) = Q_o$. Cells with settling velocities greater than $v_o$ should settle out of suspension before reaching the overflow. The probability density function for the cells in the overflow stream is obtained by substituting Eqs. (12) and (13) into Eq. (14):

$$P_o(v) = \frac{(Q_o - S(v))\, P_f(v)}{\int_0^{v_o} (Q_o - S(v))\, P_f(v)\, dv} \qquad v < v_o, \tag{15}$$

$$P_o = 0 \qquad v \geq v_o. \tag{16}$$

Similar expressions for the composition of the underflow stream are obtained by using Eqs. (7)–(9), as described by Davis et al. (104).

The theory for selective cell separations using an inclined settler was tested for both nonflocculent and flocculent cell systems. One model nonflocculent system was nondividing and dividing yeast (100). The nondividing yeast strain used was *S. cerevisiae* 378 with the plasmid pBM746. This strain has a bar1 chromosomal mutation that enables cell divisions to be stopped in the presence of $\alpha$-factor, a 13-amino peptide. It cannot grow in a $Ura^-$ medium if cells lose the plasmids. When cell divisions are stopped, the cells begin to increase their sizes and become irregular in shape. The dividing yeast strain was the same host but without the plasmid and without the addition of $\alpha$-factor during its growth phase. Size distributions of the yeast cells were measured by an Elzone 180XY particle size analyzer manufactured by Particle Data, Inc. Typical average equivalent sphere diameters are 4.5 $\mu$m for normal dividing cells and 7.4 $\mu$m for nondividing cells.

The inclined settler was fabricated from rectangular glass tubing having height $b$ = 0.5 cm, width $w$ = 4.0 cm, and length $L$ = 20 cm. After the cells were grown batchwise, their culture medium was replaced with a non-growth buffer, and the suspension was drawn through the inclined settler using a peristaltic pump. Cells that settled out of suspension formed a thin sediment layer on the upward-facing wall of the settler. These cells then slid down the wall due to gravity and were thereby returned to the reservoir. The overflow stream was recycled through a separate port. The overflow rate, $Q_o$, was varied systematically. After each adjustment in the overflow rate was made, the total recycle system was allowed to operate for at least 3 h before sampling in order to assure steady-state conditions. After steady state was reached for each overflow rate, samples were taken from both the feed reservoir and the overflow stream. The samples were analyzed by the

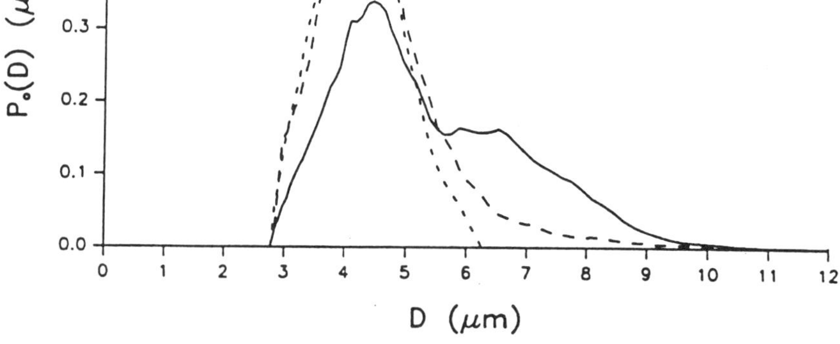


**Figure 12** Size distributions of yeast cells in the feed stream and settler overflow streams for a mixed culture of dividing and nondividing cells. (Reprinted from Ref. 100 with permission, copyright 1991 American Chemical Society.)

Elzone 180XY for cell concentrations and size distributions. The fraction of nondividing yeast cells was determined by replica plating on first a rich medium and then a defined Ura$^-$ medium.

The size distributions of yeast cells in the feed stream and the overflow stream for a typical experiment are shown in Figure 12. The feed stream consisted of 50% dividing cells (by number) and 50% nondividing cells. The partially resolved peaks for these two subpopulations are apparent in the feed distribution shown as the solid line in Figure 12. This suspension was drawn through the settler at the rate $Q_o = 0.62$ ml/min, with the settler angle from the vertical being $\theta = 45°$. These conditions correspond to a cutoff diameter of $D_o = 6.2$ μm for the largest cells predicted to reach the overflow without settling. Thus, the overflow should contain primarily the smaller, dividing cells. Indeed, the overflow was found to contain 89% dividing cells for this particular experiment. The measured size distribution in the overflow (dashed line in Fig. 12) is in very good agreement with the predictions of Eq. (15) (dotted line in Fig. 12).

Figures 13 and 14 show how the separation of a 50 : 50 mixture (by

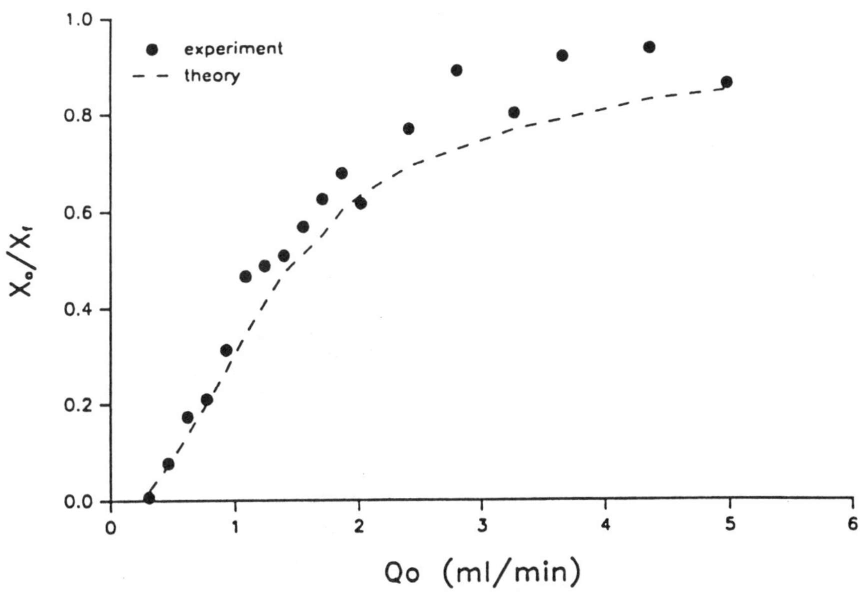


**Figure 13** Relative total cell mass concentration in the settler overflow as a function of the overflow rate for a mixed culture of dividing and nondividing yeast cells. (Reprinted from Ref. 100 with permission, copyright 1991 American Chemical Society.)

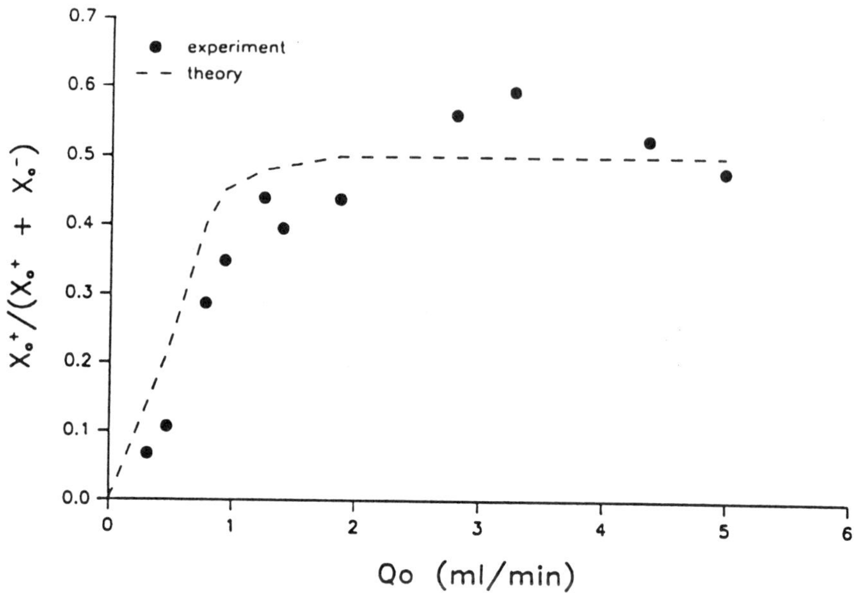


**Figure 14** Fraction of plasmid-bearing, nondividing yeast cells in the settler overflow as a function of the overflow rate for a mixed culture of dividing and nondividing yeast cells. (Reprinted from Ref. 100 with permission, copyright 1991 American Chemical Society.)

number) of dividing and nondividing yeast cells depends on the overflow rate from the settler. Figure 13 is a plot of the total cell mass concentration (dividing and nondividing cells combined) in the overflow, normalized by the feed concentration, versus the overflow rate from the settler. At low overflow rates, the cell concentration in the overflow was very small because there was sufficient holdup time in the settler for most of the cells to settle out of suspension. As the overflow rate was increased, the cell concentration in the overflow increased because there was less holdup time for settling. The experimental data follow closely the prediction of Eq. (14). Of more direct relevance for selective cell separation is Figure 14, which shows the fraction of nondividing cells in the overflow stream versus the overflow rate. In this figure, the (+) superscript refers to the nondividing strain, and the (−) superscript refers to the dividing strain. At low overflow rates, the larger cells had sufficient time to settle out of suspension, so the overflow contained primarily dividing cells. As the overflow rate was increased, a greater percentage of the larger, nondividing cells reached the settler overflow, so the selectivity of the separation was decreased. At very high overflow rates, the composition of the overflow stream approached the 50 : 50 composition of the feed stream. There is good agreement between the data and the theoretical prediction obtained by applying Eq. (14) independently for the two subpopulations.

A second nonflocculent system studied experimentally was viable and nonviable hybridomas (100,103). A mouse hybridoma cell line, AB2-143.2, derived from Sp2/0 myloma and which produces IgG2a antibodies against benzene-arsonate, was used. As measured by the Elzone 180XY particle size analyzer, the average diameter of the nonviable cells is approximately 8 $\mu$m, whereas the average diameter of the viable cells is approximately 13 $\mu$m. The selective separation experiments for hybridomas were performed in a fashion similar to the yeast experiments, except that a chemostat operation was used in which fresh nutrient medium was continuously added to the bioreactor while a product stream was removed at an equal rate. This was necessary in order to maintain the viable cell population at a nearly constant cell size distribution over the duration of the experiments. Typical results are demonstrated in Figure 15, which contains forward angle light scattering (FALS) histograms of the viable and nonviable cell subpopulations in the bioreactor (solid lines) and the overflow stream (dashed lines) made by an EPICS flow cytometer. These results are for a perfusion experiment having $Q_o$ = 72 ml/h and a viable fraction of 50% by number in the

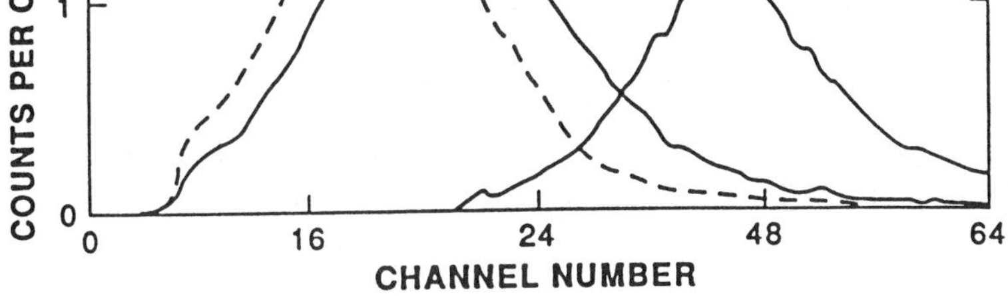


**Figure 15** FALS histograms of viable and nonviable hybridoma cell subpopulations in bioreactor (solid lines) and overflow (dashed line). The number of viable cells reaching the overflow stream was too small to be shown on the scale of this graph. (Reprinted from Ref. 100 with permission, copyright 1991 American Chemical Society.)

bioreactor. The overflow stream contained a negligible (<0.1%) quantity of viable cells, and its cell size distribution is close to that of the smaller, nonviable subpopulation in the bioreactor. As the overflow rate was increased, the viable cells also began to appear in significant amounts in the overflow, and the resolution of the separation decreased (100,103).

A systematic study (previously unpublished) of the selective separation of flocculent and nonflocculent yeast cells was also performed in the author's laboratory using an inclined settler. *S. cerevisiae* 402, a moderately flocculent strain ranging from single cells to flocs containing more than 100 cells, was used. The probability density functions of the floc-settling velocities in the feed and overflow streams were measured using a sedimentation/light extinction method (51). Cell mass concentrations were measured by drying and weighing the samples. The cell mass concentration in the overflow stream, nondimensionalized by that in the feed stream, is shown in Figure 16 as a function of the overflow rate from the inclined settler. The overflow rate is nondimensionalized by $\bar{Q} = S(\bar{v})$, where $\bar{v}$ is the median floc-settling velocity in the feed distribution. At low overflow rates, the larger flocs had sufficient time to settle out of suspension and the overflow was found to contain primarily single cells and small flocs. As the

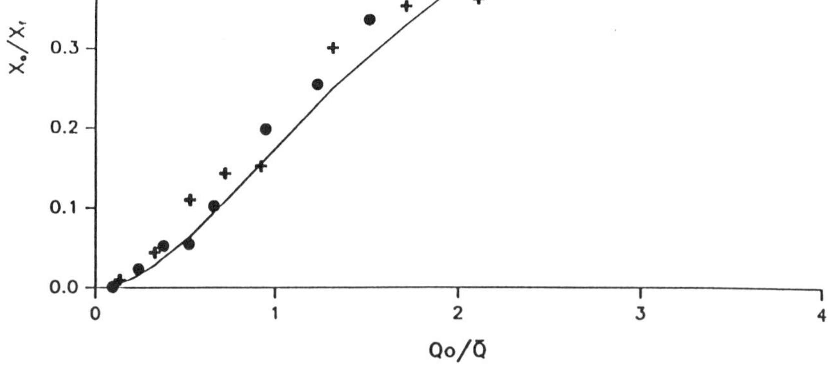


**Figure 16** Relative total cell mass concentration in an inclined settler overflow as a function of the dimensionless overflow rate for a moderately flocculent yeast strain.

overflow rate was increased, the concentration in the overflow increased owing to the reduction in the settler holdup time, which allowed larger flocs to reach the overflow. The solid line in Figure 16 is the nondimensional equivalent of Eq. (14). In this form, the nondimensional predictions are independent of the angle of inclination, and the experimental data for both $\theta = 30°$ and $\theta = 45°$ are in good agreement with the theory.

## 4.2 Applications of Sedimentation-Based Selective Cell Separations

Mixed culture systems involving two or more species or subpopulations competing for the same growth-limiting substrate are known to be unstable in continuous-flow bioreactors—the population with higher specific growth rate under the operating conditions employed will dominate while the other species or subpopulation will wash out of the system (93). There are many applications, however, for which it is desirable to tamper with this microbial version of *natural selection* so as to maintain a slower-growing subpopulation as dominant, or at least in stable coexistence with a faster-growing subpopulation. Examples include coexistence of floc-forming bacteria together with filamentous bacteria in activated sludge processes, elimination of unproductive contaminant strains that compete with productive strains during fermentation processes, and the maintenance of plasmid-bearing recombinant cells as dominant despite competition from plasmid-free segregant cells, which have higher specific growth rates due to the absence of the metabolic burden associated with the overexpression of a recombinant protein. One possible strategy for achieving the desired *unnatural selection* is to include with the bioreactor system a separator that is able to selectively separate the cell subpopulations so that the desired, slower-growing strain may be recycled to the bioreactor while the undesired, faster-growing strain is discarded. Based on the studies described in the previous section, it is clear that inclined settlers may be used to at least partially separate cell subpopulations, provided they exhibit an exploitable difference in their settling velocity distributions.

### *4.2.1 Activated Sludge Processes*

Although only occasionally using inclined or lamella settlers (93), the most long-standing application of sedimentation-based selective recycle of microorganisms is in the activated sludge process for secondary wastewater treatment. As reviewed by Sheintuch (105), the competition between floc-forming and filamentous microorganisms is affected by the substrate and dissolved oxygen levels. Since these levels may be shifted by incorporating biomass or effluent recycle, it follows that the competition outcome may be shifted by recycle. Moreover, the floc-forming microorganisms settle more rapidly than the filamentous microorganisms, so the recycle from the settler may be enriched in the faster-settling, floc-forming microorganisms. Although this effect is limited because of coflocculation of the two subpopulations, it does improve the reactor performance (105). Similar phenomena have been reported for algal cultures (106).

### 4.2.2 Contamination Control

Stephanopolous and co-workers (93,107) were the first to employ an inclined settler to selectively recycle a slower-growing species to maintain it as dominant in a continuous flow bioreactor, despite competition from a faster-growing, contaminant species. They investigated a mixed culture of *E. coli* and *S. cerevisiae* growing on a glucose-limited medium. The bacteria have a higher growth rate than the yeast and would normally dominant a chemostat culture. However, the yeast settle considerably faster because of

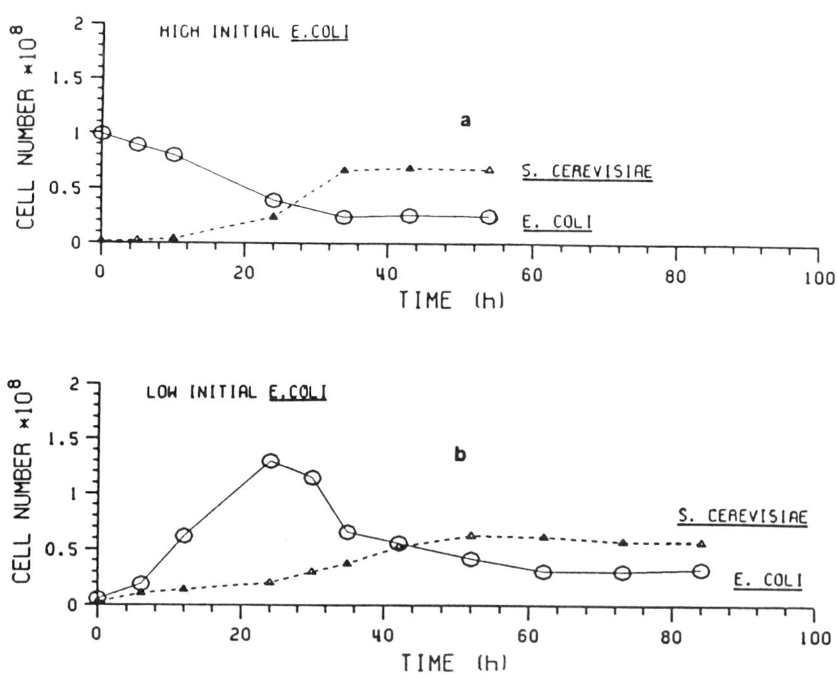


**Figure 17** Stable coexistence of *E. coli* bacteria and *S. cerevisiae* yeast at moderate dilution rates using an inclined settler sidearm to selectively recycle yeast cells. (Reproduced from Ref. 107 by permission of the American Institute of Chemical Engineers, copyright 1985 AIChE.)

their larger size. Using the design shown in Figure 10, with proper selection of the overflow rate from the settler, the yeast cells settled out of suspension in the inclined settler sidearm and were recycled by gravity to the bioreactor, whereas the bacteria did not settle appreciably and were removed from the top of the settler. At moderate dilution rates, a stable coexistence was achieved, with the yeast cells dominating (see Fig. 17). At high dilution rates, the bacteria cells were washed out of the bioreactor, whereas the yeast cells were maintained due to their selective recycle (see Fig. 18). These results have potential application in overcoming the contamination of cultures by faster-growing but slower-settling microorganisms.

### 4.2.3 Flocculent Yeast Cultures

Davis and Parnham (101) extended the use of inclined settlers for selective recycle to the case of continuous fermentations with competition between a flocculent and a nonflocculent strain of *S. cerevisiae* yeast. The flocculent

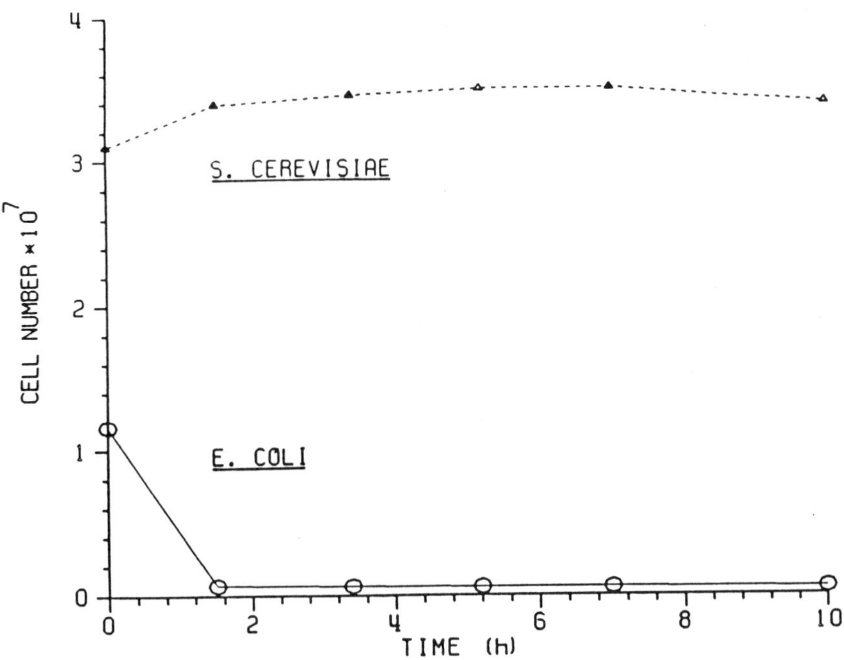


**Figure 18** Washout of *E. coli* bacteria at high dilution rates using an inclined settler sidearm to selectively recycle yeast cells. (Reproduced from Ref. 93 by permission of the American Institute of Chemical Engineers, copyright 1985 AIChE.)

strain contained the *FLO1* gene and was highly flocculent. It had a lower growth rate than the nonflocculent strain (a commercial baker's yeast selected for its powdery nature and high growth rate). This growth rate disadvantage was exacerbated by diffusional limitations of substrate to the cells in the interior of the millimeter-sized flocs. As a result, the slower-growing flocculent strain washed out of the fermentor when no recycle was used. In contrast, the inclined settler was able to selectively recycle the flocculent cells due to their higher sedimentation velocities, thereby maintaining them as dominant (see Fig. 19).

Davis and Parnham (101) also developed a theory to describe the reactor/settler system shown in Figure 20. The bioreactor is considered, in general, to have two product lines. One is a direct product line in which the cell concentrations remain unchanged. The other passes through a selective cell separator and is split into two streams, a concentrated stream that is recycled and a diluted stream that is removed from the system. They considered competition between two microbial strains, a desired (+) strain with

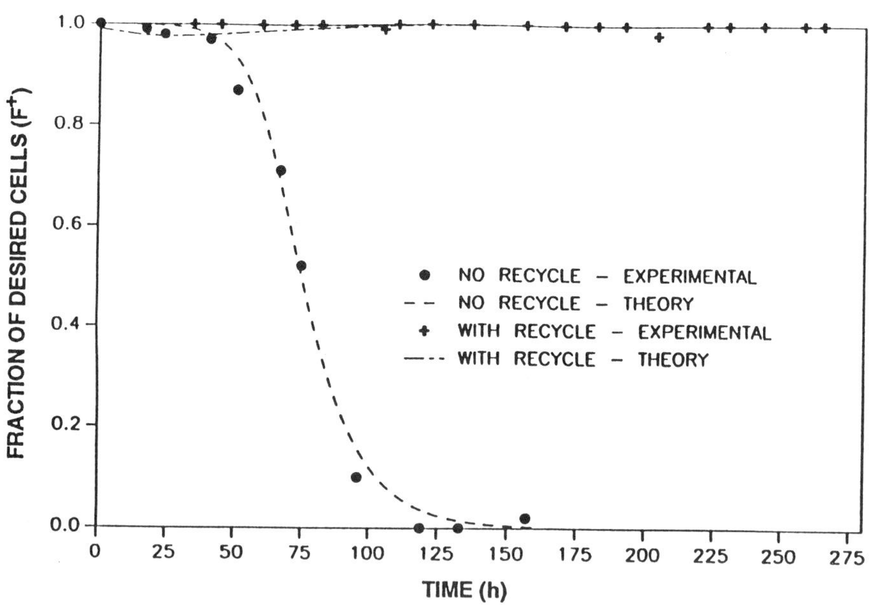


**Figure 19** Washout of a slower-growing, flocculent yeast strain by a faster-growing, nonflocculent yeast strain without recycle, and maintenance of the flocculent strain with selective recycle. (Reprinted from Ref. 101 by permission of John Wiley & Sons, Inc., copyright 1989.)

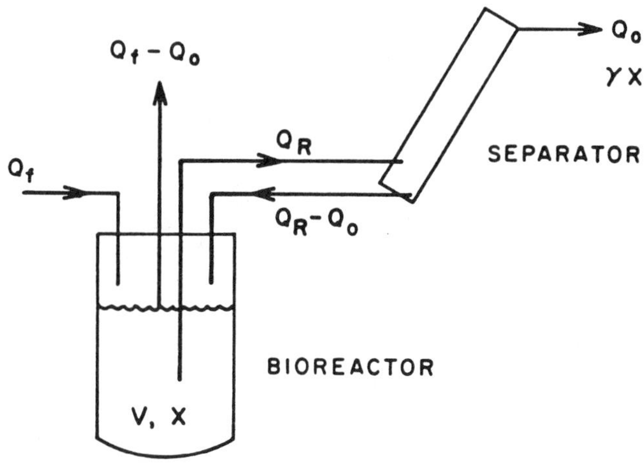


**Figure 20** Schematic of a bioreactor/separator system employing an inclined settler for selective cell recycle. (Reprinted from Ref. 101 by permission of John Wiley & Sons, Inc., copyright 1989.)

concentration $X^+$ in the fermentor and an undesired (−) strain with concentration $X^-$ in the fermentor.

Unsteady mass balances about the entire system (reactor plus cell separator) are

$$\frac{dX^+}{dt} = (\mu^+ - \beta^+ D)X^+ - \mu^+ X^+, \tag{17}$$

$$\frac{dX^-}{dt} = (\mu^- - \beta^- D)X^- + \mu^+ pX^+, \tag{18}$$

where $\mu$ is the specific growth rate, $D = Q_f/V$ is the dilution rate, $p$ is probability for segregation of the (+) strain into the (−) strain [which may be important if these refer to plasmid-bearing (+) and plasmid-free (−) strains], and $\beta$ is an effective cell dilution factor accounting for both the direct product line and the product line that passes through the separator. The factor $\beta$ is given by

$$\beta = 1 - (1 - \gamma)Q_o/Q_f. \tag{19}$$

The quantity $\gamma$ is a cell dilution factor equal to the ratio of the cell concentration in the diluted stream exiting the cell separator to that entering the cell separator, the latter being the same as the cell concentration in the well-stirred bioreactor. This quantity may either be measured directly for each subpopulation or predicted using Eq. (14). For a "perfect" separator, $\gamma^+ = 0$ and $\gamma^- = 1$, indicating that all the desired cells are concentrated and recycled, whereas the undesired cells flow through the separator and are removed without being concentrated. In deriving Eqs. (17) and (18), sterile feed, negligible cell maintenance and death terms, and negligible cell growth within the separator are assumed. The latter assumption is a good approximation when the separator volume is small compared to the fermentor volume. It is also assumed that any physiological differences between the recycled cells and those in the bioreactor, due to the shock experienced by the cells during their residence in the different environment of the settler, have a negligible effect on the overall reactor performance.

Monod kinetics on a limiting substrate are used:

$$\mu^- = \frac{\mu_m^- S}{K^- + S} \qquad \mu^+ = \frac{\mu_m^+ S}{K^+ + S}. \tag{20}$$

The mass balance on the limiting substrate is

$$\frac{dS}{dt} = D(S_o - S) - \frac{\mu^+ X^+}{Y^+} - \frac{\mu^- X^-}{Y^-}, \tag{21}$$

where $S$ is the concentration of limiting substrate in the bioreactor, $S_o$ is the limiting substrate concentration in the feed stream, and $Y^+$ and $Y^-$ are the yield coefficients for cell growth (substrate consumption for product formation is assumed to be either negligible or growth-associated and included in the yield coefficients).

When $p = 0$, which corresponds to the experimental conditions examined by Davis and Parnham (101), the steady-state solution of Eqs. (17)–(21) depends on the single parameter $G \equiv (\mu^+\beta^-)/(\mu^-\beta^+)$, called the maintenance ratio. For $G < 1$, the steady-state solution is

$$X^+ = 0 \qquad \mu^- = \beta^- D. \tag{22}$$

This corresponds to complete washout of the desired (+) strain, such as would be observed if there were no selective recycle ($\beta^+ = \beta^-$) and if the desired strain had a growth rate disadvantage ($\mu^+ < \mu^-$). On the other hand, the steady-state solution for $G > 1$ is

$$X^- = 0 \qquad \mu^+ = \beta^+ D. \tag{23}$$

This corresponds to the desired strain remaining dominant in the reactor and is possible even if there is a growth rate disadvantage, provided the selective enrichment factor, $\beta^-/\beta^+$, is sufficiently greater than $\mu^-/\mu^+$.

For the experiments shown in Figure 19, the maintenance ratio was less than unity without recycle and greater than unity with recycle, correctly predicting washout and maintenance of the desired strain, respectively

(101). Moreover, the dynamic measurements for the fraction of the flocculent strain in the bioreactor, $F^+ \equiv X^+/(X^+ + X^-)$, as a function of time are in good agreement with the theory. In further experiments, Davis and Parnham (85) showed that selective recycle could be used to recover the slower-growing desired strain after it had been allowed to be nearly washed out in the absence of recycle. They also observed a fivefold increase in the total biomass concentration when recycle was used compared to when it was not (108).

### *4.2.4 Recombinant Bacterial Fermentations*

Another potential application of selective cell separation technology is overcoming plasmid instability in continuous fermentations with recombinant microorganisms. Most recombinant products are produced by batch processes, even though continuous processes are usually more economical and efficient for the production of most chemicals. One reason for the preferred use of batch processes for recombinant DNA products is the problem of plasmid instability. Upon cell division, a daughter cell has a finite possibility of not receiving a plasmid, even though its parent cell contained multiple copies. The resulting cell is called a segregant. Although this phenomenon occurs in only a few percent or less of the new cells, the decreased metabolic burden of these plasmid-free cells results in an increased growth rate (109). This means that even if a chemostat is inoculated with 100% plasmid-bearing cells, segregants can result, and after a few generations, they can take over the reactor. Other reasons why continuous fermentations are not used extensively in industry include contamination and validation difficulties, as well as the ready availability of batch fermentation equipment.

A variety of methods have been used to reduce the problem of plasmid instability, although not without drawbacks. These methods include incorporating antibiotic-resistance genes in the plasmids and then adding antibiotics to the culture medium, using defined media that lack essential amino acids that only plasmid-bearing cells are able to synthesize, employing two reactors in series for which cell growth occurs in the first stage and product expression in the second stage, and modifying the genetics of the cells so that the segregant cells are incapable of reproducing (102). Ollis (110) was the first to propose that selective recycle could be used to overcome plasmid instability. He developed a theory to predict the resulting bioreactor behavior but did not comment extensively on a method for separating the plasmid-bearing and plasmid-free cells. Stephanopoulos et al. (93) noted that plasmid-bearing bacterial cells contain additional internal protein and so are slightly larger and more dense than the segregant cells. The author's group has attempted to exploit this difference to separate recombinant and segregant cells by differential sedimentation, but the conclusion is that the settling velocities of individual bacteria are too small for this possibility to be practical without resorting to high-speed centrifuges.

To get around the problem of small settling velocities of recombinant bacterial cells, Henry et al. (102) took advantage of the genetic control of flocculation. They used a host strain which was Pil$^-$ (lacking the gene for pilin protein) and inserted a plasmid, pORN108, which contains the *pil* operon (10). As a result, the plasmid-bearing cells synthesize type 1 pili, which leads to flocculation, whereas the plasmid-free segregant cells do not synthesize pili and are nonflocculent. They then used this strain in the reactor/settler system shown in Figure 19. When no recycle was used, the plasmid-bearing (+) cells were overtaken by plasmid-free (−) cells in only two days (see Fig. 21). This was due to the relatively high segregation probability ($p = 0.03$), which followed from the low average copy number of 3.8 plasmids per cell, and to the large growth rate differential ($\mu_m^+ = 0.72$/h vs. $\mu_m^- = 1.14$/h), which followed from the significant metabolic burden associated with the overproduction of pili. In contrast, the plasmid-bearing strain was maintained as dominant when the inclined settler was

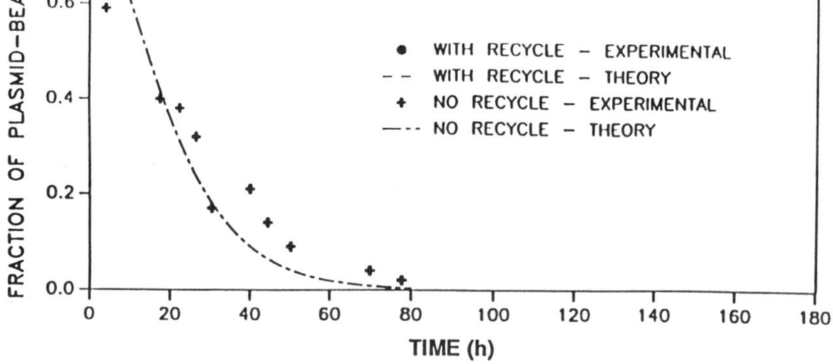


**Figure 21** Washout of a plasmid-bearing, flocculent bacterial strain by the plasmid-free, nonflocculent host strain in the absence of recycle, and maintenance of the plasmid-bearing strain with selective recycle. (Adapted from Ref. 102 with permission, copyright 1990 American Chemical Society.)

used (see Fig. 21). The inclined settler was designed so that most of the flocculent cells had time to settle out of suspension and be recycled to the bioreactor ($\gamma^+ = 0.45$), while the nonflocculent cells did not settle and were removed through the settler effluent ($\gamma^- = 1.0$).

For a segregating culture, the maintenance ratio is defined as $G \equiv \beta^-\mu^+(1 - p)/\beta^+\mu^-$ (102). The steady-state solution to Eqs. (17) and (18) is then

$$F^+ = 0 \qquad G \leq 1, \tag{24}$$

$$F^+ = \frac{G - 1}{G - 1 + p\mu^+/\mu^-} \qquad G > 1, \tag{25}$$

where $F^+ \equiv X^+/(X^+ + X^-)$ is the fraction of plasmid-bearing cells in the fermentor (95). Thus, stable maintenance is predicted for $G > 1$, and complete washout of plasmid-bearing cells is predicted for $G \leq 1$. Indeed, the values of $G = 1.33$ and $G = 0.61$ for the experiments shown in Figure 21 with and without selective recycle, respectively, confirm these predictions. Henry et al. (102) also showed that selective recycle is able to recover the fraction of plasmid-bearing cells after they had been allowed to be almost completely washed out in the absence of recycle (see Fig. 22).

In more recent work, a new plasmid was constructed that allowed for the overexpression of a model protein, $\beta$-lactamase (12). This plasmid is constructed so that pilin expression, which leads to bacterial flocculation, is under control of the *tac* operon. The plasmid-bearing cells were induced to form pili and flocculate by the addition of a chemical, IPTG, only within the inclined settler (111). In this way, cell growth and product synthesis could take place in the main fermentation vessel without the inhibiting effects of pilin production. Once again, selective recycle was shown to maintain, and even recover, the productive, plasmid-bearing cells by using their flocculent character to separate them from the nonflocculent, segregant cells in the inclined settler. As shown in Figure 23, product levels were increased more than 10-fold using this strategy (111). Moreover, the plasmid-bearing cells were observed to wash out of the fermentor after only 3 days in the absence of selective recycle, leading to the cessation of $\beta$-lactamase production, whereas high levels of $\beta$-lactamase production were maintained for more than 10 days with selective recycle. After this, the addition of IPTG to the separator was stopped. This caused the plasmid-bearing cells to no longer produce pili and flocculate, so they were quickly washed out of the bioreactor/separator system due to the normal plasmid instability phenomenon.

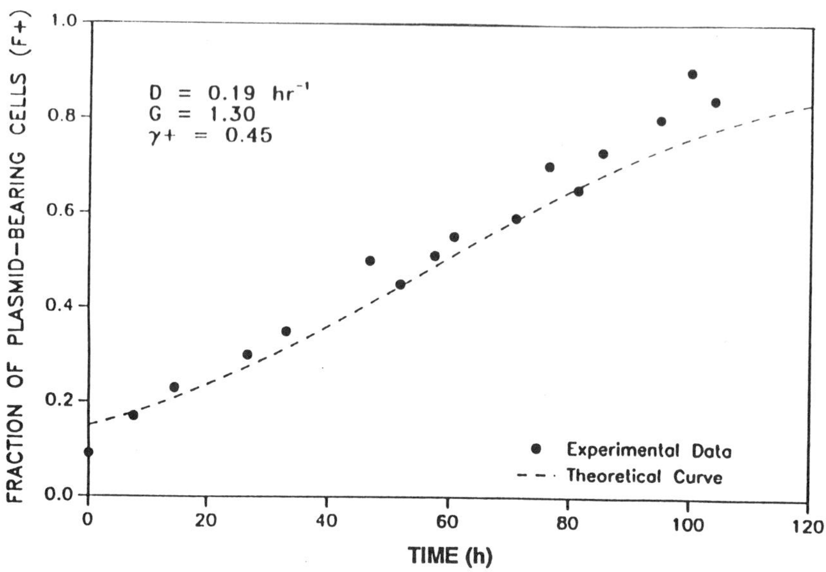


**Figure 22** Recovery of a plasmid-bearing bacterial strain using selective flocculation and recycle. (Reprinted from Ref. 102 with permission, copyright 1990 American Chemical Society.)

### *4.2.5 Mammalian Cell Cultures*

The final application of selective cell separations using inclined settlers that is discussed in this chapter, and perhaps the one with the most promise, involves mammalian cell cultures. In particular, cells that result from a hybrid fusion of a fast-growing cancerous cell and an antibody-producing immune cell are chosen for a model system. The production of monoclonal antibodies using these hybridomas is usually performed in continuous perfusion cultures because of the need to remove toxic by-products that would otherwise accumulate in the growth media. Because of their low growth rates, it is desirable to use a cell retention or recycle device that enables the bioreactors to be perfused without cell washout at dilution rates greater than the maximum specific growth rate. A typical cell retention device employs an internal spin filter, which is attached to the reactor's impeller shaft and permits the removal of cell-free culture medium (112,113). Cell recycle devices typically employ an external tangential flow filter in order to concentrate the cell suspension and recycle it back to the bioreactor (114). An alternative passes the reactor effluent through a vertical sedimen-

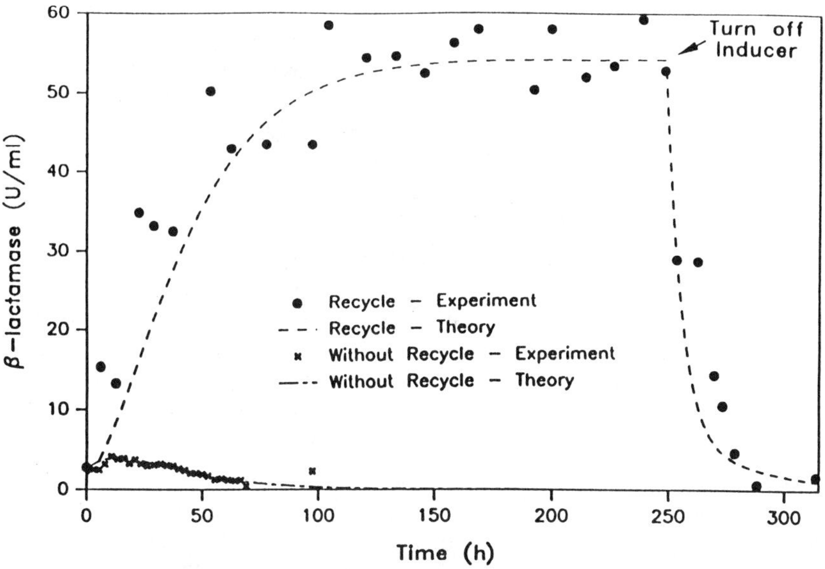


**Figure 23** Comparison of protein overexpression in a continuous bioreactor with and without selective recycle of plasmid-bearing cells. (Reprinted from Ref. 111 by permission of John Wiley & Sons, Inc., copyright 1991.)

tation column, which is sufficiently long to allow suspended cells to settle back into the bioreactor before they can be washed out (115).

Filter devices can be detrimental to long-term culture productivity because the hybridomas are subjected to excessive shear forces, resulting in higher cell death rates. Moreover, all the methods mentioned above are characterized by the continuous accumulation of unproductive dead cells in the bioreactor, which forces the intermittent removal of a culture fraction containing both viable and nonviable cells. This is especially imperative in the case of filtering devices when cell density and dilution rates become sufficiently high to cause filter clogging. The accumulation of nonviable cells and the removal of viable cells along with nonviable cells limit the culture productivity.

It is conceivable that the limitations in culture productivity may be overcome by continuously removing only nonviable hybridomas from the reactor while selectively retaining all viable cells in the reactor, provided a method may be developed for accomplishing a selective separation of these subpopulations. It is known that faster-growing, viable mammalian cells are larger in average cell volume and specific mass than slower-growing, dead cells (103). Therefore, it is expected that the desired selective cell separation may be accomplished by exploiting the different sedimentation velocities of viable and nonviable hybridomas using inclined sedimentation.

Tyo and Thilly (116) have developed a conical lamella cell separator, which they placed on top of a small spinner-flash perfusion device. Their original design was to achieve total recycle due to sedimentation in the separator, with essentially cell-free fluid being withdrawn from the top of the separator. Indeed, they were able to achieve a final cell density of $2.2 \times 10^7$ cells/ml, with a viability of about 80%, using a perfusion rate of 10 culture volumes per day. More interesting is that they found very few small cells ($D < 10$ $\mu$m) in their bioreactor when using this high perfusion rate. They attributed this to a selective separation that took place in the conical lamella settler. The smaller cells, which included predominantly nonviable cells, were preferentially removed from the system, while the larger cells, which were predominantly viable, settled out of suspension in the lamella settler and were subsequently returned to the bioreactor.

Batt et al. (103) used a rectangular inclined settler above their bioreactor to retain virtually all the viable hybridomas, while selectively removing from the culture a portion of the smaller, nonviable cells. As shown in Figure 24, both the viable and nonviable cells had sufficient time to settle out of suspension at the low dilution rate of 0.9/day used initially. As the dilution rate was gradually increased, the nonviable cells began to appear in the settler overflow due to the reduced holdup time for sedimentation. At the maximum dilution rate of 1.7/day, the nonviable cell concentration exiting the top of the inclined settler was 20% of that entering the settler. In contrast, the concentration of the larger, viable cells in the settler effluent was less than 0.1% of that entering the settler, even at the maximum dilution rate. The total cell concentration after 10 days of perfusion was $2.3 \times 10^7$ cells/ml, but the viability was only 45%. The authors concluded that this could be improved by increasing the dilution rate to remove a higher fraction of the nonviable cells while still retaining nearly all the viable cells with the inclined settler.

The antibody concentration in the reactor during the perfusion culture by Batt et al. (103) is shown in Figure 25. The antibody concentration increased from 10.4 $\mu$g/ml in the chemostat without cell recycle to 30.4 $\mu$g/ml on the final day of the perfusion culture with selective cell recycle. The increase in antibody concentration during the perfusion culture reflects not only the increased viable cell concentration, but also an increase in specific antibody productivity per viable cell as well. The variation in specific antibody productivity also is plotted in Figure 25. The specific antibody productivity nearly doubled over the duration of the perfusion culture. The total antibody productivity per culture volume increased from 8.6 $\mu$g/ml/

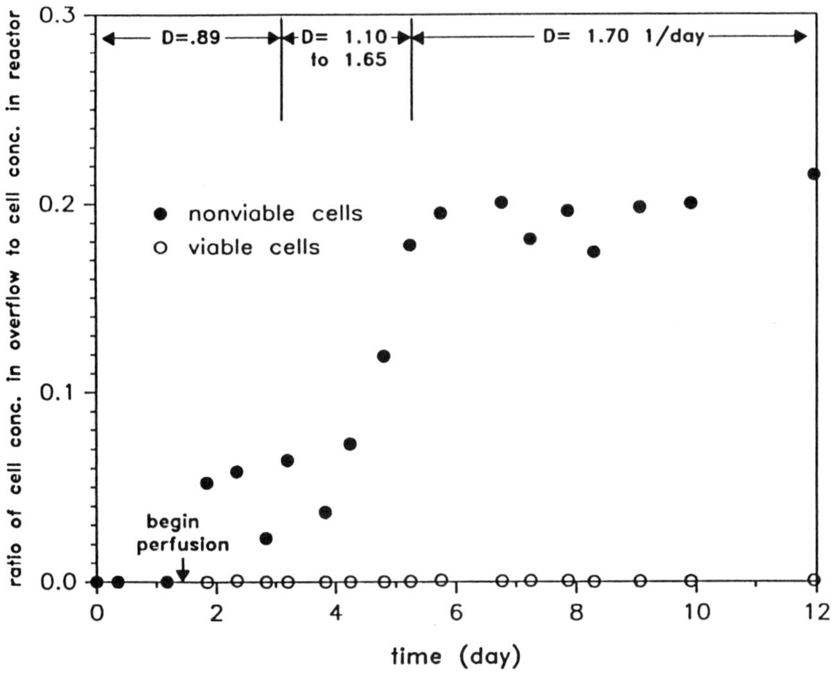


**Figure 24** Viable and nonviable hybridoma cell concentrations in the inclined settler overflow relative to those in the stream entering the settler from the bioreactor over the course of a perfusion culture. (Reprinted from Ref. 103 with permission, copyright 1990 American Chemical Society.)

day in the chemostat to 52 $\mu$g/ml/day at the time that the perfusion culture was terminated, a sixfold increase.

The results of the studies described above show that selective recycle of productive cells and removal of nonproductive cells from continuous culture is practical with inclined sedimentation, provided the productive and nonproductive cells exhibit a difference in sedimentation velocities. This may either be due to a difference in cell size and density, as is the case with hybridoma cells, or due to a difference in flocculation character, which may be controlled genetically in recombinant cells. For a given cell suspension, the degree of cell separation depends on the overflow rate through the settler. The overflow rate determines the residence time of suspended cells in the sedimentation channel. Depending on the Stokes settling velocity of a cell or floc, the allotted residence time may be sufficient for it to settle out of suspension; if not, it is washed out of the culture. Since inclined settlers

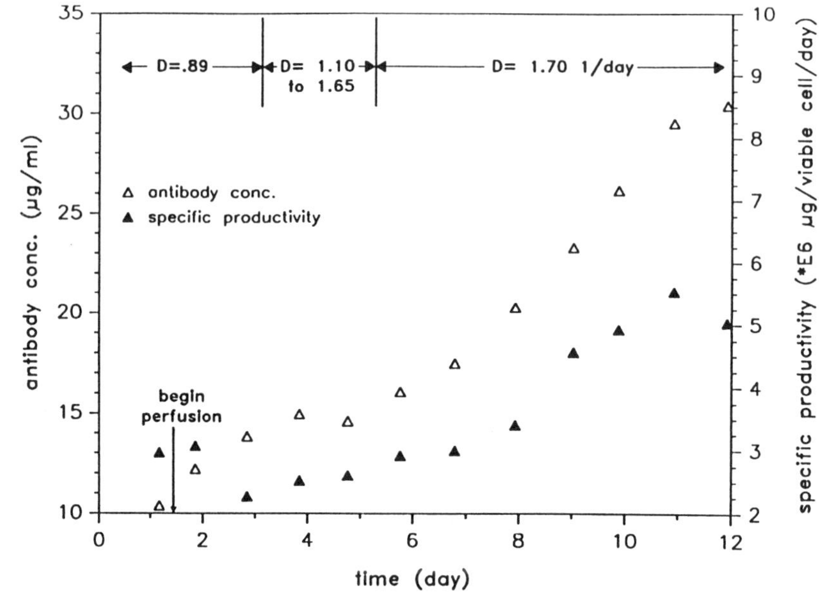


**Figure 25** Antibody concentration and specific productivity in the bioreactor over the course of a perfusion culture using an inclined settler for selective retention of viable cells. (Reprinted from Ref. 103 with permission, copyright 1990 American Chemical Society.)

are simple devices that are easily scaled up to provide greater available area for sedimentation, it is expected that they have considerable potential for being extended from laboratory studies to commercial applications.

## REFERENCES

1. Calleja, G.B. (1984). *Microbial Aggregation*, CRC Press, Boca Raton, FL.
2. Attai, Y.A., ed. (1987). *Flocculation in Biotechnology and Separation Systems*, Elsevier, Amsterdam.
3. Marshall, K.C., ed. (1984). *Microbial Adhesion and Aggregation*, Springer-Verlag, Berlin.
4. Atkinson, B. (1984). Consequences of aggregation, in *Microbial Adhesion and Aggregation* (K.C. Marshall, ed.), Springer-Verlag, Berlin, pp. 351–371.
5. Unz, R.F. (1987). Aspects of bioflocculation, in *Flocculation in Biotechnology and Separation Systems* (Y.A. Attai, ed.), Elsevier, Amsterdam, pp. 351–368.
6. Russell, I., Stewart, G.G., Reader, H.P., Johnston, J.R., and Martin, P.A.

(1980). Revised nomenclature of genes that control yeast flocculation, *J. Inst. Brew*, *86*: 120–121.
7. Stewart, G.G., and Russell, I. (1977). The identification, characterization and mapping of a gene for flocculation in *Saccharomyces* sp., *Can. J. Microbiol.*, *23*: 441–447.
8. Miki, B.L.A., Hung Poon, N., James, A.P., and Seligy, V.L. (1982). Possible mechanism for flocculation interactions governed by gene *FL01* in *Saccharomyces cerevisiae*, *J. Bact.*, *150*: 878–889.
9. Miki, B.L.A., Hung Poon, N., and Seligy, V.L. (1982). Repression and induction of flocculation interactions in *Saccharomyces cerevisiae*, *J. Bact.*, *150*: 890–899.
10. Orndorff, P.E., and Falkow, S. (1984). Identification and characterization of a gene product that regulates type 1 piliation in *Escherichia coli*, *J. Bact.*, *160*: 61–66.
11. Orndorff, P.E., and Falkow, S. (1984). Organization and expression of genes responsible for type 1 piliation in *Escherichia coli*, *J. Bact.*, *159*: 736–744.
12. Klemm, P. (1984). The *finA* gene encoding type 1 fimbrial subunit of *Escherichia coli*, *Eur. J. Biochem.*, *143*: 395–399.
13. Orndorff, P.E., and Falkow, S. (1985). Nucleotide sequence of *pil A* gene encoding the structural component of type 1 pili in *Escherichia coli*, *J. Bact.*, *162*: 454–457.
14. Ogden, K.L., and Taylor, A.L. (1991). Genetic control of flocculation in *Escherichia coli*, *J. Ind. Microbiol.*, *7*: 279–286.
15. Warne, S.R., and Bowden, C.P. (1987). Biosurface properties and their significance to primary separation: a genetic approach to the utilization of bacterial auto-aggregation, in *Separations for Biotechnology* (M.S. Verrall and M.J. Hudson, eds.), Ellis Horword, Chicester, UK, pp. 1–15.
16. Easson, B.D., Peoples, O.P., Rha, C.K., and Sinskey, A.J. (1987). Engineering of biopolymer flocculents: a recombinant DNA approach, in *Flocculation in Biotechnology and Separation Systems* (Y.A. Attai, ed.), Elsevier, Amsterdam, pp. 369–382.
17. Esser, K., Hinrichs, J., and Kues, U. (1987). Genetic control of flocculation of yeast with respect to application in biotechnology, in *Flocculation in Biotechnology and Separation Systems* (Y.A. Attai, ed.), Elsevier, Amsterdam, pp. 383–398.
18. Stewart, G.G., Russell, I., and Garrison, I.F. (1975). Some considerations of the flocculation characteristics of ale and lager yeast strains, *J. Inst. Brew.*, *81*: 248–257.
19. Amri, M.A., Bonaly, R., Duteurtre, B., and Moll, M. (1979). Interrelation between $Ca^{++}$ and $K^{+}$ ions in the flocculation of two brewer's yeast strains, *Eur. J. Appl. Microbiol. Biotechnol.*, *7*: 235–240.
20. Mill, P.J. (1964). The effect of nitrogenous substances on the time of flocculation of *Saccharomyces cerevisiae*, *J. Gen. Microbiol.*, *35*: 53–60.
21. Mill, P.J. (1964). The nature of the interactions between flocculent cells in the flocculation of *Saccharomyces cerevisiae*, *J. Gen. Microbiol.*, *35*: 61–68.

22. Jayatissa, P.M., and Rose, A.H. (1976). Role of wall phosphomannan in flocculation of *Saccharomyces cerevisiae, J. Gen. Microbiol., 96*: 165–174.
23. Beaven, M.J., Belk, M., Stewart, G.G., and Rose, A.H. (1979). Changes in electrophoretic mobility and lytic enzyme activity associated with development of flocculating ability in *S. cerevisiae, Can. J. Microbiol., 25*: 888–895.
24. Lyons, T.P., and Hough, J.S. (1970). Flocculation of Brewer's yeast, *J. Inst. Brew., 76*: 564–570.
25. Lyons, T.P., and Hough, J.S. (1970). The role of yeast cells in brewing, *Brew. Dig., 45*: 51–61.
26. Lyons, T.P., and Hough, J.S. (1971). Further evidence for the cross-bridging hypothesis for flocculation of brewer's yeast, *J. Inst. Brew., 77*: 300–305.
27. Rose, A.H. (1984). Physiology of cell aggregation: flocculation by *Saccharomyces cerevisiae* as a model system, in *Microbial Adhesion and Aggregation* (K.C. Marshall, ed.), Springer-Verlag, Berlin, pp. 323–325.
28. Calleja, G.B., and Johnson, B.F. (1977). A comparison of quantitative methods for measuring yeast flocculation, *Can. J. Microbiol., 23*: 68–74.
29. Eddy, A.A. (1955). Flocculation characteristics of yeast. II. Sugar as dispersing agents, *J. Inst. Brew., 64*: 313–317.
30. Silverstein, J., Bowman, A.M., Hess, T., Schmidt, S.K., and Howe, B. (1990). 2,4-Dinitrophenol degradation and sludge floc size, in *Env. Eng. Proc.* (C.R. O'Melia, ed.), pp. 205–212.
31. Prince, I.G., and Barford, J.P. (1982). Tower fermentation using *Zymomonas mobilis* for ethanol production, *Biotechnol. Lett., 4*: 525–530.
32. Miki, B.L., Poon, N.H., and Seligy, V.I. (1982). Repression and induction of flocculation interactions in *Saccharomyces cerevisiae, J. Bacteriol., 150*: 890–899.
33. Szlag, D.C. (1988). Factors affecting yeast flocculation, M.S. thesis, University of Colorado, Boulder.
34. Hunt, T.P. (1985). Quantitative determination of yeast flocculation, M.S. thesis, University of Colorado, Boulder.
35. Porter, A.M., and MacAulay, R.J. (1965). Studies on flocculation. I. A relationship between pH and calcium content of the growth medium, *J. Inst. Brew., 71*: 175–179.
36. Levine, S., and Friesen, W.I. (1987). Flocculation of colloid particles by water-soluble polymers, in *Flocculation in Biotechnology and Separation Systems* (Y.A. Attai, ed.), Elsevier, Amsterdam, pp. 3–20.
37. Eriksson, L.B., and Hardin, A-M. (1987). Flocculation of *E. coli* bacteria with cationic polyelectrolytes, in *Flocculation in Biotechnology and Separation Systems* (Y.A. Attai, ed.), Elsevier, Amsterdam, pp. 441–456.
38. Thompson, J.M., and Forster, C.F. (1983). Nutrient conditions in relation to bacterial flocculation, extracellular polymer production and sludge settlement, *Biotechnol. Lett., 5*: 761–766.
39. Kakii, K., Sugahara, E., Shirakashi, T., and Kuriyama, M. (1986). Isolation and characterization of $Ca^{++}$-dependent, floc-forming bacterium, *J. Ferment. Technol., 64*: 57–62.

40. Litchfield, J.H. (1984). Single-cell proteins, in *Biotechnology and Biological Frontiers* (P.H. Abelson, ed.), AAAS, Washington, D.C., pp. 269–283.
41. Aunins, J.G., and Wang, D.I.C. (1989). Induced flocculation of animal cells in suspension culture, *Biotechnol. Bioeng.*, *34*: 629–638.
42. Persson, I., and Lindman, B. (1987). Flocculation of cell debris for improved separation by centrifugation, in *Flocculation in Biotechnology and Separations Systems* (Y.A. Attai, ed.), Elsevier, Amsterdam, pp. 457–466.
43. Madsen, M., and Schundt, J. (1989). Low technology water purification by bentonite clay flocculation as performed in Sudanese villages: bacteriological examination, *Water Res.*, *23*: 873–882.
44. Garrod, D.R. (1984). Aggregation, cohesion, adhesion, phagocytosis and morphogenesis in *Dictyostelium*—mechanisms and implications, in *Microbial Adhesion and Aggregation* (K.C. Marshall, ed.), Springer-Verlag, Berlin, pp. 337–349.
45. Atkinson, B., and Daoud, I.S. (1976). Microbial flocs and flocculation in fermentation process engineering, *Adv. Biochem. Eng.*, *4*: 41–124.
46. Calleja, G.B., and Johnson, B.F. (1979). Temperature sensitivity of flocculation induction, conjugation and sporulation in fission yeast, *Antonie van Leeuwenboek J. Microbiol., Serol.*, *45*: 391–400.
47. Stahl, U., Kues, U., and Esser, K. (1983). Flocculation in yeast, an assay on the inhibition of cell aggregation, *Eur. J. Appl. Microbiol. Biotechnol.*, *17*: 199–202.
48. Mandelbrot, B. (1982). *The Fractal Geometry of Nature*, W.H. Freeman, San Francisco.
49. Weitz, D.A., Huang, J.S., Lin, M.Y., and Sung, J. (1985). Limits of the fractal dimension for irreversible kinetic aggregation of gold colloids, *Phys. Rev. Lett.*, *54*: 1416–1419.
50. Weitz, D.A., and Oliveria, M. (1984). Fractal structures formed by kinetic aggregation of gold colloids, *Phys. Rev. Lett.*, *52*: 1433–1436.
51. Davis, R.H., and Hunt, T.P. (1986). Modeling and measurement of yeast flocculation, *Biotechnol., Prog.*, *2*: 91–97.
52. Wilkinson, D.B. (1989). The fractal nature of biological aggregates, M.S. thesis, University of Arizona, Tucson.
53. Logan, B.E., and Wilkinson, D.B. (1990). Fractal dimensions and porosities of *Zoogloea ramigera* and *Saccharomyces cerevisiae*, *Biotech. Bioeng.*, *38*: 389–396.
54. Logan, B.E., and Wilkinson, D.B. (1990). Fractal geometry of marine snow and other biological aggregates, *Limunol. Oceanogr.*, *35*: 130–136.
55. Li, D-H., and Ganczarczyk, J.J. (1986). Physical characteristics of activated sludge flocs, *CRC Crit. Rev. Environ. Control*, *17*: 53–87.
56. Li, D-H., and Ganczaraczyk, J.J. (1989). Fractal geometry of particle aggregates in water and wastewater treatment processes, *Env. Sci. Technol.*, *23*: 1385–1389.
57. Li, D-H., and Ganczarczyk, J.J. (1989). Structure of activated sludge flocs, *Biotechnol. Bioeng.*, *35*: 57–65.

58. Logan, B.E., and Alldredge, A.L. (1989). The increased potential for nutrient uptake by flocculating diatoms, *Marine Biol.*, *101*: 443–450.
59. Davis, R.H., and Acrivos, A. (1985). Sedimentation of noncolloidal particles at low Reynolds numbers, *Annu. Rev. Fluid Mech.*, *17*: 91–118.
60. Richardson, J.F., and Zaki, W.N. (1954). Sedimentation and fluidization: part I, *Trans. Inst. Chem. Eng.*, *32*: 35–53.
61. Davis, R.H., and Birdsell, K.H. (1988). Hindered settling of semidilute monodisperse and polydisperse suspensions, *AIChE J.*, *34*: 123–129.
62. Li, D-H., and Ganczarczyk, J.J. (1987). Stroboscopic determination of settling velocity, size and porosity of activated sludge flocs, *Water Res.*, *21*: 257–262.
63. Alldredge, A.L., and Gottschalk, C. (1988). In situ settling behavior of marine snow, *Limnol. Oceanogr.*, *33*: 339–351.
64. Logan, B.E., and Hunt, J.R. (1988). Bioflocculation as a microbial response to substrate limitations, *Biotechnol. Bioeng.*, *31*: 91–101.
65. Stewart, G.G., and Russell, I. (1988). Yeast flocculation, in *Brewing Science* (J. Pollock, ed.), Academic Press, New York, pp. 61–92.
66. Suzzi, G., Romano, P., and Zambonelli, C. (1984). Flocculation of wine yeasts: frequency, differences and stability of the character, *Can. J. Microbiol.*, *30*: 36–39.
67. Burns, J.A. (1937). Yeast flocculation, *J. Inst. Brew (Lond.)*, *43*: 31–41.
68. Gilliland, R.B. (1957). The flocculation of brewing yeast, *Wallerstein Lab. Commun.*, *20*: 41–68.
69. Woof, J.B. (1962). An approach to yeast sedimentation characteristics, *J. Inst. Brew.*, *68*: 315–317.
70. Nishihara, H., Toraya, T., and Fukui, S. (1976). Factors affecting flocculation of brewer's yeast, *J. Ferm. Technol.*, *54*: 356–360.
71. Westermann, D.H., and Huige, N.J. (1979). Beer brewing, in *Microbial. Technology*, 2nd ed., *Vol. II*, *Fermentation Technology* (H.J. Peppler and D. Perlman, eds.), Academic Press, New York, pp. 1–37.
72. Pasteur, L. (1876). *Etudes sur la Bière*, Gauthier Villars, Paris, p. 196.
73. Kennedy, M.S., Mines, R.O., Silverstein, J., Sherrard, J.H., and Weber, A.S. (1987). Activated sludge, *J. Water Pollut. Control Fed.*, *59*: 388–395.
74. Weber, A.S., Silverstein, J., Sherrard, J.H., Mines, R.O., and Kennedy, M.S. (1988). Activated sludge, *J. Water Pollut. Control Fed.*, *60*: 816–824.
75. Mines, R.O., Silverstein, J., Sherrard, J.H., Kennedy, M.S., and Weber, A.S. (1989). Activated sludge, *J. Water Pollut. Control Fed.*, *61*: 799–807.
76. Silverstein, J., and Schroeder, E.D. (1983). Control of activated sludge settling in an SBR, *Proc. ASCE Specialty Conf. Env. Eng.*, Boulder, CO, pp. 238–246.
77. Dunlop, E.H. (1978). Characteristics of sludge from the deep shaft process, in *New Processes of Waste Water Treatment and Recovery* (G. Mattock, ed.), Ellis Horword, London, pp. 179–200.
78. Rose, A.H., ed. (1979). *Economic Microbiology*, *Vol. 4*, *Microbial Biomass*, Academic Press, London.
79. Litchfield, J.H. (1979). Production of single-cell protein for use in food or

feed, in *Microbial Technology*, 2nd ed., *Vol. I, Microbial Processes* (H.J. Peppler and D. Perlman, eds.), Academic Press, New York, pp. 93–155.
80. Benemann, J.R., Weissman, J.C., and Oswald, W.J. (1979). Algal biomass, in *Economic Microbiology, Vol. 4, Microbial Biomass* (A.H. Rose, ed.), Academic Press, London, pp. 177–206.
81. Burrows, S. (1979). Baker's yeast, in *Economic Microbiology, Vol. 4, Microbial Biomass* (A.H. Rose, ed.), Academic Press, London, pp. 315–360.
82. Hamer, G. (1979). Biomass from natural gas, in *Economic Microbiology, Vol. 4, Microbial Biomass* (A.H. Rose, ed.), Academic Press, London, pp. 315–360.
83. Crueger, W., and Crueger, A. (1984). *Biotechnology—A Textbook of Industrial Microbiology*, Science Tech., Madison, WI, pp. 267–276.
84. Boycott, A.E. (1920). Sedimentation of blood corpuscles, *Nature*, *104*: 532.
85. Ponder, E. (1925). On sedimentation and rouleaux formation, *Q. J. Exp. Physiol.*, *15*: 235–252.
86. Nakamura, N., and Kuroda, K. (1937). La cause de l'acceleration de la vitesse de sedimentation des suspensions dans les recipients inclines, *Keijo J. Med.*, *8*: 256–296.
87. Acrivos, A., and Herbolzheimer, E. (1979). Enhanced sedimentation in settling tanks with inclined walls, *J. Fluid Mech.*, *2*: 435–457.
88. Davis, R.H., Herbolzheimer, E., and Acrivos, A. (1982). The sedimentation of polydisperse suspensions in vessels having inclined walls, *Int. J. Multiphase Flow*, *8*: 571–585.
89. Walsh, T.J., and Bungay, H.R. (1979). Shallow-depth sedimentation of yeast cells, *Biotechnol. Bioeng.*, *21*: 1081–1084.
90. Bungay, H.R., and Millspaugh (1984). Cross-flow lamellar settlers for microbial cells, *Biotechnol. Bioeng.*, *26*: 640–641.
91. Davis, R.H., and Birdsell, S.A. (1985). Rapid sedimentation of microbial suspensions, *Dev. Ind. Microbiol.*, *26*: 627–633.
92. Tabera, J., and Iznaola, M.A. (1989). Design of a lamella settler for biomass recycling in continuous ethanol fermentation process, *Biotechnol. Bioeng.*, *33*: 1296–1305.
93. Stephanopoulos, G., San, K-Y., and Davison, B.H. (1985). A novel bioreactor-cell precipitator combination for high-cell density, high-flow fermentations, *Biotechnol. Prog.*, *1*: 250–259.
94. Janssens, J.H., Bernard, A., and Bailey, R.B. (1984). Ethanol from whey: continuous fermentation with cell recycle, *Biotechnol. Bioeng.*, *26*: 1–5.
95. Cysewski, G.R., and Wilke, C.R. (1978). Process design and economic studies of alternative fermentation methods for production of ethanol, *Biotechnol. Bioeng.*, *20*: 1421–1444.
96. Bull, D.N., and Young, M.D. (1981). Enhanced product formation in continuous fermentations with microbial cell recycle, *Biotechnol. Bioeng.*, *23*: 373–389.
97. Spira, W.M., and Fedorka-Cray, P.J. (1983). Euterotoxia production by *Vibrio cholerae* and *Vibrio mimicus* grown in continuous culture with microbial cell recycle, *Appl. Env. Microbiol.*, *46*: 704–709.

98. Pretlow, T.G., and Pretlow, T.P., eds. (1987). *Cell Separation—Methods and Selected Applications*, Vol. 1-5, Academic Press, New York.
99. Kompala, D.S., and Todd, P.W., eds. (1991). *ACS Symposium Series 464—Cell Separation Science and Technology*, ACS Books, Washington, D.C.
100. Davis, R.H., Lee, C-Y., Batt, B.C., and Kompala, D.S. (1991). Cell separations using differential sedimentation in inclined settlers, in *ACS Symposium Series 464—Cell Separation Science and Technology* (D.S. Kompala and P.W. Todd, eds.), American Chemical Society, Washington, D.C., pp. 113-127.
101. Davis, R.H., and Parnham, C.S. (1989). Competitive yeast fermentation with selective flocculation and recycle, *Biotechnol. Bioeng.*, *33*: 767-776.
102. Henry, K.L., Davis, R.H., and Taylor, A.L. (1990). Continuous recombinant bacteria fermentation utilizing selective flocculation and recycle, *Biotechnol. Prog.*, *6*: 7-12.
103. Batt, B.C., Davis, R.H., and Kompala, D.S. (1990). Inclined sedimentation for selective retention of viable hybridomas in a continuous suspension bioreactor, *Biotechnol. Prog.*, *6*: 458-464.
104. Davis, R.H., Zhang, X., and Agarwala, J.P. (1989). Particle classification for dilute suspensions using an inclined settler, *Ind. Eng. Chem. Res.*, *28*: 785-793.
105. Sheintuch, M. (1987). Species selection in a reactor-settler system, *Biotechnol. Bioeng.*, *30*: 598-606.
106. Weissman, J.C., and Benemann, J.R. (1979). Biomass recycling and species competition in continuous cultures, *Biotechnol. Bioeng.*, *21*: 627-648.
107. Davison, B.H., San, K-Y., and Stephanopoulos, G. (1985). Stable coexistence in a continuous fermentor with size-selective properties, *Biotechnol. Prog.*, *1*: 260-268.
108. Henry, K.L., Parnham, C.S., Davis, R.H., and Taylor, A.L. (1990). Competitive continuous fermentations with selective recycle, *Appl. Biochem. Biotechnol.*, *24/25*: 651-662.
109. Imanaka, T., and Aiba, S. (1981). A perspective on the application of genetic engineering: stability of recombinant plasmid, *Ann. NY Acad. Sci.*, *369*: 1-14.
110. Ollis, D.F. (1982). Industrial fermentations with (unstable) recombinant cultures, *Phil. Trans. R. Soc. Lond.*, *B297*: 617-629.
111. Ogden, K.L., and Davis, R.H. (1991). Plasmid maintenance and protein overproduction in selective recycle bioreactors, *Biotechnol. Bioeng.*, *37*: 325-333.
112. van Wezel, A., van der Veldon-de Groot, C., de Haan, H., van den Heuvel, N., and Schasfoort, R. (1985). Large scale animal cell cultivation for production of cellular biologicals, *Dev. Biol. Stand.*, *60*: 229-236.
113. Velez, D., Miller, L., and Macmillan, J.D. (1989). Use of tangential flow filtration in perfusion propagation of hybridoma cells for production of monoclonal antibodies, *Biotechnol. Bioeng.*, *33*: 938-940.
114. Brennan, A., Shevitz, J., and Macmillan, J. (1987). A perfusion system for

antibody production by shear-sensitive hybridoma cells in a stirred reactor, *Biotechnol. Techn.*, *1*: 169–174.

115. Takazawa, Y., Tokashiki, M., Hamamoto, K., and Murakami, H. (1988). High cell density perfusion culture of hybridoma cells recycling high molecular weight components, *Cytotechnology*, *1*: 171–178.

116. Tyo, M.A., and Thilly, W.G. (1989). Novel high density perfusion system for suspension culture metabolic studies, AIChE Annual Meeting, San Francisco, Paper 30g.

# 6
# Microbial Biofilms and Biofilm Reactors

**Brent M. Peyton**
*Pacific Northwest Laboratory, Richland, Washington*

**William G. Characklis†**
*Montana State University, Bozeman, Montana*

## 1 INTRODUCTION

A biofilm is a matrix of cells and cellular products attached to a solid surface or substratum. Biofilms can present a problem when they occur in unwanted locations such as industrial process equipment or implanted medical devices. But these microbial films can also present positive opportunities in bioremediation of hazardous and toxic substances in ground and surface water and wastewater treatment. Biofilm and other immobilized cell reactors offer significant advantages in bioprocessing, such as increased process flow rates without washing the organisms from the reactor. Engineers and scientists are just beginning to realize the significance of biofilms on process industries, natural aquatic systems, and medical technology. Intricate knowledge of microbiology, chemistry, and engineering must be combined to fully understand the processes affecting biofilm accumulation and activity. As more information concerning biofilms is made available, it has become apparent that *a biofilm investigation that does not address the microbial, chemical, and engineering aspects is incomplete*.

Biofilms typically accumulate on surfaces exposed to flowing water. The accumulation of a microbial film is most likely rate-limited by the transport

†Deceased.

of nutrients to and by-products from the bulk liquid to the biofilm. Cellular attachment to a solid surface provides a suitable environment for many types of microorganisms under varied and sometimes harsh conditions. The biofilm can provide protection from toxic substances and other detrimental environmental factors.

The goal of this chapter is to provide an in-depth introduction to the processes affecting biofilm accumulation and activity, and to provide hypotheses for future biofilm research. This chapter was written primarily for those somewhat less familiar with biofilm process engineering.

## 1.1 Relevance of Biofilms

A cross-section transmission electron microscopy (TEM) photomicrograph of a pure culture *Pseudomonas aeruginosa* biofilm is shown in Figure 1a, while scanning electron microscopy (SEM) photomicrographs are shown in Figures 1b, 1c, and 1d. These figures show the scale over which microbial processes occur. At the substratum, cells grow, reproduce, and produce extracellular polymers and other by-products. Biofilms are found in most natural and industrial aquatic systems and account for much of the overall microbial activity in these systems. In streams and rivers, a large proportion of the microbial activity occurs in attached films. Wuhrmann (1) estimated 90–99.99% of the bacterial activity in shallow streams is associated with biofilms. This microbial activity is responsible for transforming and degrading natural and man-made organic compounds in the water. In industrial process equipment, biofilms reduce heat transfer in heat exchange equipment and reduce flow capacity in pipelines, leading to increased energy consumption and increased costs. Biofilms also contribute significantly to corrosion, oil field reservoir plugging and petroleum souring, drinking water deterioration, and computer chip contamination. In industrial systems, biofilm research is usually aimed at reducing biofouling (unwanted biofilms). Many deleterious effects of biofilm formation have been reported. Biofilms cause heat exchangers and condensers (2,3) to lose their heat-transfer efficiency and oil field water-injection systems to plug with detached material. It is believed hydrogen sulfide souring of oil reserves may be caused by attached organisms, which produce hydrogen sulfide, in the oil-bearing formation. With secondary oil recovery becoming standard practice, souring of oil fields has become a major problem whereby the quality of vast oil reserves may be seriously compromised through microbial production of hydrogen sulfide. In a study of a gas storage and production facility (4), bacteria were implicated in the souring of natural gas by indigenous bacteria supplied with nutrients by the motion of water in and out of the field during injection and production. Biological plugging of oil sand reservoirs has been examined (5) with regard to the potential of the water source to induce biofilm-related plugging of the formation. The highest cell densities and extracellular polymeric substance (EPS) concentrations were found in the region where the water entered the sampled core. In the oil field, plugging around an injection well results in decreased injection rates or higher injection pressure. On the other hand, some researchers (6,7) have applied biofilm technology in an attempt to enhance oil recovery. When the larger pores were plugged with biofilm, oil remaining in the smaller pores would be exposed to more flow and more oil would be recovered. In the laboratory, selective plugging (8) of larger pores in sandstone was accomplished using a mixed population with a sucrose-mineral salts medium. The largest pore sizes were reduced from 59–69 $\mu$m before plugging to 20–38 $\mu$m after microbial plugging.

Biofouling causes an increase in fluid friction, resulting in decreased flow capacity in pipelines. Picologlou et al. (9) report the change in friction factor as a function of Reynold's number and biofilm thickness for a mixed-population biofilm. In case histories of closed conduits experiencing frictional losses as a result of biofilms, flow capacity was reduced by as much as 55%.

## 2 PROCESSES CONTRIBUTING TO BIOFILM ACCUMULATION

The progression of biofilm accumulation typically follows a sigmoidal-shaped curve in terms of biofilm mass, cell numbers, or thickness (Fig. 2). Biofilm accumulation is the net result of various processes that can be identified and quantified (Fig. 3):

1. Adsorption—the interphase accumulation of cells from the bulk liquid directly on the substratum.
2. Desorption—the reentrainment into the bulk fluid of a cell adsorbed to the substratum.
3. Attachment—the acquisition of cells from the bulk liquid by an existing biofilm.
4. Detachment—the reentrainment into the bulk fluid of cells from an existing biofilm.
5. Growth—the increase in the number of biofilm cells as a result of replication.
6. Product formation—the production of polymers and metabolic by-products in the biofilm.
7. Endogenous decay—biofilm cell metabolism of internal cellular materials.
8. Death—permanent loss of a cell's reproductive and metabolic activity.

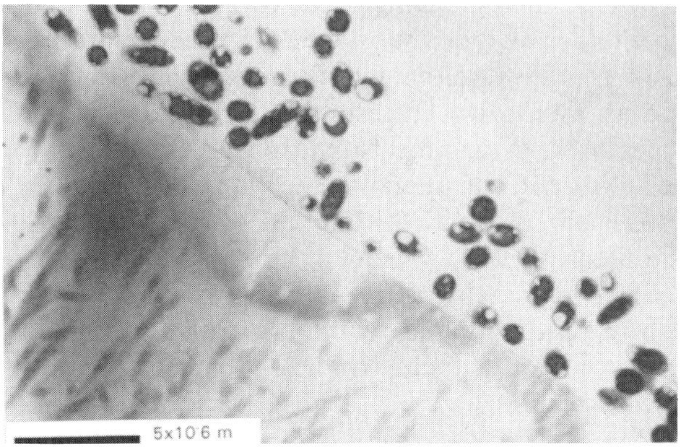


(a)

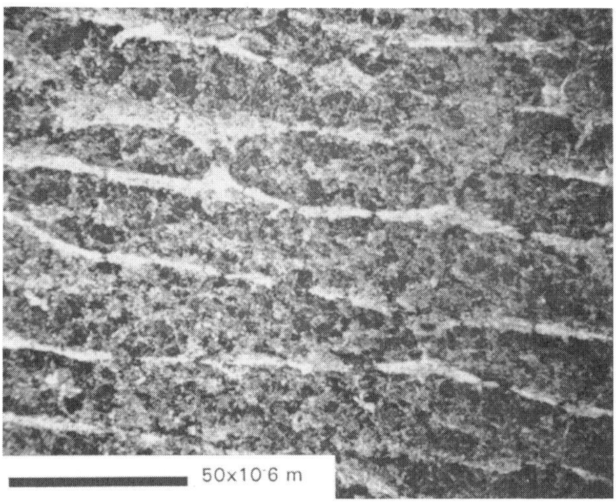


(b)

**Figure 1** Pure culture *Pseudomonas aeruginosa* biofilm. (a) TEM photomicrograph (5000×) showing cross-section of a *Ps. aeruginosa* biofilm on a polycarbonate substratum. (b) An SEM at 500× shows the macrostructure of a biofilm. (c) With an SEM at 3000×, the individual cells are vaguely apparent. Surface roughness can also be observed at this scale. (d) Magnified 20,000×, individuals cells are clearly distinguishable. Strands of polymeric material can also be distinguished; however, because of the harsh preparation before SEM the polymeric strands may only be artifacts of the true structure of EPS in the biofilm.

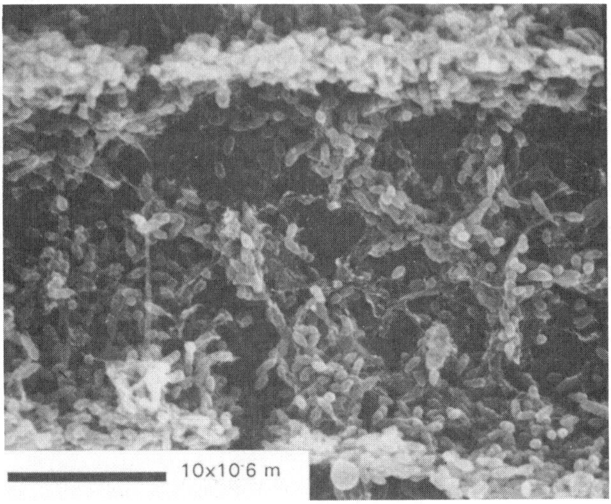


(c)

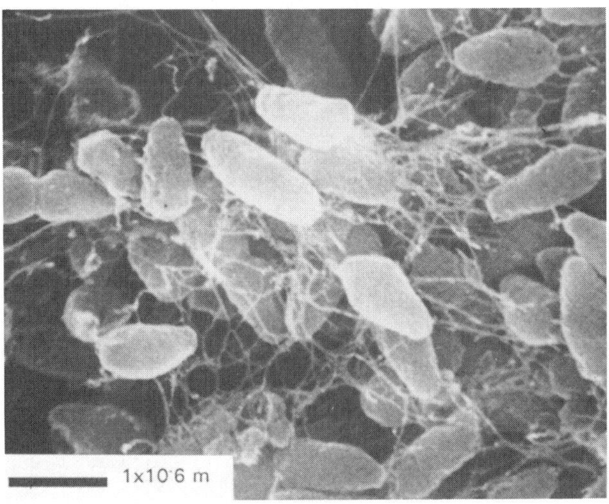


(d)

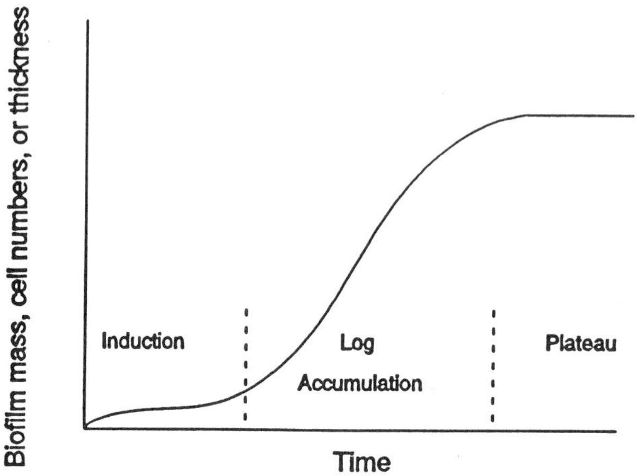


**Figure 2** Characteristic S curve of biofilm accumulation with time.

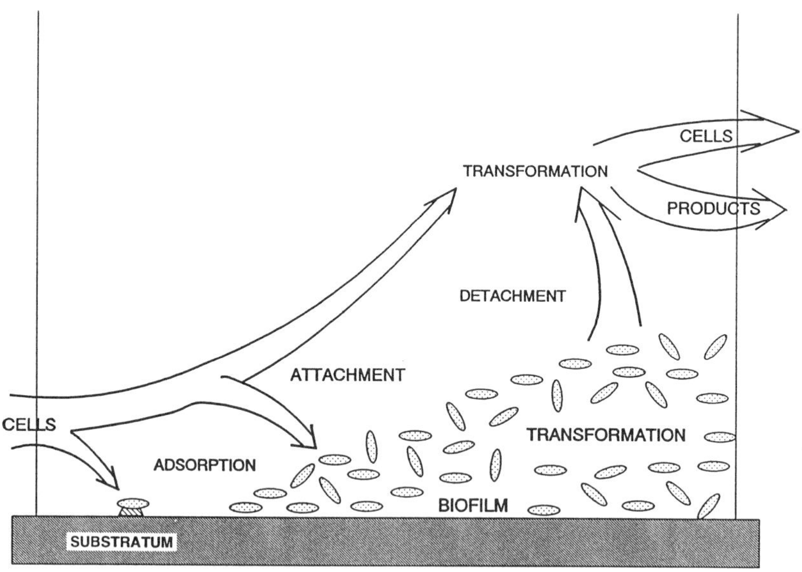


**Figure 3** Schematic illustration of the processes contributing to biofilm accumulation and activity. These processes include substrate transport and transformation in both the bulk liquid and the biofilm, cellular adsorption and attachment from the bulk liquid, and the detachment of cells and products from the biofilm.

To mathematically model the accumulation of a biofilm, a material balance is required. Although considerable differences in growth and decay rates may exist within a biofilm, the general mass balance equation for attached biomass is

$$\text{Accumulation} = \text{Input} - \text{output} + \text{transformation} \tag{1}$$

where transformation is defined as the net biomass growth rate. In the induction period, *before* a monolayer of biofilm has formed, biomass accumulation can be modeled as follows:

$$\begin{matrix}\text{Accum.}\\ \text{rate}\end{matrix} = \begin{matrix}\text{Adsorption}\\ \text{rate}\end{matrix} - \begin{matrix}\text{desorption}\\ \text{rate}\end{matrix} + \begin{matrix}\text{growth}\\ \text{rate}\end{matrix} - \begin{matrix}\text{decay}\\ \text{rate}\end{matrix} \tag{2}$$

After a monolayer of biofilm has accumulated, the processes of adsorption and desorption become negligible, when compared to attachment and detachment, so that the following equation applies:

$$\begin{matrix}\text{Accum.}\\ \text{rate}\end{matrix} = \begin{matrix}\text{Attachment}\\ \text{rate}\end{matrix} - \begin{matrix}\text{detachment}\\ \text{rate}\end{matrix} + \begin{matrix}\text{growth}\\ \text{rate}\end{matrix} - \begin{matrix}\text{decay}\\ \text{rate}\end{matrix} \tag{3}$$

In mathematical terms, Eq. (3) can be expressed as

$$\frac{dX_b}{dt} = R_a - R_d + \mu X_b - b X_b \tag{4}$$

where

$X_b$ = areal biomass concentration ($\mathrm{M\,L^{-2}}$)
$t$ = time (t)
$R_a$ = attachment rate ($\mathrm{M\,L^{-2}\,t^{-1}}$)
$R_d$ = detachment rate ($\mathrm{M\,L^{-2}\,t^{-1}}$)
$\mu$ = specific growth rate ($\mathrm{t^{-1}}$)
$b$ = endogenous decay rate coefficient ($\mathrm{t^{-1}}$)

These processes are described in more detail in the following sections.

## 2.1 Adsorption/Desorption

### *2.1.1 Adsorption of the Conditioning Film*

Immediately after a solid is immersed into water, organic molecules adsorb to the clean surface. This organic layer is called the conditioning film. Conditioning films are mainly composed of glycoproteins (10,11) and are not static, but are subject to a high turnover rate (12). Bryers (13) measured 15 mg of chemical oxygen demand (COD)/$m^2$ of organic compounds on glass. Little and Zsolnay (14) have performed similar experiments with stainless steel in seawater and after 15 min obtained up to 0.8 mg/$m^2$ of organic matter.

Deposition of proteins has been shown to affect adsorption of bacteria. Meadows (15) found that albumin inhibited adsorption, while casein and gelatin enhanced microbial adsorption rates. Fletcher (16) showed that a precoating of a mixture of albumin, gelatin, fibrinogen, and pepsin inhibited the adsorption of a pseudomonad on polystyrene. Although it has been shown that a conditioning film is deposited before cells adsorb to the substratum (17), it has not been shown to be a *prerequisite* for cell adsorption.

### 2.1.2 Adsorption of Cells to Surface

The first step in the development of a biofilm is the adsorption of a cell to a solid surface. Adsorption is defined as the interphase accumulation of cells from the bulk liquid directly on the substratum. Adsorption plays a major role in biofilm accumulation only in the initial stages of colonization. Subsequently, growth of cells is most often the dominant process. Much research has been done on the topic of adsorption; however, only recently have factors such as momentum transport been included in these studies. Shear stress has been shown to play an important role in the sticking efficiency of cells to a substratum (18–20). The sticking efficiency is defined as

$$\text{Sticking efficiency} = \frac{\text{Number of cells adsorbed to the substratum}}{\text{Number of cells transported to the substratum}} \tag{5}$$

The sticking efficiency is the ratio of the rate of cell adsorption to the rate of transport of cells to the surface and is therefore affected by the fluid-flow regime under which the adsorption occurs. In turbulent flow, turbulent bursts from the bulk fluid penetrate to the wall (21,22), and these bursts may contribute significantly to the transport of cells to the surface in turbulent flow (23). Other processes that contribute to the transport of cells to the surface are cell motility (24), gravity, and Brownian motion.

Adsorption is affected by substratum properties (25), with different materials experiencing varied rates of cellular adsorption under similar flow conditions. Roughness is also believed to strongly influence the adsorption properties of cells in flowing systems. Figure 4 shows how roughness elements on the substratum provide shelter from the shear forces exerted by fluid flow and also increase the total surface area available for colonization. For quiescent conditions, Van Haecke (26) concluded surface roughness had little effect on the adsorption kinetics of *Ps. aeruginosa* on stainless steel. Under quiescent conditions, the cell surface hydrophobicity was the major parameter affecting adsorption, and measurable adsorption occurred within 30 s. The physiological state of the cells can also play an important role in the adsorption rate. Cell adsorption per unit area was found to be

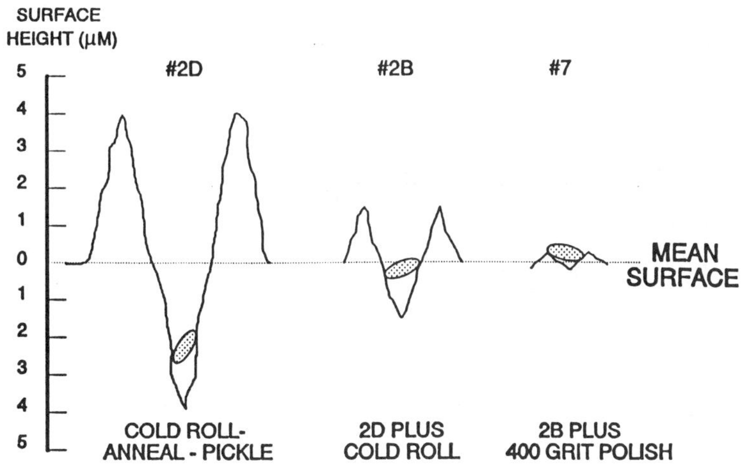


**Figure 4** Schematic comparison of the size of microroughness elements on cold-roll stainless steel with the size of a typical bacterium. (Adapted from Ref. 119.)

linear with specific growth rate history (27) in a turbulent flow system. As the specific growth rate was increased, a linear decrease in adsorbed colony-forming units was observed. Dawson (28) has shown that starved *Vibrio* were more hydrophobic than normal cells and formed polysaccharide-rich tubules that were believed to enhance adsorption.

Extracellular polymeric substances (EPS) may assist in binding microbes to a substratum and thus influence adsorption. Electron microscopy (29,30) has shown the presence of EPS adsorbed to a substratum.

### *2.1.3 Reversible/Irreversible Adsorption*

Most research has focused on irreversibly adsorbed cells, where only the cells that have "permanently" adsorbed to the substratum are included in the analysis. However, studies (20,31–33) show that some cells can adsorb to the substratum for a finite period of time, then desorb and are reentrained in the bulk liquid (Fig. 5). Factors influencing irreversible adsorption are not known.

## 2.2 Attachment/Detachment

### *2.2.1 Attachment*

Attachment is defined as cells from the bulk liquid sticking to an *existing* biofilm. Attachment of cells could play an important role in the displacement of one cell species by another. Although a method is currently being

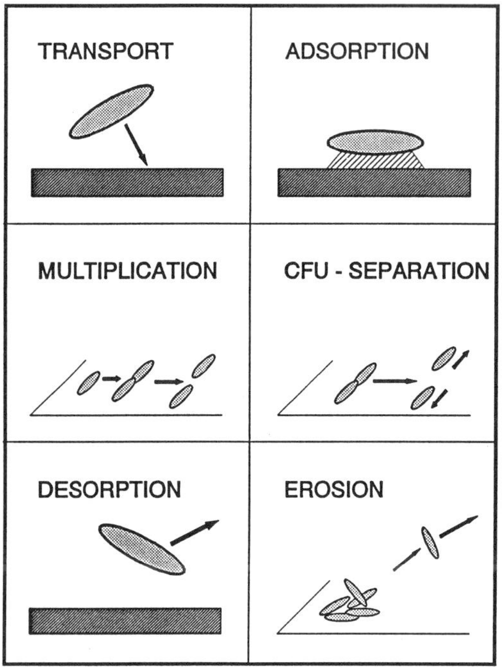


**Figure 5** Schematic diagram of the processes fundamental to the model of the initial microbial colonization of a substratum. (Adapted from Ref. 120.)

developed and tested to determine the rate of attachment of suspended cells to an existing biofilm (34), little published information is available.

### *2.2.2 Detachment*

Biofilm detachment is the process of removal and reentrainment of biomass in the bulk fluid. According to Bryers (35), portions of a biofilm can be removed in any of the following ways (Fig. 6): (1) erosion, (2) sloughing, (3) human intervention, (4) predator grazing, and (5) abrasion. Most research has been focused on the areas of erosion and sloughing. Detachment is one of the least understood processes affecting biofilm accumulation and is probably the most important process limiting both the rate and extent of biofilm accumulation.

Erosion is the removal of small portions of biofilm, thought to be a result of shear forces exerted by moving fluid in contact with the biofilm surface. Erosion is frequently treated as a continuous process, where detachment of single cells occurs evenly over the entire surface of the biofilm.

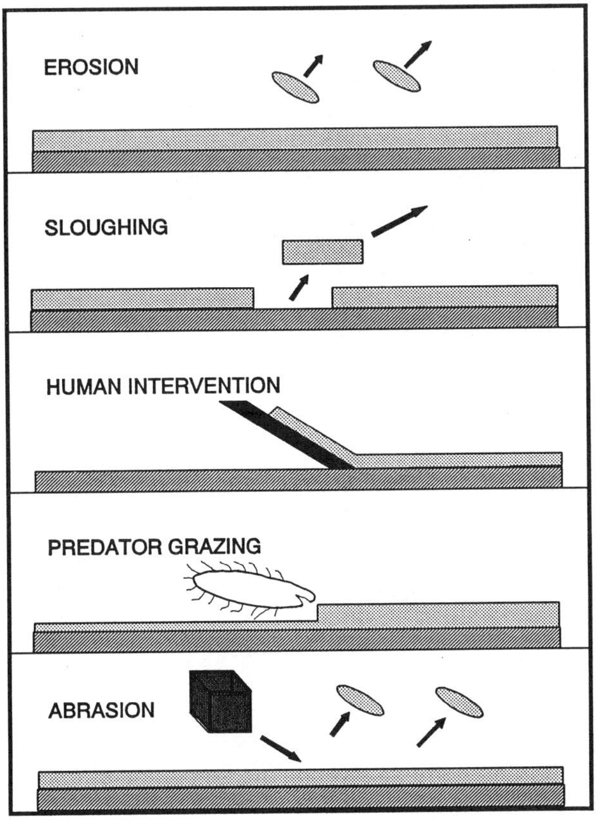


**Figure 6** Schematic diagram of the processes contributing to biofilm detachment.

Recent data, however, show that erosion is the result of a wide range of particle sizes (36) and that most biomass detaches in the form of multicellular particles. In contrast to erosion, sloughing is the detachment of large portions of a biofilm and is an apparently random, discrete process. Sloughing often occurs in older, thicker biofilms, or when environmental conditions change rapidly. Biofilms grown in high-substrate, low-shear-stress conditions are particularly susceptible to massive detachment as a result of sloughing. Few researchers have focused on detachment as a result of sloughing, and most have treated it as an annoying phenomenon that distorts and complicates biofilm research. Howell and Atkinson (37) were the first to formally attempt to model sloughing events. In the model, oxygen depletion in the depth of the biofilm triggers a sloughing event.

Sloughing has also been theorized to be the result of substrate depletion deep in a biofilm and subsequent decay of the cell and polymer matrix that holds the biofilm to the substratum (38,39). Sloughing has also been associated with nitrogen bubbles forming in denitrifying biofilms (40). Spontaneous sloughing of biofilm has been observed after step increases in lactose and especially lactate concentrations, but step changes in glucose had no significant effect on biofilm sloughing (41). In a related paper, a model of bacterial flocs (42) predicts that an increase in substrate concentration causes a breakup of the floc. This is consistent with observations at activated sludge facilities.

Sloughing has been observed after substrate concentration decreases. Bott and Miller (43) observed a 50% decrease in biofilm weight in aluminum tubes at a liquid velocity of 0.5 m/s within 4 days after stopping the nutrient supply of 4 mg/L glucose. However, at a liquid velocity of 2.0 m/s, the decrease in biofilm weight was insignificant. These data indicate that the fluid conditions under which the biofilm was grown do have an effect on the sloughing properties. Sloughing has been induced (44) by adding ethylene glycol-*bis*(β-aminoethyl ether)-*N*,*N*-tetraacetic acid (EGTA), a calcium-specific chelant. The divalent calcium ions give tertiary structure to the polymer matrix and may play a major role in determining biofilm strength. Thus, removing calcium causes a partial collapse of the biofilm structure.

Other causes of biofilm detachment include human intervention, predator grazing, and abrasion. Detachment via human intervention is the removal of biofilm as a result of chemical or physical means, such as adding biocides or scraping with brushes or other abrasive materials. Most of the research in this area is empirical and related to the control of biofouling and corrosion in industrial systems. Some more fundamental research has been accomplished (45) by searching for mechanisms and direct effects within the biofilm, rather than percent kill for a given biocide. Predator grazing occurs when biofilm mass is consumed by larger organisms, such as protozoa. This process of biofilm removal does not fit directly under the definition of biofilm detachment, since biomass is not reentrained in the bulk liquid, but is included here for completeness. Active protozoan predation of a biofilm community was observed by Fenchel (46), who suggests predation may benefit a wastewater treatment biofilm community by maintaining the bacteria in a more active physiological state. Higher organisms such as snails, worms, and insects also feed on biofilms (47). The detachment of biomass resulting from collisions of solid particles with the biofilm is termed abrasion. In fluidized bed bioreactors, abrasion can be significant, e.g., removal of biofilm in a sand filter during backwash.

### *2.2.3 Detachment Models*

The importance of biofilm detachment can be illustrated by considering an ideal biofilm. At steady state, the accumulation term of Eq. (4) is zero and can be rearranged to give

$$R_d = R_a + (\mu - b)\,X_b \qquad (6)$$

where

$R_d$ = detachment rate ($M\,L^{-2}\,t^{-1}$)
$R_a$ = attachment rate ($M\,L^{-2}\,t^{-1}$)
$\mu$ = specific growth rate ($t^{-1}$)
$b$ = endogenous decay rate ($t^{-1}$)
$X_b$ = areal biomass concentration ($M\,L^{-2}$)

For steady-state biofilms, the detachment rate is equal to the sum of the biomass attachment rate, the decay rate, and the growth rate. For systems with low suspended cell concentrations, typical of a vessel with a short residence time, the attachment rate is low such that

$$R_a \ll (\mu - b)\,X_b \qquad (7)$$

and Eq. (6) reduces to

$$R_d = (\mu - b)\,X_b \qquad (8)$$

This is a simplified model of a biofilm because the specific growth rate will not be the same throughout the entire biofilm due to diffusional resistances. But Eq. (8) points out the important concept that at steady state, the net rate of growth in the biofilm, $(\mu - b)\,X_b$, must be equal to the total rate of biofilm detachment. This first-order detachment equation as a function of total biomass is a commonly used approach (48–50) in biofilm research.

An expression for detachment as a first-order function of shear stress and biomass concentration is given by Bakke et al. (51), although little data is available to support the equation:

$$R_d = K_s\,\tau\,X_b \qquad (9)$$

where

$R_d$ = biofilm detachment rate ($M\,L^{-2}\,t^{-1}$)
$K_s$ = shear stress detachment rate coefficient ($L\,t\,M^{-1}$)
$\tau$ = shear stress at biofilm surface ($M\,t^{-2}\,L^{-1}$)
$X_b$ = areal biomass concentration ($M\,L^{-2}$)

In a turbulent flow pipeline system, this equation can be solved to explicitly include bulk-fluid velocity by substituting Blausius's law of friction in a pipe [Eq. (10)] into the definition of the friction factor (52) [Eq. (12)]:

$$f = 0.0791\, Re^{-0.25} \tag{10}$$

where

$$Re = \frac{V d}{\nu} \tag{11}$$

$$f \equiv \frac{2\tau}{\rho\, \nu^2} \tag{12}$$

where

$f$ = friction factor (unitless)
$Re$ = Reynold's number (unitless)
$V$ = average fluid velocity ($\mathrm{L\, t^{-1}}$)
$d$ = pipe diameter (L)
$\nu$ = kinematic viscosity ($\mathrm{L^2\, t^{-1}}$)
$\rho$ = fluid density ($\mathrm{M\, L^{-3}}$)

The final form for the detachment in a turbulent flow pipeline as a function of hydrodynamic shear stress is given in Eq. (13).

$$R_d = K_{tp}\, \nu^{1.75}\, X_b \tag{13}$$

Other models employ a second-order detachment rate with regard to biofilm mass (53,54):

$$R_d = K_2\, X_b^2 \tag{14}$$

where $K_2$ = detachment rate coefficient ($\mathrm{M^{-1}\, L^2\, t^{-1}}$)

In a multispecies biofilm model, a second-order (in biofilm thickness) detachment rate expression was used (55).

$$R_d = K_3\, L_f^2\, \rho_f \tag{15}$$

where $K_3$ = detachment rate coefficient ($\mathrm{L^{-1}\, t^{-1}}$)

Equation (15) has incorporated two important biofilm properties:

1. Hydrodynamic erosion is probably related to biofilm thickness.
2. Biofilm strength may be related to biofilm density.

Speitel and DiGiano (58) propose an expression for the detachment rate that includes a first-order function in biofilm mass with the addition of a growth-related detachment term, $K_g'$, to give the following detachment rate expression:

$$R_d = L_f\, (K_g + K_g'\, \mu) \tag{16}$$

This equation does not explicitly depend on shear stress and was shown to fit experimental data better than a simple first-order detachment expression during periods of high growth rates. The growth-related detachment was incorporated to account for the high biofilm loss rates observed during periods of high biodegradation rate. The experiments were performed in granular activated carbon columns using effluent suspended biomass as the measure of biofilm shear losses. The growth-associated erosion coefficient, $K_g'$, appears to be substrate-specific.

In an experimental apparatus similar to that of Speitel and DiGiano, Chang and Rittmann (56) observed high detachment during periods of high growth rate. It was noted that during periods of rapid biofilm growth, the biofilm erosion rate increased in a manner that could not be fit by a first-order detachment function. Trulear and Characklis (57) show biofilm detachment rate as a function of biofilm mass, but the data from two experiments were graphically combined and interpreted with one curve. In light of the data presented by Speitel and DiGiano (58) concerning the relationship between growth rate and biofilm detachment, the data given by Trulear and Characklis indicate growth-rate-related detachment. In Figure 7, the curves are plotted separately. In a qualitative sense, at the higher substrate loading rate, the growth rate is higher and therefore the detachment rate should be higher for a given amount of biomass. Rittmann (59) also uses the data of Trulear and Characklis to calculate an erosion loss coefficient. However, Rittmann also neglects the fact that the data were collected under two different substrate loading rates.

A recent study (60) shows that biofilm detachment rates correlate best with observed data when cellular metabolic activity is incorporated into the detachment model. Peyton and Characklis determined that Eq. (17) gave better correlation than other published detachment expressions for both pure and mixed-culture biofilm data sets.

$$R_d = k_d \frac{Q}{A} (S_i - S) \, Y_{\frac{x}{s}} L_f \tag{17}$$

where

$R_d$ = cellular detachment rate ($M\,L^{-2}\,t^{-1}$)
$k_d$ = detachment coefficient ($L^{-1}$)
$Q$ = volumetric flow rate through the bioreactor ($L^3\,t^{-1}$)
$A$ = area available for biofilm growth ($L^2$)
$S$ = effluent substrate concentration ($M\,L^{-3}$)
$S_i$ = influent substrate concentration ($M\,L^{-3}$)
$Y_{x/s}$ = observed cellular yield ($M_x\,M_s^{-1}$)
$L_f$ = biofilm thickness (L)

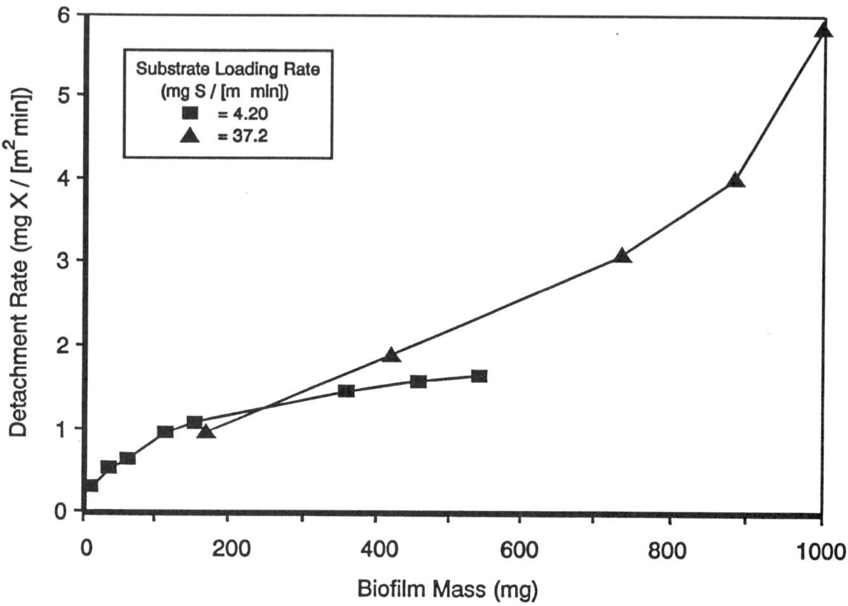


**Figure 7** Comparison of biofilm detachment rates as a function of total biofilm mass in identical reactors at two substrate loading rates. For a given amount of biomass, the detachment rate will increase with an increase in the amount of available substrate.

In a related paper (61), a theoretical model for growth-associated biofilm detachment is given, which, for certain assumptions, reduces to Eq. (17). This model gives good quantitative correlation of the data generated by Trulear and Characklis. The most usable form of the detachment expression derived from the theoretical model is

$$R_d = k_{d1}\,\mu\,\rho_f\,a^2 + k_{d2}\,\rho_f\,L_f^2 \tag{18}$$

where

$a$ = the thickness of the actively growing biofilm (L)
$\mu$ = average specific growth rate in the actively growing zone ($t^{-1}$)
$\rho_f$ = average biofilm cell density ($M_x\,L^3$)

The research of Chang and Rittmann (56) compares the effects of surface irregularities on biofilm shear losses and accumulation in packed columns of granular activated carbon (GAC). By using spherical and irregularly shaped GAC, values for a first-order erosion coefficient were estimated. During the induction period, the irregularly shaped GAC showed greater biofilm attachment and no erosion. However, as the biofilm began to accumulate, the shear loss coefficient for both types of GAC began to increase. Finally, at steady state, the erosion coefficients for both the spherical and irregularly surfaced GAC approached the same values. Thus, surface roughness may play an important role in initial events, but becomes less important to detachment rates after the biofilm completely covers the surface irregularities. It has not been shown how surface roughness affects the rate and amount of sloughing.

## 2.3 Growth

### 2.3.1 *Exponential Growth (Growth Limited)*

Growth is defined as an increase in microbial cell numbers or microbial mass as a result of cell replication, and the rate of growth is quantitatively expressed as the change in cell numbers or mass per unit time. Under the proper environmental conditions (temperature, concentration of electron acceptor or electron donor), bacteria will grow and multiply. In some cases, a bacterial cell can divide quite rapidly. For example, *Escherichia coli* in an appropriate nutrient solution can divide every 20 min. The highest rate at which a bacterial cell can divide is limited by physiological processes, such as DNA replication rates. This rate is commonly called the maximum specific growth rate, $\mu_{max}$. The maximum specific growth rate varies from one species to another under identical culture conditions and changes for a given species with different substrates, temperatures, and other environmental conditions. Under conditions of balanced growth and a nonlimiting nutrient supply, the rate of microbial growth, $r_x$, is proportional to the amount of microbial cells present. The rate equation that describes this type of growth is the following:

$$r_x = \frac{dX}{dt} = \mu_{max} X \tag{19}$$

where $X$ = biomass concentration ($M_x \, L^3$)

This equation is appropriate only while the environmental conditions remain constant. However, when the concentration of a substrate is reduced to a growth-limiting level, substrate-limited growth occurs, and Eq. (19) is no longer valid without incorporating the effects of substrate limitations.

### 2.3.2 *Substrate-Limited Growth (Monod Expression)*

In many biofilm applications, the substrate concentration is below that which is required for maximum growth. This is substrate-limited growth. For suspended microbial populations, the specific growth rate is described by the Monod equation (62,63):

$$\mu = \frac{\mu_m S}{K_s + S} \tag{20}$$

Equation (20) should be substituted for the maximum specific growth rate in Eq. (19) to account for the effects of nutrient depletion on the organism's growth kinetics.

The growth of biofilm organisms also obeys Monod kinetics (64). However, interpretation of growth kinetics in biofilms is complicated by diffusional resistance between the bulk liquid and the biofilm. Diffusional resistance extends throughout the biofilm so that the deeper layers of biofilm are exposed to substrate concentrations that are lower than in the bulk liquid.

*2.3.2.1 Transport-Limited Growth* The microenvironment of a biofilm is characterized by steep gradients in substrate concentration. The gradients are the result of the diffusion of substrates from higher to lower concentrations and the chemical reactions required for cells to maintain themselves and reproduce. Traditional batch culture studies do not encounter the problem of diffusional resistance because of high substrate concentrations and vigorous and constant stirring. With more researchers attempting in situ studies and improved techniques for field analysis comes an awareness of the important role that mass, heat, and momentum transport play in biofilm accumulation and activity. Advances in microsensors have contributed significantly to understanding of the microscale environment that exists in biofilms. Microsensor techniques (65) have helped determine photosynthetic activity and its influences on oxygen, hydrogen sulfide, and pH profiles in a cyanobacterial biofilm. In a nonphotosynthetic system, Lewandowski et al. (66) used a fiber optic and dissolved oxygen combination microprobe to locate the biofilm-substratum and biofilm–bulk liquid interface. In addition, this method also gives the dissolved oxygen profile and the decrease in oxygen concentration across the mass transfer boundary layer (Fig. 8). Profiles attained by microprobes can be used to determine in situ kinetic rate coefficients for substrates and products in biofilms.

While attempting to determine kinetic coefficients using *bulk liquid* measurements, care should be taken to remove all possible substrate transport limitations to the biofilm. If transport limitations are significant, the mass transfer rate rather than the reaction rate will be measured. The importance of distinguishing between reaction rates and mass transfer rates has been addressed elsewhere (67,68).

### 2.3.3 Substratum Effects on Growth

Biofilms will form on almost any immersed surface or interface. The definition of a substratum needs to be expanded at this point to include liquid

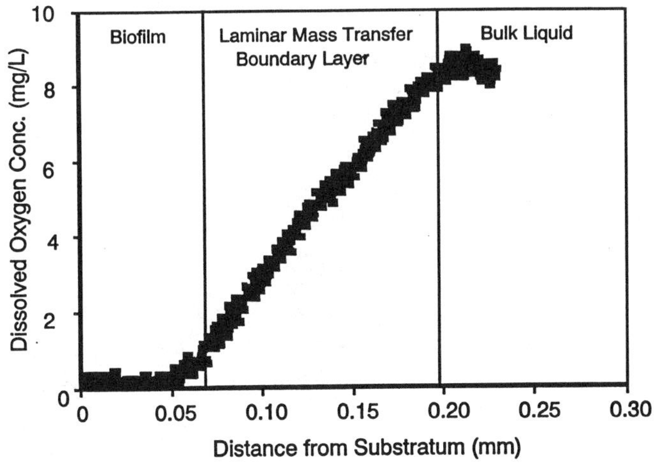


**Figure 8** Dissolved oxygen concentration as measured in a biofilm in situ with a combination dissolved oxygen and light transmission microprobe.

substrata such as oil drops in water. Some substrata can be metabolized for part of the cell's nutritional requirements, whereas other substrata are toxic to the biofilm.

Rosenberg (69) has shown *Acinetobacter calcoaceticus* to adsorb at an oil-water interface, with hydrophobic thin fimbriae as the major organelle responsible for the adsorption. Mutants that lacked the thin fimbriae did not adsorb to the hydrocarbon droplet. After a short lag phase, the cells began to use the hydrocarbon as a carbon source. Other substrata that can also be used by the cells as a growth substrate include cellulose, lignin, and protein.

A substratum can have an inhibitory effect on biofilm cell growth. Copper was found (70) to inhibit cell replication when no more than a monolayer of cells was adsorbed to the surface. If the exposure time to a cell suspension was long enough for a monolayer to form, replication of cells in the second and higher cell layers was observed.

### 2.3.4 Product Formation

Many species of bacteria produce extracellular polymeric substances (EPS). These polymers are believed to be important to biofilm formation and stability, in addition to providing a protective layer from oxidizing chemicals. The kinetics of product formation have been described by Luedeking and Piret (71). In this model, originally developed for the production of lactic acid by *Lactobacillus delbrueckii*, the product formation rate is both growth- and non-growth-dependent. That is, the rate of product formation depends on the rate at which the biomass is growing and amount of biomass present.

$$r_p = k_p \mu X + k_p' X \tag{21}$$

Weiss and Ollis (72) describe a method to determine the kinetic coefficients, $k_p$ and $k_p'$, used in Eq. (21). Robinson et al. (73) determined the polymer production rate coefficients for *Ps. aeruginosa* in continuous culture. In both studies, steady state was required to evaluate the coefficients. However, the inherent differences in batch stationary phase and continuous culture steady state, such as the mean cell age, were not addressed in this research. Experimentally determined EPS production rate coefficients for *Pseudomonas* species are given in Table 1.

### 2.3.5 Maintenance

A minimum amount of substrate is required by microorganisms to survive. Energy can be obtained from environmental sources or the cell can begin mass- and energy-conserving measures that include metabolism of internal compounds, or cell decay. The decay rate is usually characterized as first order in biomass, with decay rate coefficient, *b*. Rittmann and McCarty (74) calculated the theoretical minimum substrate concentration, $S_{min}$, re-

**Table 1** Experimentally Determined EPS Production Rate Coefficients

| Microbial species | Coefficients and ref. |
|---|---|
| *Pseudomonas* sp. | Kp = 0.25 g product $ODU^{-1}\ l^{-1}$<br>K′p = 0.0037 g product $ODU^{-1}\ h^{-1}$<br>Weiss and Ollis (72) |
| *Ps. aeruginosa* | Kp = 0.3 g polymer C $(g\ cell\ C)^{-1}$<br>K′p = 0.04 g polymer C $(g\ cell\ C)^{-1}\ h^{-1}$<br>Robinson et al. (73) |
| *Ps. aeruginosa*[a] | Kp = 2.1 g polymer C $(g\ cell\ C)^{-1}$<br>K′p = 0 g polymer C $(g\ cell\ C)^{-1}\ h^{-1}$<br>Turakhia (122) |
| *Ps. aeruginosa*[b] | Kp = 3.3 g polymer C $(g\ cell\ C)^{-1}$<br>K′p = 0.1 g polymer C $(g\ cell\ C)^{-1}\ h^{-1}$<br>Mian et al. (121) |

[a]Biofilm.
[b]Nitrogen-limited system.

quired to sustain a steady-state biofilm [Eq. (22)], although $S_{min}$ calculated using the $\mu_m$ and $K_s$ for balanced growth are probably not the same as for starvation conditions.

$$S_{min} = \frac{K_s\, b}{\mu_m - b} \tag{22}$$

### 2.3.6 Death and/or Lysis

Bacteria in a state of starvation typically go through physiological changes, including miniaturization. Turnover of proteins increases as the cell enters the stationary phase. This recycling of proteins provides the cell with amino acids needed for producing new stationary-phase-specific proteins. After a period of starvation or other environmental stress, a cell may completely lose its ability to reproduce or metabolize substrates; i.e., the cell is dead. In most cases, death is followed by lysis (rupturing of the cell wall and loss of the cell's internal contents). Cell lysis can contribute enough substrate to provide maintenance energy for other cells in the population.

## 3 METABOLISM AND PHYSIOLOGY IN BIOFILMS

The influence of bacterial adhesion on cell physiology has not been systematically and consistently defined. It appears as though surfaces do not alter bacterial metabolism and physiology, although this alteration is rarely separated into direct and indirect influences [after a review by van Loosdrecht et al. (75)]. Direct influences are defined as changes in microbial activities resulting from the presence of a surface, e.g., changes in the composition or structure of the cell as a direct result of the adhesion to a surface. These changes may or may not be permanent. Indirect influences include changes in cell activity as a result of (1) changes in the composition of the medium because of substrate adsorption or desorption phenomena at the solid interface, (2) heterogeneity of the space immediately around an attached cell as a result of specific geometry or flow conditions, (3) the fact that cells remain in a particular area after colonizing a surface. Examples of indirect influences are substrate availability or pH changes near the surface. The review concludes that surfaces do influence substrate utilization and yields in both the positive and negative direction; however, the changes are a result of indirect effects, and no conclusive evidence exists to show that adhesion directly affects microbial metabolism or physiology.

Many authors do not appear to evaluate the way the environment surrounding a cell may change as a result of the proximity of a solid surface. Instead, differences are attributed to possible changes in cell physiology, with little evidence to support the hypothesis. A notable exception is presented in two papers by Doran and Bailey (76,77) where careful consideration of mass transfer limitations was accompanied by thorough physiologically relevant measurements of cellular composition. Doran and Bailey found that for immobilized yeast, ethanol production was 40–50% greater, glucose consumption was twice as fast, and the specific growth rate was reduced by 45% as compared to suspended organisms. In addition to these parameters, cellular composition was different in immobilized organisms since measurement of intracellular polysaccharide levels showed larger quantities of reserve and structural carbohydrates when compared to suspended organisms. The most interesting data are flow cytrometry comparisons of DNA, RNA, and protein frequency functions. These data show that immobilized cells had a different physiological composition (higher ploidy) than suspended cells when extreme care was taken to minimize differences in the environmental conditions between the immobilized and suspended cultures.

The review by van Loosdrecht et al. (75) and the research of Doran and Bailey (76,77) indicate that much more remains to be investigated about the effects of cell attachment to surfaces. Before an observed effect can be attributed to physiological differences, direct physiological evidence should be obtained and all environmental factors should be ruled out.

With this information in mind, the data in the following sections is a combination of results from suspended cell and biofilm cultures. Bakke et al. (64) successfully used stoichiometry and yield coefficients obtained from suspended culture to develop a kinetic model of a biofilm system. Still, care should be taken when stoichiometry and yield coefficients from suspended cultures are applied to biofilm systems with no experimental confirmation, since significant differences may exist between immobilized and suspended cells.

## 3.1 Stoichiometry

The stoichiometry of a chemical reaction describes the exact amount of each reactant required to form a given quantity of each product. The overall reaction stoichiometry is a result of the energy reaction and the biomass synthesis reaction. Microbial stoichiometry permits all reactant or product concentrations to be calculated from measuring one reactant or product, given that initial concentrations of all components are known. This information is critical to the analysis of microbial processes. It has been shown that stoichiometric coefficients determined in suspended cultures can be applied to biofilm systems (64). In a well-defined system, Busch (78) determined the stoichiometry for the aerobic conversion of glucose to energy and biomass, and combined with an experimentally determined yield coefficient, the result of these two equations is the overall stoichiometry. Equation (23) represents the conversion of glucose to carbon dioxide and water and the amount of useful work that can be accomplished by the cell.

Energy:

$$C_6H_{12}O_6 + 6O_2 \Rightarrow 6CO_2 + 6H_2O \qquad [-120.11\ \text{kJ}\ (\text{mol e}^-)^{-1}] \tag{23}$$

Biomass synthesis:

$$C_6H_{12}O_6 + 1.2NH_3 \Rightarrow 6CH_{1.4}O_{0.4}N_{0.2} + 3.6H_2O \tag{24}$$

Overall stoichiometry:

$$C_6H_{12}O_6 + 2.5O_2 + 0.7NH_3 \Rightarrow 3.5CH_{1.4}O_{0.4}N_{0.2} + 2.5CO_2 + 4.6H_2O \tag{25}$$

Equation (24) is the synthesis reaction and represents the amount of biomass that could be produced from a mole of glucose *if no energy were required by the cell.* Equation (25) is the overall stoichiometry that would be observed in a biological reactor and can also be used to calculate all reactant and product concentrations from measuring one reactant or product, given that initial concentrations of all components are known.

The stoichiometry of other substrates and electron acceptors has also been determined using the experimentally determined biomass yield of Traore et al. (79). Sulfate-reducing bacteria can use lactate or acetate as the carbon and energy source while using sulfate as the electron acceptor. The stoichiometric equation for the conversion of lactate is the following:
Energy:

$$CH_3CHOHCOOH + 1.5H_2SO_4 \Rightarrow 1.5H_2S + 3H_2CO_3 \qquad [-11.67\ \text{kJ}\ (\text{mol e}^-)^{-1}] \tag{26}$$

Synthesis:

$$CH_3CHOHCOOH + 0.6NH_3 \Rightarrow 3CH_{1.4}O_{0.4}N_{0.2} + 1.8H_2O \tag{27}$$

Overall stoichiometry:

$$CH_3CHOHCOOH + 1.41H_2SO_4 + 0.0036NH_3 \Rightarrow 0.18CH_{1.4}O_{0.4}N_{0.2} + 1.41H_2S + 2.82CO_2 + 2.93H_2O \tag{28}$$

An overall stoichiometric equation for biomass and acetate production from lactate is proposed (80):

$$CH_3CHOHCOOH + 0.98H_2SO_4 + 0.02NH_3 \Rightarrow 0.1CH_{1.4}N_{0.2}O_{0.4} + 0.94CH_3COOH + 0.98H_2S + 1.96CO_2 + 1.08H_2O \tag{29}$$

Sulfate-reducing bacteria can also use acetate in the following manner:
Energy:

$$CH_3COOH + H_2SO_4 \Rightarrow H_2S + 2H_2CO_3 \qquad [-6.36\ \text{kJ}\ (\text{mol e}^-)^{-1}] \tag{30}$$

Synthesis:

$$5CH_3COOH + 2NH_3 \Rightarrow 10CH_{1.4}O_{0.4}N_{0.2} + 6H_2O \tag{31}$$

Overall stoichiometry:

$$CH_3COOH + 0.915H_2SO_4 + 0.034NH_3 \Rightarrow 0.17CH_{1.4}O_{0.4}N_{0.2} + 0.915H_2S + 1.83CO_2 + 1.93H_2O \tag{32}$$

## 3.2 Yield

The amount of cellular material produced per unit measure of substrate consumed is defined as the yield.

$$Y_{\frac{x}{s}} = \frac{\text{g biomass formed}}{\text{g nutrient used}} \tag{33}$$

For example, if 0.3 g of cellular carbon was produced by metabolism of 1.0 g glucose carbon, the cellular yield coefficient would be 0.3 g cellular C/1.0 g glucose C. Yield coefficients can be calculated from stoichiometric equations; conversely, stoichiometric equations are evaluated from yield information. Yield coefficients can be applied to both suspended and attached bacterial cultures. Experimentally determined biomass yields with glucose and oxygen are given in Table 2.

**Table 2** Biomass Yields from Glucose with Oxygen as Electron Acceptor

| Organism | $Y^{x/s}$ ($gC_x\, gC_s^{-1}$) | Ref. |
|---|---|---|
| *Ps. aeruginosa* | 0.33[a] | Turakhia (122) |
| *Ps. aeruginosa* | | Robinson et al. (73) |
| Total biomass | 0.54[b] | |
| Cell mass | 0.30[b] | |
| *Ps. fluorescens* | 0.38[c] | Nagai (123) |
| *Rhodopseudomonas speroides* | 0.45[c] | Nagai (123) |
| *Saccharomyces cerevisae* | 0.50[c] | Nagai (123) |
| Mixed populations | 0.44[c] | Busch (78) |
| | 0.40[c] | Servisi and Bogan (124) |

[a]Biofilm reactor.
[b]Chemostat.
[c]Batch reactor.
*Source*: Adapted from Ref. 125.

**Table 3** Elemental Composition of Microorganisms

| Organism | Composition (%) C | H | N | O | Empirical formula | Ref. |
|---|---|---|---|---|---|---|
| *C. utilis* | 50.4 | 7.7 | 5.8 | 36.1 | $CH_{1.83}N_{0.10}O_{0.54}$ | Herbert (81) |
| *K. aerogenes* | 50.6 | 7.4 | 13.0 | 29.0 | $CH_{1.75}N_{0.22}O_{0.43}$ | Herbert (81) |
| *S. cerevisiae* | 49.3 | 6.7 | 9.8 | 34.2 | $CH_{1.64}N_{0.16}O_{0.52}$ | Harrison (135) |
| *Ps. denitrificans* | 48.4 | 7.3 | 11.4 | 32.9 | $CH_{1.81}N_{0.20}O_{0.51}$ | Stouthamer (136) |
| *E. coli* | 48.1 | 7.1 | 13.4 | 31.4 | $CH_{1.77}N_{0.24}O_{0.49}$ | Bauer and Ziv (137) |
| *A. aerogenes* | 45.9 | 7.0 | 13.4 | 33.7 | $CH_{1.83}N_{0.25}O_{0.55}$ | Mayberry et al. (138) |
| *Ps. aeruginosa* | 55.3 | 7.9 | 14.6 | 22.2 | $CH_{1.73}N_{0.23}O_{0.30}$ | Turakhia (122) |

# 4 PROPERTIES OF BIOFILMS

## 4.1 Elemental Composition

Microorganisms are composed largely of carbon, hydrogen, nitrogen, and oxygen. Herbert (81) encourages using formulae that contain one gram-atom of carbon since (1) the carbon content changes very little with growth conditions and limiting nutrient, (2) carbon is the most abundant element in the cell, so changes in other elements have little effect on formula weight, and (3) a carbon balance is usually of greater importance than other elemental balances. The compositions of some organisms are given in Table 3.

## 4.2 Physical Properties

### *4.2.1 Density*

Biofilm density is a measure of the amount of biomass that exists in a given volume of biofilm and is generally reported on a basis of dry weight per unit wet volume (Table 4). Biofilm density is an important parameter in mathematical modeling of biofilm processes, since biomass concentration is often related to the activity of a biofilm. Typically, biofilm density is measured by determining the average thickness of the biofilm and then scraping a known area of biofilm into a preweighed drying dish. The biofilm is dried and the dry weight determined. Density can then be directly calculated. Rittmann and McCarty (82,83) calculated a biofilm density assuming biofilm dry weight is 50% carbon; however, the results are lower than those typically obtained with the more direct method of measurement. In a specialized high-pressure immobilized cell reactor (84), cell densities as high as 850 g dry weight per liter were measured in aggregates of starved *E. coli* grown in pressures up to nine atmospheres. The research demonstrated that *E. coli* can exert pressures up to three atmospheres on their environ-

**Table 4** Biofilm Density and Thickness (Dry Mass per Unit Wet Volume)

| Biofilm type[a] | Density ($kg\ m^{-3}$) | Thickness ($\mu m$) | Ref. |
|---|---|---|---|
| A | 66–130 | 160–210 | Kornegay and Andrews (87) |
| A | 50[b] | 100 | Rittmann and McCarty (126) |
| B | 20–105 | 30–1300 | Hoehn and Ray (127) |
| B | 5[c] | 119–126 | Herbert (81) |
| B | 5[c] | 0–125 | Rittmann and McCarty (83) |
| B | 10–65 | 10–124 | Trulear and Characklis (57) |
| C | 42–109 | 150–580 | Williamson and McCarty (128) |
| D | 17–47 | 36–47[d] | Trulear (108) |
| D | 27 | 0–60 | Bakke (129) |

[a]A, Steady state, heterotrophic mixed population; B, heterotrophic mixed population; C, steady state, nitrifying mixed population; D, steady state, *Ps. aeruginosa*.
[b]Calculated assuming biofilm is 80% volatile solids.
[c]Calculated assuming biofilm is 50% carbon.
[d]Calculated from measured thickness corrected for refractive index of biofilm.

ment. The only data available on the variation of biofilm density with biofilm depth are by Masuda et al. (85) given in Table 5. A "microslicer" was used to section the biofilm. The density was then determined for each biofilm section.

Biofilm density does appear to correlate with shear stress under some conditions. A study by Characklis (86) showed biofilm density increased with increasing shear stress at low substrate loading rates. This is somewhat in contrast to the findings of Kornegay and Andrews (87), who showed that at high substrate loading rates, no change in biofilm density occurred with changes in shear stress. The relationship between substrate loading rate and shear stress on biofilm density must be determined for more accurate computer simulation of biofilm systems.

**Table 5** Variation of Biofilm Density with Biofilm Depth

| | Density ($kg\ m^{-3}$) | Depth from water-biofilm interface ($\mu m$) | Thickness of biofilm section ($\mu m$) |
|---|---|---|---|
| Surface film | 37 | 0–400 | 400 |
| Intermediate | 98 | 400–600 | 200 |
| Base film | 102 | 600–730 | 130 |

*Source*: Ref. 139.

A difficulty in characterizing biofilm density is that the density can vary with depth. This has led to the artificial division of a biofilm, based on morphology, into a base and a surface film. The surface film is defined as the heterogeneous biofilm region near the biofilm–bulk liquid interface (Fig. 9). Advective transport dominates the movement of nutrients in the surface film. The surface film is characterized by a rough, viscoelastic surface and may contribute significantly to fluid friction. The surface film is also believed to influence particle capture in biofilms. Suschka (88) postulated the existence of a surface film with a different porosity than that of the base film. A threefold difference in experimentally measured and theoretically calculated residence times in trickling filters led to the conclusion that two liquid film layers existed: one layer of "free" liquid flowing over the top of the biofilm and another, "captured" layer flowing within the surface portion of the biofilm. Some implications of a rougher surface film on oxygen transfer are addressed. Detailed study of this rougher surface film has not been addressed except to acknowledge its existence. The base film is a relatively continuous accumulation of cells and polymer and is more dense than the surface film. The density and continuity provide a structure that may limit the amount of advective transport within the base film; therefore, mass transport of nutrients through the base film is dominated by molecular diffusion.

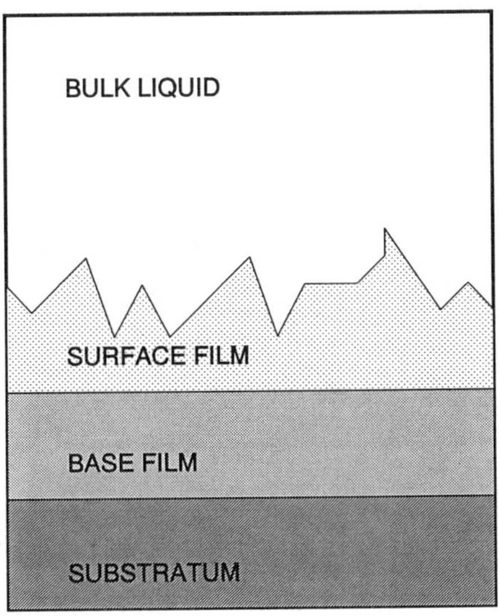


**Figure 9** Schematic diagram of the base and surface film.

### 4.2.2 Transport and Rheological Properties

The transport properties of interest in most biofilm systems are the coefficients that relate the transport rates of heat and mass through a biofilm to the environmental conditions that exist outside the biofilm. The properties of significant interest are the apparent molecular diffusivity of soluble components in a biofilm and the thermal conductivity for heat transfer through a biofilm.

The molecular diffusivity for soluble substrates in the biofilm matrix is essential to accurate biofilm modeling since calculating substrate concentration profiles within a biofilm requires a value for the molecular diffusivity. A few researchers have measured the molecular diffusivity coefficient within a biofilm. The reported (89) diffusivity coefficients in biofilm range from 8% to 100% of that in water, and are given in Table 6. Typical values for apparent molecular diffusivity in biofilm are 80% of the component's molecular diffusivity in water.

When compared to the diffusivity of various compounds through a biofilm, little research has been performed on the thermal conductivity measurements or the rheological properties of biofilms. Thermal conductivity is defined as the proportionality constant relating the heat flow per unit

**Table 6** Experimentally Determined Diffusivities in Biofilms and Bioflocs

| Diffusing species | Microbial aggregate | Diffusivity, $D_f$ ($10^{-10}$ $m^2$ $s^{-1}$) | $D_f/D$ | Ref. |
|---|---|---|---|---|
| Oxygen | Film | 15 | 0.70 | Tomlinson and Snaddon (130) |
| Oxygen | Film | 22 | 0.9 | Williamson and McCarty (128) |
| Oxygen | Film | | 0.95 | Bungay et al. (131) |
| Glucose | Film | | ≈0.5 | Siegrist and Gujer (132) |
| Sodium | Film | | ≈0.6 | Siegrist and Gujer (132) |
| Bromide | Film | | ≈0.6 | Siegrist and Gujer (132) |
| Ammonia | Film | 13 | 0.8 | Williamson and McCarty (128) |
| Nitrate | Film | 14 | 0.9 | Williamson and McCarty (128) |
| Oxygen | Floc | 4–20 | 0.2–1.0 | Matson and Characklis (89) |
| Glucose | Floc | 0.6–2.1 | 0.1–0.3 | Matson and Characklis (89) |
| Glucose | Floc | 0.48 | 0.08 | Baillod and Boyle (133) |
| Glucose | Floc | 0.6–6.0 | 0.1–1.0 | Pipes (134) |

area to the temperature change per unit distance. This property is important when computing the fouling resistance of a biofilm on a heat transfer surface. Little data is available on the thermal conductivity of biofilms; however, Characklis et al. (90) measured this parameter and found the thermal conductivity of biofilm to be insignificantly different from that of water. This is not unexpected since biofilms have a high water content. Because of the water content and the gel-like composite of polymers and cells, biofilms are not rigid structures. Characklis (86), using a Weissenburg rheogoniometer in situ on a mixed culture biofilm, showed that the biofilm is viscoelastic. The elastic (storage) modulus and the viscous (loss) modulus were determined to be 59.5 and 118 $N\ m^{-2}$, respectively.

## 5 ARTIFICIALLY IMMOBILIZED ORGANISMS

Whole microbial cells may be immobilized as a method for deploying the microbes in a reactor system. Immobilization in a bioreactor allows a higher throughput of reactants without washing out the organisms performing the specific bioconversion. Immobilization conserves expensive "designer" organisms that may be needed to produce highly specific biochemicals. Artificially immobilized biocatalysts have been used to produce the following compounds: hydrogen, methane, antimicrobials, enzymes, amino acids, organic acids, ethanol, and wine (91). On an industrial scale, immobilization is used to produce isosyrup for food use, amino acids, proinsulin, and penicinallic acid. In addition to chemical production, biodegradation of undesired compounds has been accomplished. Some of the degraded compounds include phenolics, nitrate, urea, and malic acid. Immobilization techniques can be divided into three categories: (1) entrapment within a support, (2) adsorption to a support, and (3) covalent bonding to a support. Each method has its advantages and disadvantages, and in most cases no perfect method exists. A number of thorough reviews are available in the area of artificially immobilized biocatalysts (91–94). The following sections give information pertinent to cell entrapment and will not cover information regarding cell adsorption or covalent bonding.

### 5.1 Entrapment

Entrapment is one of the least disruptive methods for artificially immobilizing whole cells. The organisms are generally entrained in a membrane or gel matrix. The membrane or gel surrounding the cell allows nutrients and cell products to diffuse to and from the cell while providing a "cage" to keep the cell entrapped. Although diffusion occurs, poor mass transfer in the matrix can decrease the biocatalytic activity drastically when compared to the activity of freely suspended organisms.

### 5.1.1 Membrane Entrapment

A membrane reactor of particular importance in whole cell immobilization is the hollow fiber (HF) reactor. The HF reactor has found wide acceptance in the entrapment of eukaryotic cells, but has also been used for prokaryotic immobilization (95). In most cases, the cells are introduced on the shell side of the fiber bundle, with nutrients supplied to the tube (or lumen) side. Nutrients must diffuse through the lumen before reaching the cells. Diffusion limitations can cause significant cell density variations throughout the reactor (96).

### 5.1.2 Immobilization Gels

Gels provide a three-dimensional matrix for cell immobilization. Many methods of gel entrapment have been tried with varying degrees of success. Some of the gel entrapment polymers impose stressful conditions on the cells during matrix polymerization, while other polymers do not have the structural integrity to withstand processing conditions.

Polyacrylamide gel is the most widely used matrix for whole cell entrapment. Linear chains of polyacrylamide are produced with acrylamide monomers in a free radical polymerization reaction. This reaction is carried out in an isotonic, phosphate-buffered aqueous solution that contains suspended organisms. A crosslinking agent, usually methylene *bis* acrylamide, is added to give the polymer a three-dimensional structure. Pellet porosity of the polyacrylamide gel is determined by the extent of crosslinking achieved during polymerization. This method has the disadvantage that the acrylamide monomer is toxic to the cells, and the free radical reaction can cause enzyme deactivation. Polymerization has been shown to damage bacterial cell walls (97), but the damage can be reduced by adding 0.2 M magnesium ions to the cell suspension before acrylamide monomer addition. The final shape of the gel is determined by the vessel in which the polymerization reaction takes place, but the solidified gel containing microorganisms can be easily granulated to serve as column packing, fluidized bed fill, and so forth. The polymerization reaction is exothermic, so casting the gel in a large block can create detrimental temperatures up to 75°C in the block's core (98). To reduce damage to the organisms, the reaction can be carried out in a thin film on a cooled plate.

The first developed gel for whole cell entrapment was ionic crosslinking of alginic acid with polyvalent metal ions (99). Metal ions that have been used as crosslinking agents include the following: calcium, iron III, aluminum, barium, and strontium, although barium and strontium are toxic to eukaryotes. Calcium is the most commonly used divalent metal ion for this technique. To form a calcium alginate gel, a suspension of microorganisms in a sodium alginate solution is added dropwise to a calcium chloride solution. The drops precipitate in the calcium chloride solution to form spherical pellets of high porosity. This method is simple to perform and is nontoxic to microorganisms. The dominant limitation with using calcium alginate is its susceptibility to calcium removal by chelating agents such as phosphates. This problem has been addressed by Stenroos et al. (100) with immobilized *Lactobacillus*, and by Lewandowski et al. (101) with autotrophic nitrifiers and denitrifiers. The addition of insoluble calcium carbonate in the cell-alginate mixture prevents the beads from falling apart, since the protons produced during microbial respiration react with the calcium carbonate particles to release calcium ion. The release of the calcium ions replaces the calcium lost to the bulk solution.

Another noncovalent gel is a seaweed-derived polysaccharide, K-carrageenan. This polysaccharide has proven useful for whole-cell entrapment since it gels at room temperature and in the presence of metal ions, ammonium ions, or water-miscible organic solvents (102–104). Most K-carrageenan gelling methods employ 0.3M potassium chloride as the gelling agent. The K-carrageenan gel structure is too open to entrap enzymes, but when whole cells are used, this property gives the gel higher effective diffusion coefficients for nutrients and oxygen when compared to other entrapment gels.

## 6 BIOFILM REACTORS

### 6.1 Continuous Stirred Tank Reactor (CSTR)

A CSTR is the most common reactor for performing biofilm research. The word "continuous" indicates that liquid is continuously flowing into and out of the reactor; hence, the reactor is an open system. Biofilms are most typically found in open systems. With the continuous flow of liquid into the reactor, a constant low concentration of nutrients can be maintained. A CSTR is well mixed. This implies, in the ideal case, the bulk liquid concentration of all dissolved and suspended materials is constant throughout the reactor. The advantages of uniform liquid-phase concentrations provide for easy sampling of the bulk liquid components. A characteristic of open systems (CSTR), as compared to closed systems (batch), is that open systems can reach a steady state with regard to viable biomass and substrate concentrations.

### 6.2 Batch Reactor

The traditional microbiologist's tool is the shake flash (a batch reactor). This type of reactor is easy to operate and maintain. All nutrients are added at the beginning with the inoculum. This mixture is then shaken or stirred continuously for a specific period of time, during which time samples can be procured until the system reaches a desired end-point, usually a stationary phase. Batch reactors are very useful as a screening tool in biodegradation studies or to obtain kinetic parameters. However, these closed systems simulate very few natural systems and do not favor the formation of a biofilm. Batch systems do not reach a steady state in which viable organisms exist; i.e., there is no steady state except the final thermodynamic equilibrium.

## 6.3 Pipe Flow Reactor

Pipe flow reactors have been used extensively in biofilm research since many industrial biofouling problems occur in pipe systems. Much of the impetus for biofilm research has come from the fouling of heat exchangers and condenser tubing in many of the process industries. To simulate these industrial processes, laboratory experimental systems may use tubing of the same composition and diameter as the field system in question. In the laboratory, most pipe flow reactors are operated in a CSTR fashion by operating at a high recycle rate (Fig. 10). Recycle operation reduces the amount of liquid and nutrients required to maintain flow in an open system. At high recycle rates, no axial gradients occur within a short pipe section, so the pipe process analysis is simplified to CSTR analysis. When the recycle ratio is low, pipe reactors operated in the turbulent flow regime can be characterized by a plug flow approximation. However, ideal plug flow theory does not account for gradients in the radial direction that develop as a result of substrate depletion by the biofilm.

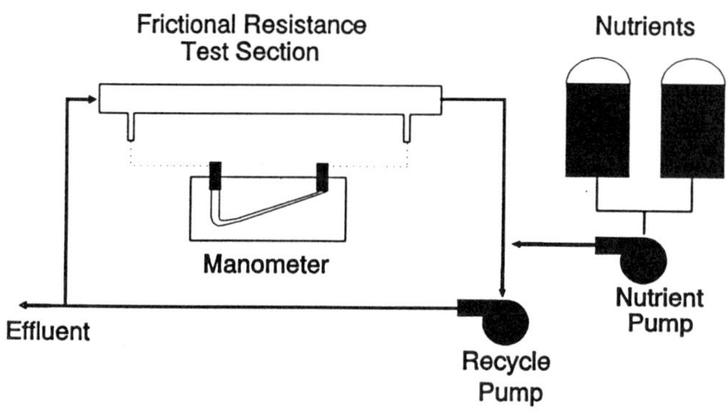


**Figure 10** Schematic diagram of a typical laboratory pipeline reactor operated in recycle mode. The nutrient flow determines the reactor residence time, and the recycle pump can be used to control the desired fluid velocity and shear stress.

In a pipe flow reactor, the shear stress depends on the pipe diameter, the friction factor, and the volumetric flow rate. This implies that as the friction factor increases because of biofilm accumulation, the shear stress will also increase. Therefore, to perform experiments at a constant value of shear stress, the volumetric flow rate must be adjusted continuously throughout the experiment. This last point is particularly important if the pipe reactor is to be operated as a once-through system since changing the volumetric flow rate also changes the substrate loading rate. For example, doubling the volumetric flow rate to achieve a higher shear stress also doubles the amount of substrate entering the reactor (per unit pipe area). Even though the concentration may be the same, the total amount of substrate available for biofilm growth will be doubled. A pressure drop across a tubing test section can be measured with a manometer or pressure transducer and can be used to calculate the fluid frictional resistance if the flow rate is known. A mathematical model for biofilm accumulation in a turbulent flow pipe has been published (105). The model was used to determine heat and mass transfer in heat exchanger or condenser tubes in the presence of a biofilm.

## 6.4 Rotating Annular Reactor (RAR)

The RAR (106) reactor design has proved to be the most versatile for biofilm research. This reactor type has several characteristics that make it superior to other biofilm reactors. The first of these characteristics is independent control of shear stress. The inner cylinder can be rotated at any speed, and the torque can be measured (107) to determine average shear stress on the biofilm (Fig. 11). This reactor also has a large surface area to volume ratio, which means a large amount of biomass can be produced with relatively small volumetric flow rates while still maintaining a short residence time to prohibit growth of suspended organisms. Twelve removable coupons allow the biofilm to be sampled without taking the reactor off-line. Draft tubes inside the inner cylinder completely mix the reactor contents, giving it the mixing characteristics of a CSTR (108).

## 6.5 Open Channel

The open channel reactor is generally used when easy access to the biofilm is necessary for in situ analyses such as microelectrode measurements, or when simulation of a streambed or a culvert is desired. Open channel reactors can be operated to give a high surface area exposed to a gas phase. The flow regime is usually laminar "plus ripples" in these reactor configurations; true laminar flow is difficult to produce.

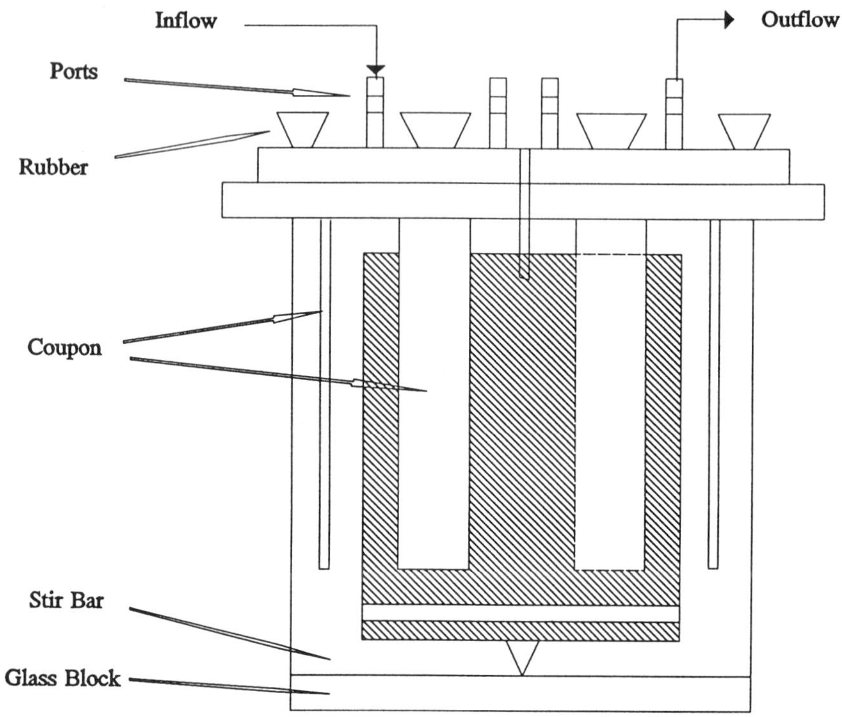


**Figure 11** Simplified schematic diagram of a rotating annular reactor. The reactor consists of an inner rotating drum inside a stationary outer cylinder. Draft tubes (not shown) keep the liquid in the annular space completely mixed.

## 6.6 Rotating Biological Contractor

The rotating biological contractor (RBC) is probably the most common biofilm reactor used for industrial wastewater treatment applications. This reactor is characterized by a series of parallel rotating disks half-submerged in liquid with the other half exposed to the atmosphere (Fig. 12). Atmospheric exposure and rotation (peripheral velocity $\approx$ 0.3 m $s^{-1}$) gives good oxygen mass transfer to the biofilm. The RBC can also be run completely submerged to operate as an anaerobic biofilm reactor. The most common disk materials are high-density polypropylene or polyethylene.

The RBC has many advantages in full-scale industrial operations, including the following: (1) low power requirements for mixing as compared to activated sludge system, (2) less vulnerability than trickling filters to the detrimental effects of low-flow conditions, (3) flexibility in oxygen transfer

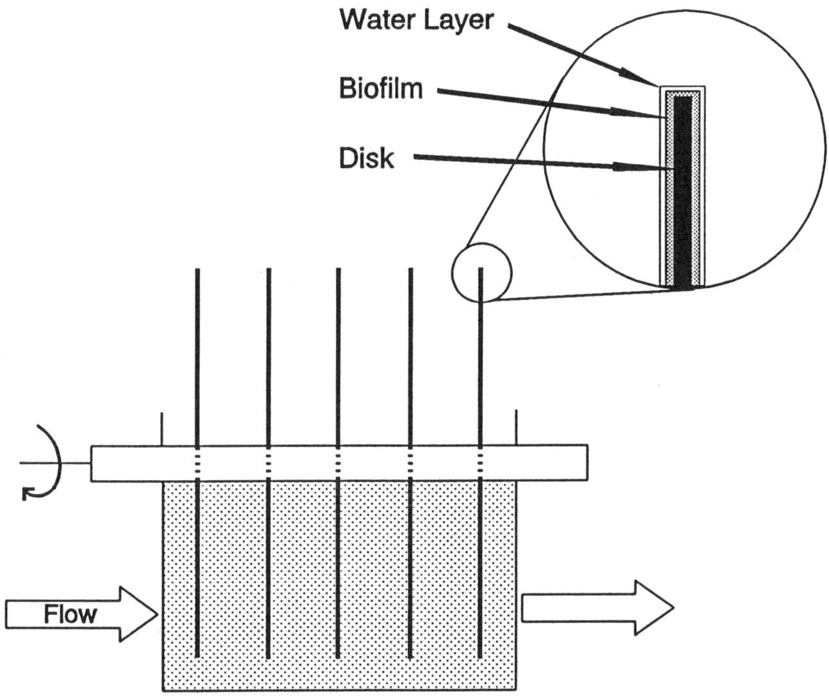


**Figure 12** Schematic diagram of a rotating biological contactor showing disks mounted on a rotating shaft. As the shaft rotates, a given area of biofilm will be submerged and then exposed to the atmosphere. (Inset) Biofilm with a water layer. Oxygen from the atmosphere must diffuse through this water layer to reach the biofilm cells.

and biofilm control by rotational speed, and (4) low susceptibility to shock loadings of toxic materials. If higher oxygen transfer rates are needed, oxygen can be added to the bulk liquid, or additional RBCs can be placed in series, to increase the oxygen capacity of the system. Watanabe et al. (109,110) have presented models on RBC nitrification, and a review by Grady (111) includes information on RBC.

## 6.7 Rotating Disk

A few researchers (112–114) have chosen to use a submerged rotating disk operated in the laminar flow regime. Shear stress can be expressed as a function of radial distance from the center to allow correlation with observed biofilm accumulation. A characteristic of this reactor design is a constant mass-transfer boundary-layer thickness across the entire disk; therefore, mass transfer does not vary with radial position along the disk.

### 6.8 Packed and Fluidized Bed

Packed beds have a very high surface-to-volume ratio permitting intimate contact of fluid elements and biomass. Typically, this reactor performs in a plug-flow manner whereby the fluid elements that are just entering the reactor do not mix with the fluid that is already inside the reactor. A packed bed operated in a downflow mode at a low flow rate is sometimes called a trickling or percolating filter. Trickling filters are usually characterized by a large amount of airspace. Models for packed bed performance are common (115), and many are listed as references in a paper by Skowlund and Kirmse (116). Because the substrate concentration is higher at the inlet end of the packed bed, biomass tends to concentrate at this end of the reactor. This may lead to detrimental channeling or plugging. Reports of packed bed simulation of biodegradation of toxic compounds in groundwater have been numerous in the literature, although many studies do not characterize the biomass concentration responsible for ameliorating the toxic compounds.

A fluidized bed is a bed of small particles (0.2–2.0 mm in diameter) freely suspended in an upward flow of water, or a water and gas mixture. Fluidized beds have the advantage of being well mixed, allowing equal distribution of biomass throughout the entire reactor. This prevents plugging and also enhances mass transfer to the biofilm. Cooper and Atkinson (117) collected a considerable quantity of information on the research and development of fluidized bed reactors. Mulcahy and Shieh (118) report that the Richardson-Zaki correlation for rigid solid particles applies well to the fluidization of bioparticles. Based on this and other correlations given in the paper, an iterative design procedure was developed for a fluidized bed biofilm reactor.

## 7 SUMMARY

Scientists and engineers have realized the industrial and environmental significance of biofilm accumulation and activity. The ability to predict and control biofilm formation has led to less fouling and corrosion in industrial systems and a better understanding of biofilm importance in natural aquatic systems. Understanding the fundamental processes contributing to biofilm formation is beneficial to anyone involved with natural or industrial systems where biofilms may play a significant role in determining variables such as bulk water quality, toxic compound biodegradation, or product quality.

## 8 NOMENCLATURE

| | |
|---|---|
| $b$ | Endogenous decay rate coefficient ($t^{-1}$). |
| $d$ | Pipe diameter (L). |
| $f$ | Fanning friction factor (unitless). |
| $K_g$ | Growth-associated detachment coefficient ($M\ L^{-1}$). |
| $K'_g$ | Non-growth-associated detachment coefficient ($M\ L^{-1}\ t^{-1}$). |
| $k_p$ | Growth-associated product formation coefficient (unitless). |
| $k'_p$ | Non-growth-associated product formation coefficient ($t^{-1}$). |
| $K_s$ | Monod half-saturation coefficient ($M\ L^{-3}$). |
| $K_{sd}$ | Shear stress dependent detachment rate coefficient ($L\ t\ M^{-1}$). |
| $K_{tp}$ | Velocity dependent detachment rate coefficient for a turbulent pipe ($t^{0.75}\ L^{-1.75}$). |
| $K_2$ | Second-order detachment rate coefficient ($L^2\ M^{-1}\ t^{-1}$). |
| $K_3$ | Mixed detachment rate coefficient ($L^{-1}\ t^{-1}$). |
| $L_f$ | Biofilm thickness (L). |
| $R_a$ | Attachment rate ($M\ L^{-2}\ t^{-1}$). |
| $R_d$ | Detachment rate ($M\ L^{-2}\ t^{-1}$). |
| $Re$ | Reynold's number (unitless). |
| $r_x$ | Cellular accumulation rate ($M\ L^{-3}\ t^{-1}$). |
| $S$ | Substrate concentration ($M\ L^{-3}$). |
| $t$ | Time (t). |
| $v$ | Average fluid velocity ($L\ t^{-1}$). |
| $X$ | Biomass concentration ($M\ L^{-3}$). |
| $X_b$ | Areal biomass concentration ($M\ L^{-2}$). |
| $Y_{x/s}$ | Biomass yield ($M_x\ M_s^{-1}$). |
| $\mu$ | Specific growth rate ($t^{-1}$). |
| $\mu_m$ | Maximum specific growth rate ($t^{-1}$). |
| $\nu$ | Kinematic viscosity ($L^2\ t^{-1}$). |
| $\rho$ | Fluid density ($M\ L^{-3}$). |
| $\rho_f$ | Biofilm volumetric density ($M\ L^{-3}$). |
| $\tau$ | Shear stress at biofilm surface ($M\ t^{-2}\ L^{-1}$). |

## ACKNOWLEDGMENTS

This work was supported in part by the NSF Center for Biofilm Engineering, College of Engineering, Montana State University, and by the U.S. Department of Energy as part of Pacific Northwest Laboratory's Laboratory Directed Research and Development. Pacific Northwest Laboratory is operated for the U.S. Department of Energy by Battelle Memorial Institute under Contract DE-AC06-76RLO 1830.

## REFERENCES


1. Wuhrmann, K. (1971). Stream purification, in *Water Pollution Microbiology* (R. Mitchell, ed.), Wiley, New York, p. 119.
2. Characklis, W.G., Bryers, J.D., Trulear, M.G., and Zelver, N. (1980). Biofouling film development and its effects on energy losses: a laboratory study, in *Condenser Biofouling Control* (J.F. Garey et al., eds.), Ann Arbor Science Publishers, Ann Arbor, MI, p. 49.
3. Characklis, W.G., Turakhia, M.H., and Zelver, N. (1984). Biofouling of heat transfer surfaces—part II, *Corrosion Maintenance*, July–Sept., p. 213.
4. Dziewulski, D.M., Fraleigh, S.P., Pope, D.H., Thomas, S., Puente, P., and Annexstad, G. (1990). Microbial production of hydrogen sulfide in gas storage and production fields: field studies, preliminary modeling and control, *Corrosion '90*, April 23–27, 1990, Las Vegas, NV, paper 35.
5. Geesey, G.G., Mittelman, M.W., and Lieu, V.T. (1987). Evaluation of slime-producing bacteria in oil field core flood experiments, *Appl. Environ. Microbiol.*, *53*: 278.
6. Crawford, P.B. (1966). Water technology: continual changes observed in bacterial stratification rectification, *Prod. Mon.*, *26*: 12.
7. Lappin-Scott, H.M., Cusack, F., and Costerton, J.W. (1988). Nutrient resuscitation and growth of starved cells in sandstone cores: a novel approach to enhanced oil recovery, *Appl. Environ. Microbiol.*, *54*: 1373.
8. Torbati, H.M., Raiders, R.A., Donaldson, E.C., McInerney, M.J., Jenneman, G.E., and Knapp, R.M. (1986). Effect of microbial growth on pore entrance size distribution in sandstone cores, *Soc. Ind. Microbiol.*, *1*: 227.
9. Picologlou, B.F., Zelver, N., and Characklis, W.G. (1980). Biofilm growth and hydraulic performance, *J. Hydraul. Div.*, ASCE, *106*: 733.
10. Baier, R.E. (1980). Substrata influences on adhesion of microorganisms and their resultant new surface properties, in *Adsorption of Microorganisms to Surfaces* (G. Bitton and K.C. Marshall, eds.), Wiley, New York, p. 59.
11. Baier, R.E., and Weiss, L. (1975). Demonstrations of the involvement of adsorbed proteins in cell adhesion and cell growth on solid surfaces, in *Applied Chemistry at Protein Interfaces* (R.E. Baier, ed.), Advances in Chemistry Series, 145, American Chemical Soc., Washington, D.C., p. 300.
12. Brash, J.L., and Samak, Q.M. (1978). Dynamics of interactions between human albumin and a polyethylene surface, *J. Colloid Int. Sci.*, *65*: 495.
13. Bryers, J.D. (1980). Dynamics of early biofilm formation in a turbulent flow system, Ph.D. thesis, Rice University, Houston, TX.
14. Little, B.J., and Zsolnay, A. (1985). Chemical fingerprinting of adsorbed organic materials on metal surfaces, *J. Colloid. Int. Sci.*, *104*: 79.
15. Meadows, P.S. (1971). Attachment of bacteria to solid surfaces, *Arch. Mikrobiol.*, *75*: 374.
16. Fletcher, M. (1976). The effects of proteins on bacterial attachment to polystyrene, *J. Gen. Microbiol.*, *94*: 400.
17. Marshall, K.C. (1979). Strategies of microbial life in extreme environments,

in *Dahlem Konferenzen Life Sciences Research Report* 13 (M. Shilo, ed.), Verlag Chemie, Weinheim, p. 281.

18. Hermanowicz, S.W., Danielson, R.E., and Cooper, R.C. (1989). Bacterial deposition on and detachment from surfaces in turbulent flow, *Biotech. Bioeng.*, *33*: 157.
19. Powell, M.S., and Slater, N.K.H. (1983). The deposition of bacterial cells from laminar flow onto solid surfaces, *Biotechnol. Bioeng.*, *25*: 891.
20. Escher, A. (1986). Colonization of a smooth surface by *Pseudomonas aeruginosa*: image analysis methods, Ph.D. thesis, Montana State University, Bozeman.
21. Cleaver, J.W., and Yates, B. (1975). A sublayer model for the deposition of particles from a turbulent flow, *Chem. Eng. Sci.*, *30*: 983.
22. Cleaver, J.W., and Yates, B. (1976). The effect of re-entrainment on particle deposition, *Chem. Eng. Sci.*, *31*: 147.
23. Escher, A., and Characklis, W.G. (1990). Modelling initial events in biofilm accumulation, in *Biofilms* (W.G. Characklis and K.C. Marshall, eds.), Wiley, New York, p. 445.
24. Jang, L.K., and Yen, T.F. (1985). A theoretical model of diffusion of motile and nonmotile bacteria toward solid surfaces, in *Microbes and Oil Recovery, 1* (J.E. Zajic and E.C. Donaldson, eds.), *Int. Bioresour. J.*, Vol 1, p. 226.
25. Mueller, R.F. (1990). Characterization of initial events of bacterial colonization at solid-water interfaces using image analysis, Master's thesis, Montana State University, Bozeman.
26. Van Haecke, E., Remon, J.P., Moors, M., Reas, F., DeRudder, D., and Van Peteghem, A. (1990). Kinetics of *Pseudomonas aeruginosa* adhesion to 304 and 316-L stainless steel: role of cell surface hydrophobicity, *Appl. Environ. Microbiol.*, *56*: 788.
27. Nelson, C.H., Robinson, J.A., and Characklis, W.G. (1985). Bacterial adsorption to smooth surfaces: rate, extent, and spatial pattern, *Biotechnol. Bioeng.*, *27*: 1662.
28. Dawson, M.P. (1981). The effect of starvation on the cell surface characteristics of marine bacteria, Bachelor's honors thesis, University of New South Wales.
29. Fletcher, M., and Floodgate, G.D. (1973). An electron-microscopic demonstration of an acidic polysaccharide involved in the adhesion of a marine bacterium to solid surfaces, *J. Gen. Microbiol.*, *74*: 365.
30. Marshall, K.C., and Cruickshank, R.H. (1973). Cell surface hydrophobicity and the orientation of certain bacteria at interfaces, *Arck. Mikrobiol.*, *91*: 29.
31. Zobel, C.E. (1943). The effect of solid surfaces on bacterial activity, *J. Bacteriol.*, *46*: 39.
32. Marshall, K.C., Stout, R., and Mitchell, R. (1971). Mechanism of initial events in the sorption of marine bacteria to surfaces, *J. Gen. Microbiol.*, *68*: 337.
33. Powell, M.S., and Slater, N.K.H. (1983). The deposition of bacterial cells from laminar flows onto solid surfaces, *Biotechnol. Bioeng.*, *25*: 891.

34. Gunawan, C. (1990). The rate of cellular attachment to an established biofilm, Master's thesis, Montana State University, Bozeman.
35. Bryers, J.D. (1987). Biologically active surfaces: processes governing the formation and persistence of biofilms, *Biotechnol. Prog.*, *3*: 57.
36. Stewart, P.S., Peyton, B.M., Drury, W.J., and Murga, R. (1993). Quantitative observations of heterogeneities in *Pseudomonas aeruginosa* biofilms, *Appl. Environ. Microbiol.*, *59*: 327.
37. Howell, J.A., and Atkinson, B. (1976). Sloughing of microbial film in trickling filters, *Water Res.*, *10*: 307.
38. Heukelelian, H., and Crosby, E.S. (1956). Slime formation in polluted waters, *Sewage Industrial Wastes*, *28*: 78.
39. Heukelelian, H., and Crosby, E.S. (1956). Slime formation in sewage, *Sewage Industrial Wastes*, *28*: 206.
40. Jansen, J., and Kristensen, G.H. (1980). Fixed film kinetics: denitrification in fixed films, Report 80-59, Department of Sanitary Engineering, Technical University of Denmark.
41. Bakke, R. (1983). Dynamics of biofilm processes: substrate load variations, Master's thesis, Montana State University, Bozeman.
42. Lee, S.S., Jackman, A.P., and Schroeder, E.D. (1975). The role of flocculation in transient microbial growth, *Water Res.*, *9*: 491.
43. Bott, T.R., and Miller, P.C. (1983). Mechanisms of biofilm formation on aluminum tubes, *J. Chem. Tech. Biotechnol.*, *33B*: 177.
44. Turakhia, M.H., Cooksey, K.E., and Characklis, W.G. (1983). Influence of a calcium-specific chelant on biofilm removal, *Appl. Environ. Microbiol.*, *46*: 1236.
45. Marshall, P.A., Loeb, G.I., Cowan, M.M., and Fletcher, M. (1989). Response of microbial adhesives and biofilm matrix polymers to chemical treatments as determined by interference reflection microscopy and light section microscopy, *Appl. Environ. Microbiol.*, *55*: 2827.
46. Fenchel, T. (1986). The ecology of heterotrophic microflagellates, *Adv. Microbial Ecol.*, *9*: 57.
47. Metcalf and Eddy, Inc. (1979). *Wastewater Engineering* (G. Tchobanoglous, ed.), McGraw-Hill, New York, p. 445.
48. Rittman, B.E. (1982). The effect of shear stress on biofilm loss rate, *Biotechnol. Bioeng.*, *24*: 501.
49. Kreikenbohm, R., and Stephan, W. (1985). Application of a two compartment model to the wall growth of *Pelobacter acidigallici* under continuous culture conditions, *Biotechnol. Bioeng.*, *27*: 296.
50. Chang, H.T., and Rittmann, B.E. (1987). A comparative study of biofilm on activated carbon, *J. WPCF*, *60*: 362.
51. Bakke, R., Characklis, W.G., Turakhia, M.H., and Yeh, A. (1990). Modeling a monopopulation biofilm system: *Pseudomonas aeruginosa*, in *Biofilms* (W.G. Characklis and K.C. Marshall, eds.), Wiley, New York, p. 487.
52. Bird, R.B., Stewart, W.E., and Lightfoot, E.N. (1960). *Transport Phenomena*, Wiley, New York, p. 656.

53. Bryers, J.D. (1984). Biofilm formation and chemostat dynamics: pure and mixed culture considerations, *Biotechnol. Bioeng.*, *26*: 948.
54. Crawford, D.J. (1987). Hydraulic effects of biofilm accumulation in simulated porous media flow system, Master's thesis, Montana State University, Bozeman.
55. Wanner, O., and Gujer, W. (1986). A multi-species biofilm model, *Biotechnol. Bioeng.*, *28*: 314.
56. Chang, H.T., and Rittmann, B.E. (1988). Comparative study of biofilm shear loss on different adsorptive media, *J. WPCF*, *60*: 362.
57. Trulear, M.G., and Characklis, W.G. (1982). Dynamics of biofilm processes, *J. WPCF*, *54*: 1288.
58. Speitel, G.E., Jr., and DiGiano, F.A. (1987). Biofilm shearing under dynamic conditions, *J. Environ. Eng.*, *113*: 464.
59. Rittmann, B.E. (1982). The effect of shear stress on biofilm loss rate, *Biotechnol. Bioeng.*, *24*: 501.
60. Peyton, B.M., and Characklis, W.G. (1993). A statistical analysis of the effects of substrate utilization and shear stress on the kinetics of biofilm detachment, *Biotechnol. Bioeng.*, *41*: 728.
61. Stewart, P.S. (1993). A model of biofilm detachment, *Biotechnol. Bioeng.*, *41*: 111.
62. Monod, J. (1942). *Recherches sur la Croissance des Cultures Bacteriennes*, Hermann, Paris.
63. Monod, J. (1949). The growth of bacterial cultures, *Annu. Rev. Microbiol.*, *3*: 371.
64. Bakke, R., Trulear, M.G., Robinson, J.A., and Characklis, W.G. (1984). Activity of *Pesudomonas aeruginosa* in biofilms: steady state, *Biotechnol. Bioeng.*, *26*: 1418.
65. Jorgensen, B.B., and Des Marais, D.J. (1986). Competition for sulfide among colorless and purple sulfur bacteria in cyanobacterial mats, *FEMS Microbiol. Ecol.*, *38*: 179.
66. Lewandowski, Z., Walser, G., Larsen, R., Peyton, B.M., and Characklis, W.G. (1990). Biofilm surface positioning, Proceedings 1990, EE Div/ASCE, Arlington, VA, July 8–11, 1990.
67. Characklis, W.G. (1978). Microbial reaction rate expressions, *J. Environ. Eng. Div. ASCE*, *104*: 531.
68. Atkinson, B. (1974). *Biochemical Reactors*, Pion Ltd., London, p. 40.
69. Rosenberg, E. (1989). Biofilms on water-insoluble substrates, in *Structure and Function of Biofilms* (W.G. Characklis and P.A. Wilderer, eds.), Wiley, New York, p. 59.
70. Mueller, R.F. (1990). Characterization of initial events of bacterial colonization at solid-water interfaces using image analysis, Master's thesis, Montana State University, Bozeman.
71. Luedeking, R., and Piret, E.L. (1959). *J. Biochem. Microbiol. Technol. Eng.*, *1*: 393.
72. Weiss, R.M., and Ollis, D.F. (1980). Extracellular microbial polysaccharides.

I. Substrate, biomass, and product kinetic equations for batch xanthan gum fermentation, *Biotechnol. Bioeng.*, *22*: 859.
73. Robinson, J.A., Trulear, M.G., and Characklis, W.G. (1984). Cellular reproduction and extracellular polymer formation by *Pseudomonas aeruginosa* in continuous culture, *Biotechnol. Bioeng.*, *26*: 1409.
74. Rittmann, B.E., and McCarty, P.L. (1980). Evaluation of steady state biofilm kinetics, *Biotechnol. Bioeng.*, *22*: 2358.
75. van Loosdrecht, M.C.M., Lyklema, J., Norde, W., and Zehnder, A.J.B. (1990). Influences of interfaces on microbial activity, *Microbiol. Rev.*, *54*: 75.
76. Doran, P.M., and Bailey, J.E. (1986). Effects of immobilization on growth, fermentation properties, and macromolecular composition of *Saccharomyces cerevisiae* attached to gelatin, *Biotechnol. Bioeng.*, *28*: 73.
77. Doran, P.M., and Bailey, J.E. (1987). Effects of immobilization on the nature of glycolytic oscillations in yeast, *Biotechnol. Bioeng.*, *29*: 892.
78. Busch, A.W. (1971). *Aerobic Biological Treatment of Wastewaters*, Gulf Publishing Co., Houston, TX.
79. Traore, A.S., Hatchikian, C.E., Legall, J., and Belaich, J.P. (1982). *J. Bacteriol.*, *149*: 606.
80. Okabe, S. (1990). Unpublished results, Center for Interfacial Microbial Process Engineering, Montana State University, Bozeman.
81. Herbert, D. (1976). Stoichiometric aspects of microbial growth, in *Continuous Culture 6: Applications and New Fields* (A.C.R. Dean, D.C. Ellwood, C.G.T. Evans, and J. Melling, eds.), Ellis Horwood Ltd., Chichester, pp. 1–30.
82. Rittmann, B.E., and McCarty, P.L. (1980). Model of steady state biofilm kinetics, *Biotechnol. Bioeng.*, *22*: 2343.
83. Rittmann, B.E., and McCarty, P.E. (1981). Substrate flux into biofilms of any thickness, *J. Env. Eng. Div. ASCE.*, *107*: 831.
84. Stewart, P.S., and Robertson, C.R. (1989). Microbial growth in a fixed volume: studies with entrapped *Escherichia coli*, *Appl. Microbiol. Biotechnol.*, *30*: 34.
85. Masuda, S., Watanabe, Y., and Ishiguro, M. (1991). Biofilm properties and simultaneous nitrification and denitrification in aerobic rotating biological contractors, *Water Sci. Technol.*, *23*: 1355.
86. Characklis, W.G. (1980). Biofilm development and destruction, Final report, EPRI CS-1554, Project RP902-1, Electric Power Research Institute, Palo Alto, CA.
87. Kornegay, B.H., and Andrews, J.F. (1967). Characteristics and kinetics of fixed-film biological reactors, Final report, Grant WP-01181, Federal Water Pollution Control Administration, U.S. GPO, Washington, D.C.
88. Suschka, J. (1987). Hydraulic performance of percolating biological filters and consideration of oxygen transfer, *Water Res.*, *21*: 865.
89. Matson, J.V., and Characklis, W.G. (1976). Diffusion into microbial aggregates, *Water Res.*, *10*: 877.
90. Characklis, W.G., Nimmons, M.J., and Picologlou, B.F. (1981). *Heat Transfer Eng.*, *3*: 23.

91. Kennedy, J.F., and Cabral, J.M.S. (1983). *Appl. Biochem. Bioeng.*, *4*: 189.
92. Kennedy, J.F. (1979). *ACS Symp. Ser.*, *106*: 119.
93. Rosebear, A. (1984). Immobilized biocatalysts—a critical review, *J. Chem. Tech. Biotechnol.*, *34b*: 127.
94. Scott, C.D. (1987). *Enzyme Microb. Technol.*, *9*: 66.
95. Inloes, D.S., Smith, W.J., Taylor, D.P., Cohen, S.N., Michaels, A.S., and Robertson, C.R. (1983). *Biotechnol. Bioeng.*, *25*: 2653.
96. Ku, K., Kuo, M.J., Delente, J., Wildi, B.S., and Feder, J. (1981). Development of a hollow-fiber system for large-scale culture of mammalian cells, *Biotechnol. Bioeng.*, *23*: 79.
97. Starostina, N.G., Lusta, K.A., and Fikhte, B.A. (1983). Morphological and physiological changes in bacterial cells treated with acrylamide, *Eur. J. Appl. Microbiol. Biotechnol.*, *18*: 264.
98. Wheatley, M.A., and Phillips, C.R. (1983). Temperature effects during polymerization of polyacrylamide gels used for bacterial cell immobilization, *Biotechnol. Bioeng.*, *25*: 623.
99. Hackel, V., Klein, J., Megnet, R., and Wagner, F. (1975). Immobilization of microbial cells in polymeric matrices, *Eur. J. Appl. Microbiol.*, *1*: 291.
100. Stenroos, S.L., Linko, Y.Y., and Linko, P. (1982). *Biotechnol. Lett.*, *4*: 159.
101. Lewandowski, Z., Bakke, R., and Characklis, W.G. (1987). *Water Sci. Technol.*, *19*: 175.
102. Wada, M., Kato, J., and Chibata, I. (1979). Immobilized growing cells using carrageenan gel and their properties, *Eur. J. Appl. Microbiol. Biotechnol.*, *8*: 241.
103. Tosa, T., Sato, T., Mori, T., Yamamoto, K., Takata, I., Nishida, Y., and Chibata, I. (1979). Immobilization of enzymes and microbial cells using carrageenan as matrix, *Biotechnol. Bioeng.*, *21*: 1697.
104. Chibata, I. (1980). *Enzyme Eng.*, *5*: 393.
105. Kirkpatrick, J.P., McIntire, L.V., and Characklis, W.G. (1980). Mass and heat transfer in a circular tube with biofouling, *Water Res.*, *14*: 117.
106. Characklis, W.G. (1990). Biofilm processes, in *Biofilms* (W.G. Characklis and K.C. Marshall, eds.), Wiley, New York, p. 55.
107. Characklis, W.G., and Roe, F.L. (1984). Monitoring buildup of fouling deposits on surfaces of fluid handling systems, U.S. Patent 4,485,450, Nov. 27, 1984.
108. Trulear, M.G. (1983). Cellular reproduction and extracellular polymer formation in the development of biofilms, Ph.D. thesis, Montana State University, Bozeman.
109. Watanabe, Y., Nishidome, K., Thananteseth, C., and Ishiguro, M. (1982). Kinetics and simulation of nitrification in a RBC, in *Proc. 1st International Conference on Fixed Film Biological Processes* (Y.C. Wu, E.D. Smith, R.D. Miller, and E.J. Opatken, eds.), Kings Island, OH, p. 309.
110. Watanabe, Y., Masuda, S., Nishidome, K., and Wantawin, C. (1984). *Water Sci. Technol.*, *17*: 385.
111. Grady, C.P.L. (1982). Modelling of biological fixed films: a state-of-the-art review, in *Proc. 1st International Conference on Fixed Film Biological Pro-*

*cesses* (Y.C. Wu, E.D. Smith, R.D. Miller, and E.J. Opatken, eds.), Kings Island, OH, p. 344.

112. Loeb, G.I., Laster, D., and Gracik, T. (1984). The influence of microbial fouling films on hydrodynamic drag of rotating disks, in *Marine Biodeterioration: an Interdiscipline Study* (J.D. Costlow and R.C. Tipper, eds.), Naval Institute Press, Annapolis, MD, p. 88.
113. Abbot, A., Rutter, P.R., and Berkeley, R.C.W. (1983). The influence of ionic strength, pH, and a protein layer on the interaction between *Streptococcus mutans* and glass surfaces, *J. Gen. Microbiol.*, *129*: 439.
114. Gulevich, W., Renn, C.E., and Liebman, J.C. (1968). *Env. Sci. Technol.*, *2*: 113.
115. Logan, B.E., and Hermanowicz, S.W. (1987). Application of the penetration theory to oxygen transfer to biofilms, *Biotechnol. Bioeng.*, *29*: 762.
116. Skowlund, C.T., and Kirmse, D.W. (1989). Simplified models for packed-bed biofilm reactors, *Biotechnol. Bioeng.*, *33*: 164.
117. Cooper, P.F., and Atkinson, B. (eds.) (1981). *Biological Fluidized Bed Treatment of Water and Wastewater*, Ellis-Horwood, Chichester.
118. Mulcahy, L.T., and Shieh, W.K. (1987). Fluidization and reactor biomass characteristics of the denitrification fluidized bed biofilm reactor, *Water Res.*, *21*: 451.
119. Characklis, W.G. (1990). Biofilm processes, in *Biofilms* (W.G. Characklis and K.C. Marshall, eds.), Wiley, New York, p. 195.
120. Escher, A., and Characklis, W.G. (1990). Modeling the initial events in biofilm accumulation, in *Biofilms* (W.G. Characklis and K.C. Marshall, eds.), Wiley, New York, p. 445.
121. Mian, F.A., Jarman, T.R., and Righelato, R.C. (1978). Biosynthesis of exopolysaccharide by *Pseudomonas aeruginosa*, *J. Bacteriol.*, *134*(2): 418.
122. Turakhia, M. H. (1986). The influence of calcium on biofilm processes, Ph.D. thesis, Montana State University, Bozeman.
123. Nagai, S. (1979). Mass and energy balances for microbial growth kinetics, in *Advances in Biochemical Engineering* (T.K. Ghose, A. Fletcher, and N. Blakebrough, eds.), Springer-Verlag, New York, p. 53.
124. Servisi, J.A., and Bogan, R.H. (1963). *J. San. Eng. Div., Proc. ASCE*, *89*: 17.
125. Characklis, W.G. (1990). Energetics and stoichiometry, in *Biofilms* (W.G. Characklis and K.C. Marshall, eds.), Wiley, New York, p. 161.
126. Rittmann, B.E., and McCarty, P.L. (1978). Variable-order model of bacterial film kinetics, *J. Env. Eng. Div., ASCE*, *104*(EE5): 889.
127. Hoen, R.C., and Ray, A.D. (1973). *J. Water Pollut. Contr. Fed.*, *46*: 2302.
128. Williamson, K., and McCarty, P.L. (1976). Verification studies of the biofilm model for bacterial substrate utilization, *J. Water Pollut. Contr. Fed.*, *48*: 281.
129. Bakke, R. (1986). Biofilm detachment, Ph.D. thesis, Montana State University, Bozeman.
130. Tomlinson, T.G., and Snaddon, D.M. (1966). *Int. J. Air Water Pollut.*, *10*: 865.

131. Bungay, H.R., Whalen, W.J., and Saunders, W.M. (1969). Microprobe techniques for determining diffusivities and respiration rates in microbial slime systems, *Biotechnol. Bioeng.*, *11*: 765.
132. Siegrist, H., and Gujer, W. (1985). Mass transfer mechanisms in a heterotrophic biofilm, *Water Res.*, *19*(11): 1369.
133. Baillod, R.C., and Boyle, W.C. (1970). *J. San. Eng. Div.*, *Proc. ASCE*, *96*(SA4): 525.
134. Pipes, D.M. (1974). Variations on glucose diffusion coefficients through biological floc, Master's thesis, Rice University, Houston, TX.
135. Harrison, J.S. (1967). *Process Biochem.*, *2*: 41.
136. Stouthamer, A.H. (1977). Theoretical calculations on the influence of the inorganic nitrogen source on parameters for aerobic growth of microorganisms, *Antonie van Leeuwenhoek*, *43*: 351.
137. Bauer, S., and Ziv, E. (1976). Dense growth of aerobic bacteria in a bench-scale fermentor, *Biotechnol. Bioeng.*, *18*: 81.
138. Mayberry, W.R., Prochazka, G.L., and Payne, W.J. (1968). Factors derived from studies of aerobic growth in minimal media, *J. Bacteriol.*, *96*(4): 1424.
139. Masuda, S., Watanabe, Y., and Ishiguro, M. (1991). Biofilm properties and simultaneous nitrification and denitrification in aerobic rotating biological contactors, *Water Sci. Tech.*, *23*: 1355.

# 7

# Matrices and Activation Methods for Cell Adhesion/Immobilization Studies

**William H. Scouten**
*Utah State University, Logan, Utah*

## 1 INTRODUCTION

This chapter will be limited to a basic overview of matrices and activation processes employed in cell studies. We will not attempt to consider the applications of the resulting immobilized cells; these are considered in other chapters. For more detail concerning matrices for cell immobilizations, we refer readers to the review by Kennedy and Cabral (1) or, alternatively, to a summary of enzyme coupling methods by this author in *Methods in Enzymology* (2). For application specifically to cell immobilization, the reader may refer to the two-volume series on immobilized cells and organelles by Bo Mattiasson (3) or the three-volume series in *Methods in Enzymology* edited by Klaus Mosbach (4).

### 1.1 Matrix Choice

Three factors are involved in the choice of an appropriate matrix for any solid-phase biochemical procedure, whether it be affinity chromatography, enzyme immobilization, or, as in the case at hand, the immobilization of whole cells. Those three factors are the mechanical, chemical, and biological properties of the matrix. Each needs to be considered individually.

Mechanical considerations in matrix choice consist chiefly of the availability of the desired matrix in the form desired for the ultimate application, as well as the mechanical durability of the matrix during use. For example, nylon is an appropriate matrix for fibrous materials which are used to great advantage in cell purification. Moreover, it is also readily available as membranes and in a wide variety of solid shapes and sizes. However, for those applications where large surface areas and high capacities are needed, there are few examples of porous nylon having a high surface area for a given volume.

The chemical considerations in choice of matrix are quite simple. The question asked is "Can this material be modified to yield the surface needed for proper interaction with biomolecules, and can it be activated to covalently bond to them?" For example, a hydroxylic matrix can be readily activated to immobilize proteins and cell components but, unfortunately, is not easily used in organic solvents. Similarly, polyesters would not be the best materials to use where exposure to pH extremes or esterases would tend to hydrolyze the matrix backbone. Likewise, polystyrene is an exceptionally useful matrix from a mechanical standpoint, but the possibilities for chemical activation are very limited.

Finally, and most important, are the biological considerations. The first and foremost of these is that the matrix does not contain anything that would be toxic to cells which are often needed in a viable form. Second, but by no means least, is the need for the matrix to have little or no nonspecific binding of undesired materials. This is often the greatest problem that is faced by the investigator. Materials having appropriate mechanical strength, chemically defined surfaces, and good biological properties often have such nonspecific adsorption of biomolecules as to be simply intolerable for the end-uses of the system. For example, we may desire to purify cells, using immobilized lectins to select cells having specific sugar moieties on their surface. In this case, it would be extremely important that the immobilized lectin not nonspecifically adsorb proteins that might secondarily adsorb cell types other than the ones desired. If that were to happen, the cells, which we isolate via lectin binding, would be contaminated with cells having affinity for the nonspecifically bound proteins.

It can be seen from this, as well as from the list of factors involved in matrix choice in Table 1, that the choice of a matrix is not easy, nor is the search for the best matrix ever complete. This choice always requires compromise between conflicting excellence in some properties and inferiority in others, exhibited by virtually all available matrices. The search for "ideal" matrices will certainly continue, although the bulk of applications will continue to be performed using the matrix that is easiest to obtain and/or that for which the most articles have been published. In cell adhesion and immobilization a somewhat broader group of matrices has been studied

**Table 1** Factors in Matrix Choice

| Mechanical | Biological | Chemical |
|---|---|---|
| Dimensionally stable | Nontoxic | Stable to normal pH values |
| Resists fracture | No undesirable leakage | Nonionogenic |
| | Low nonspecific adsorption | Noncharged |
| | | Polar (or, at least, wettable) |

with reasonable diligence (5). Unfortunately, as in most of solid-phase biochemistry, beaded agarose, owing to its wide commercial availability, the excellent commercial support it has received, the advertising by its producers, and the number of applications in which it has already been employed, will be the matrix used by most investigators. Table 2 lists most (but not necessarily all, as new offerings of new matrices and/or new commercial sources are constantly appearing) commercially available matrices and their suppliers, and Table 3 provides an example of cell purification using affinity techniques. The latter includes an excellent series of protocols to aid the researcher in designing new systems.

## 2 MATRICES

Matrices for all aspects of solid-phase biochemistry are most easily considered when we divide them into various categories with similar properties. This author has found it useful to consider naturally occurring matrices separately from synthetic ones. Naturally occurring matrices have certain properties in common, for example their availability and natural interaction with biological molecules, that make them distinct from most synthetic ones. Therefore, we will first consider natural matrices composed of either organic or inorganic materials, followed by a similar discussion of those that are man-made.

### 2.1 Naturally Occurring Matrices

Naturally occurring matrices can further be divided into two subclasses, those of organic origin, often consisting of various polysaccharides, and those that are inorganic, consisting of various minerals that have been processed to yield the proper shape and size.

#### *2.1.1 Hydroxylic Matrices*

The most widely used matrix, as already mentioned, is beaded agarose. Agarose is composed of a repeating dimer of galactose and anhydrogalactose in a long polysaccharide chain which, above its melt temperature, exists

**Table 2** Selected Commercial Sources for Matrices/Compounds Used in Enzyme and Cell Immobilization

| Matrix/ compound | Commercial name | Derivative | Most reactive function | Source |
|---|---|---|---|---|
| Agarose | BioGel A-150m, A-50m, A-15m, A-5m, A-1.5m, A-0.5m | Underivatized, various percent agarose, inclusion size of 150 $\times 10^6$ to 0.5 $\times 10^6$ | $-OH$ | Bio-Rad |
| | Affi-Gel 101 Affi-Gel 102 | Aminoalkyl, different spacers | $-NH_2$ | Bio-Rad |
| | Affi-Gel 201 Affi-Gel 202 | Aminoalkyl carboxylic acid | $-COOH$ | Bio-Rad |
| | Affi-Gel 401 | Thiol alkyl | $-SH$ | Bio-Rad |
| | Affi-Gel 501 | Organic mercurial | $-Hg^+$ | Bio-Rad |
| | Affi-Gel 10 | Carboxylic acid *N*-hydroxy-succinimide ester | | Bio-Rad |
| | Ultrogel A | Underivatized, various pore sizes | $-OH$ | LKB-IBF |
| | Ultrogel AcA | Polyacrylamide agarose, copolymer, various new sizes | $-OH$ | LKB-IBF |
| | Magnogel A YR | Magnetic particle | $-OH$ | LKB-IBF |
| | Magnogel AcA | Magnetic particle and polyacrylamide copolymer | $-OH$ | LKB-IBF |
| | Act-Ultrogel AcA | Agarose, polyacrylamide with glutaraldehyde | $-\overset{O}{\overset{\parallel}{C}}-H$, others | LKB-IBF |
| | AC-Ultrogel AcA | Carboxylate | $-COOH$ | LKG-IBF |
| | AC-Magnogel AcA | As above with magnetic particle | $-COOH$ | LKB-IBF |
| | HMD-Ultrogel AcA | Aminoalkyl | $-NH_2$ | LKB-IBF |
| | HME-Magnogel AcA | As above with magnetic particles | $-NH_2$ | LKB-IBF |
| | Sea Plaque | Alkylated | $-OH$ | Marine Colloids/FMC |
| | Sea Kem | To be low melting various gel temperatures available | | |

| | | | |
|---|---|---|---|
| *N*-Methylpyridinium Trisacryl | *N*-Methylpyridinol | $-CH_2-O-$ | Bioprobe Intl. |
| PL-Agarose | Poly-L-lysyl- | $-NH_2$ | Miles |
| PAL-Agarose | Poly(DL-alanine)polylysine | $-NH_2$ | Miles |
| SPL-Agarose | Succinylpolylysyl | $-COOH$ | Miles |
| SPAL-Agarose | Succinyl-poly(DL-alanine)-polylysyl | $-COOH$ | Miles |
| Adipoyl | $-C(=O)-NHNH_2$ | $-C(=O)NHNH_2$ | Miles |
| Succinyl-adipoyl- | $-C(=O)-N-N-C(=O)-CH_2-CH_2-COOH$ | $-COOH$ | Miles |
| Sepharose 2B, 4B, 6B | Underivatized 2%, 4%, 6% | $-OH$ | Pharmacia |
| Sepharose AH | Aminohexyl | $-NH_2$ | Pharmacia |
| Sepharose CH | Aminohexanoic acid | $-COOH$ | Pharmacia |
| CNBr-Sepharose | Cyanoester | $-O-C\ N$ | Pharmacia |
| Active thio- | Glutathione-2-mercapto- | $-S-S-$ | Pharmacia |
| Activated Sepharose-CH | Aminohexanoic acid *N*-hydroxysuccinimide ester | $-C-O-N$ | Pharmacia |
| Epoxy-activated Sepharose | Bisoxirane | | Pharmacia |

**Table 2** Continued

| Matrix/ compound | Commercial name | Derivative | Most reactive function | Source |
|---|---|---|---|---|
| | Tresyl-activated Sepharose | Tresyl ester | $-CH_2-O-S(=O)_2-CH_2-CF_3$ | Pharmacia |
| | CDAP | 1-Cyano-4-dimethyllaminopyridinium tetrafluoroborate | | Research Organics |
| Cellulose | Avicel | Unsubstituted | $-OH$ | Asahi Kasei |
| | Cellex | Unsubstituted | $-OH$ | Bio-Rad |
| | CMR cellulose | Carboxymethyl | $-COOH$ | Bio-Rad |
| | Cellex-CM | Carboxymethyl | $-COOH$ | Bio-Rad |
| | Cellex-PAB | *p*-Aminobenzyl | (diazotizable) | Bio-Rad |
| | | Unsubstituted | $-OH$ | Merck |
| | | Aminohexyl | $-NH_2$ | Merck |
| | | Aminodecyl | $-NH_2$ | Merck |
| | Aminohexyl succinyl | Succinyl | $-COOH$ | Merck |
| | CM-cellulose hy-drazide | Hydrazide | $-C(=O)-NHNH_2$ | Merck |
| | APT-paper/ ABM-paper | Arlyamine | | Schleicher and Schuell |
| | Nitrocellulose | Nitrate | Unknown adsorption | Schleicher and Schuell |

| | | | | |
|---|---|---|---|---|
| | CM-cellulose | Carboxymethyl | $-\overset{O}{\overset{\|}{C}}-OH$ | Schleicher and Schuell |
| | | Unsubstituted | $-OH$ | |
| | Various forms | Underivatized | $-OH$ | Whatman |
| Collagen | | Underivatized | $-NH_2$ | Yellow Springs Instrument Co. |
| Dextran | Sephadex | Unsubstituted, various protein exclusion limits, ±1500–±200,000 | $-OH$ | Pharmacia |
| | DEAE-Sephadex | Diethylaminoethyl | $-CH_2CH_2-\overset{C_2H_5}{\overset{\|}{N}}-C_2H_5$ | Pharmacia |
| | QAE-Sephadex | Diethylamino-2-hydroxypropyl | Chiefly quaternary amine, 2°$-OH$ | Pharmacia |
| | CM-Sephadex | Carboxymethyl | $-COOH$ | Pharmacia |
| | SP-Sephadex | Sulfopropyl | $-SO_3H$ | Pharmacia |
| Diatomaceous earth | Celite | | $-Si-OH$ | Johns-Manville |
| Glass | CPG | Underivatized | $-SIOH$ | Electro-Nucleonics |
| | | Aminopropyl | $-NH_2$ | Electro-Nucleonics |
| | | Carboxypropyl | $-COOH$ | Electro-Nucleonics |
| | CPG | Underivatized | $-Si-OH$ | Pierce-Corning |
| | CPG-dextran | Aminopropyl CPG coupled to dextran | $-CH-OH$ | Pierce-Corning |
| | CPG-glycerol | Glycerolpropyl silane | $-OH$ | Pierce-Corning |
| | Glycidoxy CPG | Clycidoxyl propyl silane | $-O-CH_2-\overbrace{CH-CH_2}^{O}$ | Pierce-Corning |
| | CPG-phenylhydrazine | Aminopropyl-*p*-benzoylhydrazine | | Pierce-Corning |
| | CPG-hydroxysuccinimide | Aminopropylsuccinimidyl-*N*-Hydroxysuccinimide | | Pierce-Corning |

**Table 2** Continued

| Matrix/ compound | Commercial name | Derivative | Most reactive function | Source |
|---|---|---|---|---|
| | Carboxy-CPG-CPG-thiol | Carboxylate | $-COOH$ | Pierce-Corning |
| | | Aminopropylcysteine | $-SH$ | Pierce-Corning |
| | CPG-aminoaryl | Aminopropyl-*p*-aminobenzoyl | | Pierce-Corning |
| | CPG-lipoamide | Lipoyl | | Pierce-Corning |
| Magnetic iron particles | Ferrofluid | None | Used as magnetic inclusion for gel particle formation | Ferrofluids |
| | Iron oxide-maganese oxide particles | None | As above | La Radiotechnique |
| Nylon | Nylon tube | Underivatized | $-C(=O)-N(H)-$ | Portex, Hythe, Kent, U.K. |
| Polyacrylamide | BioGel P | Underivatized various exclusion limits from $2 \times 10^3$ to $300 \times 10^3$ | $-OH$ | Bio-Rad |
| | BioGel P hydrazide | Hydrazide | $-C(=O)-N(H)-NH_2$ | Bio-Rad |
| | A-E-BioGel-P | Aminoethyl | $-NH_2$ | Bio-Rad |
| | BioGel-CM | Carboxymethyl | $-COOH$ | Bio-Rad |
| | Enzacryl-AA | Arylamine | | Koch-Light |
| | Enzacryl-AH | Alkylhydrazide | $-C(=O)-N(H)-NH_2$ | Koch-Light |

| | | | | |
|---|---|---|---|---|
| | Enzacryl-PT | Cysteinyl | $-SH$ | Koch-Light |
| | Enzacryl-PTL | Cysteinyl thioacetyl | | Koch-Light |
| Polyacrylamide | Enzafix P-HZ | Alkylhydrazide | $-C(=O)-NH-NH_2$ | Wake |
| | Enzafix P-AB | Arylamine | | Wake |
| | Sepheron | Hydroxymethyl | OH | Lacema |
| | Trisacryl | Trishydroxymethyl | $-CH_2OH$ | LKB-IBF |
| Polyester | Insolmer | | Proprietary crosslinker | Accurate Chemical Co. |
| Polyethylene | | Underivatized | $-CH_2CH_2$ | Aldrich |
| Polystyrene | Polysep P | None, various pore sizes, forms | | Polysciences |
| Polyvinylalcohol | PVA | None | $-OH$ | Aldrich |
| Silica | Unisil | Underivatized | SiOH | Clarkson Chemical |
| | Zirconia-coated controlled pore glass | Amyloglucosidase | | |
| Polyvinyl | Fractogel | None, various pore sizes | $-OH$ | E. Merck-Pierce |
| | Fractogel CDI | Carbiimidazole | | |

**Table 3** Examples of the Use of Cellular Affinity Chromatography

| Substances isolated | Affinity ligands | Solid support or immobilized ligands | Ref. |
|---|---|---|---|
| *Cells, cell membranes, and organelles* | | | |
| Adipose cells | Insulin | Sepharose 4B | 6 |
| Antigen-binding cells from spleens of mice | Antigens | Nylon fibres | 7,8 |
| Antigen-reactive cells from rabbit bone marrow | Antigens | Antigen-coated glass beads | 9 |
| Anti-hapten plaque-forming cells | δ-*N*-2,4-dinitrophenyl-L-ornithine, HCl | Sepharose 2B | 10 |
| Anti-hapten specific lymphocytes | Axophenyl-β-lactoside hapten groups | Bio-Gel P-6 with histamine | 11 |
| Avian antigen-binding cells | Antigen-coated beads | Polyacrylamide | 12,13 |
| Blood platelets | Sepharose 2B | Sepharose 2B | 14 |
| Cells carrying IgG and blood group antigens | Antigen-coated beads | Glass of plastic beads | 15 |
| Cells (erythrocytes, thymocytes) | Concanavalin A | Nylon fibres | 16 |
| Spleen | Dinitrophenylated serum albumin | Gelatin fibres | 16 |
| Populations of specific antibody-producing and specific memory cells | Human or bovine serum albumin or hen ovalbumin | Glass beads or Degalan, V 26 | 12 |
| Producing antihapten antibodies | Azophenyl-β-glycosides | Bio-Gel P-6 with histamine and hydrazine | 17 |
| Cholinergic receptor enriched membranes | α,ω,-bis-4-trimethyl-ammonium polyethylene oxide | Dextran | 18,19 |
| Cultured tumor cells | Isoproteronol, corticotropin (ACTH), triiodothyronine | Glass or Sepharose | 20 |
| Cultured leukemic cells (L-1210) | Concanavalin A | Nylon | 21 |

| | | | |
|---|---|---|---|
| Erythrocytes | Anti-erythrocyte antibodies | Reticulated polyester polyurethane foam | 22,23 |
| Erythrocytes | Concanavalin A | Sepharose 6B | 24 |
| Fat cells | Insulin | Sepharose 4B | 6 |
| Flagellae | Anti-Hb globulins | Sepharose 4B | 25 |
| Human lymphocytes | *Lens culinaris* lectin | Sepharose 2B | 26 |
| Human T lymphocytes | Wheat germ agglutinin | Sepharose 4B | 27 |
| Human reticulocytes | Transferrin | Sepharose 4B | 28 |
| Immunoglobulin-positive cells | Anti-fluorescein antibody | Anti-fluorescein antibody | 29 |
| Immune cells | Antigen (serum albumin or ovalbumin) | Glass and plastic beads coated with antigenic protein molecules | 30 |
| Immunogenic tumor cells | Concanavalin A | Nylon fibres | 31 |
| Immunoglobulin-beraing lymphocytes | Anti-immunoglobulin antibodies | Plastic beads coated with anti-immunoglobulin antibodies | 32 |
| Immunological memory cells | Antigen-coated glass or plastic beads | Glass or plastic beads | 33 |
| Immunospecific precursor cells from unimmunized mice | Azophenyl-$\beta$-lactoside | Bio-Gel P-6 with hydrazine and histamine | 34 |
| Inside-out membrane vesicles from pig lymphocytes | Concanavalin A Anti-lymphocytic serum | Sepharose 4B | 35 |
| Inside-out plasma membranes from erythrocytes | Polylysine-coated glass beads | Polylysine-coated glass beads | 36,37 |
| Inside-out vesicles from rat liver mitochrondria | Cytochrome *c* | Sepharose 4B | 38 |
| Lymphocytes | $\beta$-Lactoside haptens | Acrylamide | 39 |
| Anti-hapten specific | Phytomitogens | Sepharose 4B | 40 |
| Specific immunocompetent T | Anti-idiotyic antibodies | Degalan beads coated with antibodies | 41 |

**Table 3** Continued

| Substances isolated | Affinity ligands | Solid support or immobilized ligands | Ref. |
|---|---|---|---|
| (T and B cells) from spleens of mouse | Antigens (hapten-bovine serum albumin conjugates, *Limulus* haemocyanine or concanavalin A) | Nylon fibers | 42 |
| | Human immunoglobulin (HGG) after treatment with anti-HGG antisera | Surface of tissue culture grade plastic ware | 43 |
| (T and B cells) from rat thoracic duct lymph | Anti-rat F $(ab')_2$ antibody | Sephadex G-200 | 44 |
| Lymphocyte membrane vesicles | Concanavalin A | Coliphage K 29 | 45,46 |
| Lymphocyte plasma membrane | Concanavalin A | Sepharose 4B | 47,48 |
| Lymphoid cells | Bovine serum albumin or its derivatives | Nylon fibres | 49 |
| | Antigen-coated beads | Glass or plastic | 50 |
| Lymphocytes from rat spleen and thymus and mouse spleen | Aggregated rat immunoglobulin | Sepharose 4B | 51 |
| Lymph node cells from guinea-pigs | Dinitrophenyl | Polyacrylamide beads | 52 |
| Membranes from eukaryotic cells | Polylysine-coated beads | Polyacrylamide | 37 |
| Mouse bone marrow cells | Wheat germ agglutinin | Sepharose 6MB | 53 |
| Murine cytotoxic T lymphocytes | Lectin | Sepharose | 54 |
| Mouse spleen cells (IgG bearing) | Protein A of *Staphylococcus aureus* (sp A) | Sepharose 6MB | 55 |
| Neural cell separation | Normal rabbit globulin | Sepharose 6MB | 56 |
| Ovalbumin-synthesizing polysomes in complex with anti-albumin antibody | Ovalbumin | Ovalbumin cross linked with glutaraldehyde | 57 |
| Peripheral blood T lymphocytes | *Helix pomatia* lectin | Sepharose | 58 |
| Plasma membrane from erythrocytes | Polylysine-coated beads | Glass | 59,60 |

| | | | |
|---|---|---|---|
| Plasma membranes from HeLa cells | Polylysine-coated glass beads | Polylysine-coated polyacrylamide beads | 61 |
| Plasma membranes from pig lymph node cells | Concanavalin A | K 29 Coliphage | 62 |
| Plasma membrane vesicles from eukaryotic cells | Concanavalin A | Sepharose 4B | 63 |
| Plasma membrane vesicles from thymocytes | Concanavalin A | Sepharose 4B | 64–66 |
| Polyribosomes from mouse plasmacytoma producing IgG 1 immunoglobulin type X | Complex of mouse immunoglobulin with rabbit antibodies | Amino-cellulose | 67 |
| Polysomes (albumin-synthesizing) | Antigen-antibody complex | Amino-cellulose | 68,69 |
| Polysomes (galactosidase-specific) | *p*-Aminophenyl-D-thio-galactopyranoside | Sepharose with 3-aminosuccinyl-1,6-diaminohexane | 70 |
| Polysomes | Antibody to specific protein | Agarose | 71 |
| | Pyridoxamine phosphate | Sepharose 4B with ethylenediamine and succinic anhydride | 72 |
| Proliferated cells | Concanavalin A | Sepharose 4B | 73 |
| Rat basophilic leukemia cells | Lentin-lectin | Sepharose 4B | 74 |
| Reticulocyte ribosomes | Polyuridylic acid | Sepharose 4B | 75 |
| Ribosomes from *E. coli* | Streptomycin or gentamicin | Indubiose 4A | 76 |
| Ribosomes synthesizing tyrosine aminotransferase from hepatoma tissue culture cells | Pryidoxamine phosphate | Sepharose 4B | 77 |
| Splenocytes | *Helix pomatia* A hemagglutinin | Sepharose | 78 |
| Spleen lymphocytes | Insulin, Con A | Sepharose 4B | 79 |
| Sympathetic ganglion neurones | $\alpha$-Bungarotoxin | Sepharose 6MB | 80 |

**Table 3** Continued

| Substances isolated | Affinity ligands | Solid support or immobilized ligands | Ref. |
|---|---|---|---|
| Synaptic plasma membrane fraction from guinea pig brain | Concanavalin A | Sepharose 4B | 81 |
| Synaptosomes | Concanavalin A | Sepharose 4B | 82,83 |
| Synaptic vesicles | Sepharose 6MB | Sepharose 6MB | 84 |
| Tissue culture cells (HeLa, SV 3T3) | Lens culinaris lectin | Sepharose 2B | 85 |
| Thymocytes | Anti-thymocyte globulin | Sepharose 4B | 86 |
| Thymocytes | Anti-peanut agglutinin | Sepharose | 87 |
| Translating ribosomes | Periodate-oxidized polyuridylic acid | CM-Cellulose with dihyrazide of dithioglycolic acid | 88 |
| *Phages and viruses* | | | |
| Bacteriophage SKV 1 | *Shigella sonnei* liposaccharide | CNBr-activated Sepharose 4B | 89 |
| T-4 Phage | Poly DL-lysine | Sepharose 2B | 90 |
| Aleutian mink disease virus | Antibody-Sepharose | Antibody-Sepharose 4B | 91 |
| | Immunoglobulin from chronically infected mink | Immunoglobulin from chronically infected mink | 92 |
| Foot-and-mouth disease virus (FMDV) | FNDV antibodies | Sepharose 4B | 93 |
| Influenza virus | γ-Globulin of rabbits immunized with influenza virus | Disulphide-linked γ-globulin with *N*-acetylhomocysteine thiolactone | 94 |
| Murine type-C virus p30 precurose protein | Single-stranded DNA | Cellulose | 95 |
| Plant virus | Antibodies | Antibodies crosslinked by glutaraldehyde | 96 |
| Semliki forest virus | Antibody-Sepharose | Antibody Sepharose | 97 |
| Tobacco mosaic virus | Anti-tobacco mosaic virus antibodies | Antibody Sepharose *p*-Aminobenzyl cellulose | 98 |

*Source:* Reprinted with permission of the authors, from Ref. 5.

in free solution as a soluble saccharide (99). When it is gelled, however, a triple helix of three polysaccharide chains forms in an interwoven mesh (100). This mesh is interwoven at the corners of a series of pentagonal holes. The size of these pentagonal holes is dependent on the percentage of agarose in the material. Normally concentrations only slightly below 1% will not gel, and even at 2% agarose, the mechanical strength of the resulting gel leaves much to be desired. Most agarose, therefore, is utilized as beaded agarose with concentrations between 4 and 6%. To achieve additional mechanical and chemical strength, it is often crosslinked by treatment with epichlorohydrin or similar reagents (101). Treatment with the crosslinking agent somewhat decreases the porosity of the agarose and thus limits the size of biomolecules that can penetrate the matrix. In addition, some added hydrophobicity results from crosslinking. It is important to note that agarose, though hydrophilic, is not *as* hydrophilic as one would imagine when looking at its basic structural unit. This is because most of the hydroxyl groups are hydrogen-bonded to other hydroxyls to form the helical backbone of the gel. The current thought is that most of the "free" hydroxyls in gelled agarose occur chiefly at the corners of the pentagonal array where the triple helix loses its high degree of structure. It is perhaps for that reason that biomolecules enclosed in agarose often behave as if they were in a more hydrophobic solution than the same material in bulk water. Dyes, for example, which are normally soluble in water, will adsorb to agarose. One could view agarose in this case as a macro-sized cyclodextran, since the hydrocarbon backbone is what biomolecules entering the pentagonal holes encounter. It must be understood that agarose is one of the most hydrophilic of all the matrices available, even though it exhibits a small, but unexpected, degree of hydrophobicity.

Likewise, agarose often has some ionic character that might not be expected. This is because agarose, like any naturally occurring material, does not consist solely of galactose and anhydrogalactose moieties. Rather, it contains small amounts of other saccharides that are sulfated or carboxylated and, therefore, impart anionic character to the gel. Most commercial preparations minimize this by using a reduction process that removes most, if not all, of the sulfate groups (102). Nevertheless, the investigator is cautioned not to accept a naïve view of the composition of agarose, nor, for that matter, of any of the matrices employed in solid-phase biochemical systems.

Chitin, and its water-soluble derivative chitosan, are among the world's most ubiquitous organic molecules, second only, perhaps, to cellulose. Chitin consists chiefly of (1-4)-linked 2-acetoamido-2-deoxy-beta-D glucose moieties with about 15% of the glucose hydroxyl residues nonacetylated (103). Chitosan is created by treating the very insoluble chitin with concentrated alkali to produce a deacetylated water-soluble polymer (104). Chitosan is deacetylated to various degrees, depending on the length and severity of alkali treatment, and, therefore, can have varying degrees of intrinsic hydrophobicity. Chitin would appear to be an extremely useful support because of its ready availability, as it is found in insect exoskeletons and, in particular, the exoskeletons of lobster, shrimp, and similar materials. It can, therefore, be obtained in large quantities at relatively low cost. By appropriate treatment with epichlorohydrin, similar to that employed with agarose, a beaded, crosslinked chitin derivative can be prepared.

Cellulose is a ubiquitous, inexpensive, world-wide product, unlike agarose, which is isolated only from seaweed and requires substantial purification. Cellulose has had, however, a poor reputation created by its earlier availability in inferior grades for use in ion exchange chromatography. Indeed, DEAE and CM cellulose are among the world's basic ion exchangers and are the backbone of any protein purification laboratory (105). Cellulose in the fibrous form that is most commonly available is heterogeneous and contains both highly crystalline and amorphous regions (106). This permits biomolecules to penetrate some areas of the matrix and not others. Moreover, its fibrous nature does not yield good chromatographic procedures; particularly, flow rate is slow and compressibility is high. However, relatively recently, beaded cellulose, which has the desired flow properties, has become available commercially. Beaded cellulose (107) is created by dissolving cellulose in one of several systems, most commonly in cupric ammoniacal nitrate or as a xanthate created by treatment with carbon disulfide in base. Each method leaves residual solvent, which must be carefully removed from commercial bead preparations before use. This caution is not, however, restricted to cellulose, as, for example, agarose beads also need careful washing before use, because various preservatives and similar materials will most certainly be present.

Beaded cellulose has almost as wide a range of porosities and strengths as the more widely employed agarose. It is this investigator's belief that beaded cellulose should be considered by anyone embarking on a study of any aspect of solid-phase biochemistry, including cell affinity chromatography, cell adhesion, or cell immobilization, as it may prove to be the best, cheapest, and most reliable of all commercially available supports.

Calcium alginate is another frequently used polysaccharide derivative. Alginate is widely used in cell immobilization (108–110). Alginate is soluble in water as a sodium salt and consists of predominantly mannuronic and glucuronic acid residues. Several types of alginate, as shown in Figure 1, exist depending on the length of the polysaccharide, the percent of mannuronic and glucuronic acid, and the length of blocks of these repeating units.

When sodium alginate solutions are mixed with calcium salts, the result-

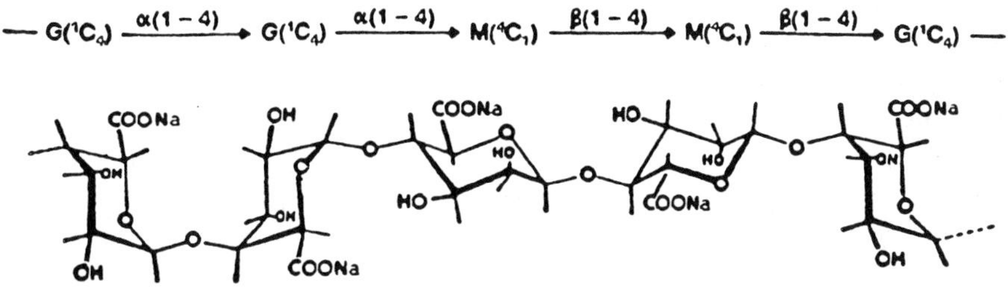


**Figure 1** Chemical structure of $Na^+$ alginate. (Reprinted with permission of the authors, Dr. Olav Smidsrod and Dr. Gudmund Skjak-Braek, *Trends in Biotechnology, 8:* 72, 1990.)

**Table 4** Preparation of Alginate as an Immobilization Matrix for Cells

Preparation of the matrix requires the following steps:

Preparation of alginate solution

A 2–4% (w/v) aqueous solution of $Na^+$ alginate is prepared by suspending the polymer in distilled water or buffer. The suspension should be stirred for 6 h by a magnetic stirrer or left overnight on a rotary shaker at room temperature.

Sterilization

Alginate solutions can be sterilized by autoclaving or sterile filtration. However, since high temperatures will cause some depolymerization of the chains, sterile filtration is recommended. If autoclaving is chosen, the pH should be adjusted to 7–8 before the temperature is raised. Highly purified alginate can be filtered directly through a 0.22-$\mu$m membrane filter. Other qualities should be passed through a series of membrane filters of pore sizes 1.2 $\mu$m, 0.8 $\mu$m, 0.45 $\mu$m, and 0.22 $\mu$m. This procedure also removes some contaminants such as proteins and polyphenols, and the final solution forms highly transparent beads.

Mixing with cells

Sterile alginate solution is mixed with an equal volume of suspended cells. The cell suspension may be in an isotonic buffer solution. Phosphate, citrate, EDTA, and divalent cations should be avoided.

Immobilization

The gel beads are formed by dripping the alginate-cell suspension (syringe inner diameter 0.22–1.0 mm, about 20 cm from the $CaCl_2$ surface) into a solution containing 20–100 mM $Ca^{2+}$ ions. The beads should be left to harden in $CaCl_2$ for 5–30 min, depending on their diameter.

Dissolving of beads

The entrapped cells can be recovered easily by a gentle dissolution of the gel beads. This is done simply by immersing the beads in a solution containing phosphate or citrate, which will sequester calcium ions (50 mM $Na^+$ citrate or phosphate buffer at pH 7.0). If a high-G alginate is used, this process takes several hours, compared with 10–30 min for a low-G alginate.

*Source:* Reprinted with permission of the authors, Dr. Olav Smidsrod and Dr. Gudmund Skjak-Braek, *Trends in Biotechnology*, *8*, p. 72 (1990).

ing calcium alginate immediately gels. To prepare beaded calcium alginate, an aqueous sodium alginate solution is dropped into a solution containing calcium ions (±200 mM). This results in a fairly soft gel with a high degree of porosity. The gelling process is readily reversed by the addition of calcium chelating agents such as EDTA or sodium citrate. Very small beads may be formed by using an ultrasonically vibrating nozzle to force the cell/sodium alginate solution into the calcium hardening bath. Once beads have formed, the mechanical and chemical stability of the alginate can be increased by crosslinking with gluteraldehyde, either alone or in combination with a carrier protein such as serum albumin. This makes a very stable bead that can be further employed in the absence of stabilizing calcium ions. An alternative hardening procedure, which can be employed if the biological system will allow it, is to dry the alginate beads. This produces an irreversible shrinking of the bead, further entrapping the cells and imparting additional stability to the matrix material.

A variety of other naturally occurring polysaccharides and proteins can be used in preparing matrices for solid-phase systems (111). One widely used nonsaccharide is collagen, or its hot-water-soluble component gelatin. Both can be obtained as membranes as well as in beaded form and have wide applicability. Often cells find a collagen or gelatin surface very desirable for growth, which accounts for its wide popularity in the area of cell immobilization and study (112). Synthetic polylysine also is widely used for membrane isolation, as shown in Figure 2.

A beaded form for all the above matrices has long been considered to be the preferable format. Only recently have membranous materials been prepared that possess the needed chemical, biological, and mechanical properties for widespread use. For the preparation of beads, a well-stirred, two-phase system is created in which beads of one component form an emulsion in the other solvent (113). For example, a melted solution of agarose may be poured into a rapidly stirred toluene bath, and as cooling takes place, well-formed beads develop. This is a very simple method, although, in large-scale manufacturing, special vessels containing appropriate invaginations with stirring at a determined rate are necessary for the most economical operation. Even with such equipment, beads of various sizes form and must subsequently be separated, often through a series of sieves or by being allowed to settle in large cylinders for various periods of time. Such size separation contributes to the high cost of beaded materials.

The processes to form membranes are not simpler. Several methods are available, the most common of which, in the research laboratory, is to evaporate the solvent in which the desired membrane materials are dissolved, leaving behind a membrane of the desired thickness and size. Unfor-

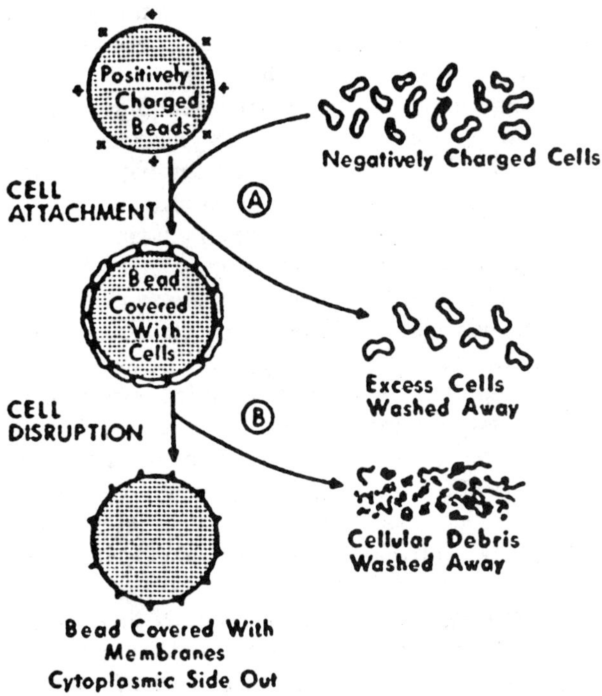


**Figure 2** Schematic illustration of membrane isolation on polylysine-coated glass beads. (Reprinted with permission of the authors, Dr. S. K. Sharma and Dr. P. P. Mahendroo, *Journal of Chromatography, 184*, p. 471, 1980.)

tunately, this is a very crude process, and preparation of appropriate membrane materials should, almost always, be left to our industrial colleagues.

### *2.1.2 Inorganic Matrices*

Inorganic materials have frequently been used in small-scale experiments by the researcher, but only a few have ever found their way to wide-scale application in a variety of research systems. The simplest inorganic matrix that one could use is "dirt," which is, of course, a simple, but highly impure, silica-containing, mineral substance. Such materials have been used, and "brick dust" has been reported as a basis for several investigations (114). However, controlled pore glass and sintered silica are the only materials commercially available that have had wide-scale use reported in the research literature. In addition, diatomaceous earth of particularly useful porosity and size is available (Celite, Celite Co., Lompoc, CA) (115).

Controlled pore glass (116–118) is formed by holding borosilicate glass at an elevated temperature for a prolonged period of time. This glass, in a semimelted form, separates into boronate-rich phases and silicate-rich phases. Boronate is rather soluble in acid, whereas the silica phase is not. Therefore, borosilicate glass that has been treated to effect phase separation is subsequently etched with acid to yield a high-silica-content material containing pores of approximately 40 Å in size. This is known as "thirsty Vycor," and if it is heated to cause collapse of the pores, high-silica Vycor glass results. For controlled pore glass, however, the nominally 40-Å pore size is expanded by treatment with base to further etch away silica. This results in a material containing pores of 150–1500 Å. Quite understandably, the larger the pore size, the more fragile the glass, as the lower the silica content, the thinner the supporting network. Controlled pore glass is very expensive because this process requires considerable time and skill. It is, nevertheless, a very useful material, since the porosity of the glass can be precisely controlled such that a 600-Å pore glass might vary in pore size by less than 10%. (For some applications, however, such as gel permeation chromatography, such tightly controlled pores may actually be detrimental!)

The real problem with controlled pore glass, however, is that, *in its virgin state*, it has an *extremely* high nonspecific adsorption of protein. This contrasts with agarose, which generally has little nonspecific adsorption *until* it has been derivatized or further treated, after which it may have properties that are worse than those of virgin porous glass. Nevertheless, the nonspecific binding properties of controlled pore glass are such that it is not widely used in any area other than high-performance liquid chromatography, where its dimensional stability is essential. Controlled pore glass has several other difficulties, including the fact that there is a variable amount of boronate on the glass surface, which makes complete coating with silane derivatives, as described below, somewhat difficult. To obviate these problems, the industry has produced porous silica (119). This is considerably less costly to produce and is much purer than controlled pore glass, but has a wide porosity range. There are two ways of producing porous silica. The first is a sintered silica in which highly purified silica is ground to a fine powder that is compressed and heated to the point where the particles melt or fuse together. When the particles have fused, they create a single network containing pores the size of which depends on the properties (size, etc.) of the silica particle used to form the product.

Another type of porous silica matrix can be formed by the precipitation of silica from water-soluble silicates (120). Most of these are proprietary materials that are not readily formed in the research laboratory.

Other metal oxides have been employed for the immobilization of proteins, cells, saccharides, and other biomolecules. The most important of these are magnetic oxides, which are frequently used to impart magnetic properties to other matrices, whether organic or inorganic. Using these, magnetic affinity chromatography has significant potential in cell isolation (121).

## 2.2 Synthetic Organic Matrices

There are almost as many synthetic matrices as there are natural ones. Nearly all the synthetic matrices are organic, and, as with the naturally occurring matrices, only a few of them have been utilized with any particular regularity. Certainly the most important class of polymers used for the immobilization of biomolecules are polyacrylamide gels. There are many different polyacrylamides and their derivatives, including the Enzacryls (122), which are a group of thiol amine, aryl amine, and aldehyde-containing acrylate polymers, and Eupergit (123), an oxirane-containing polyacrylamide that is "preactivated" but with a limited shelf life. Acrylamides can be used in much the same fashion as agarose and similar matrices. Polyacrylamides tend to have poor mechanical strength and limited porosity. This generalized statement, however, is subject to many exceptions, which depend on the monomer used to form the acrylamide. Simple polyacrylamide has all the negative characteristics described above. Hydroxyethyl derivatives, however, are quite hydrophilic and excellent for protein and cell immobilization. Hydrophobic materials can also be formulated, as well as hydrophilic ones. One matrix in particular, Trisacryl (124), formed from tris aminohydroxymethane, is exceptionally hydrophilic.

Polyesters (125), on the other hand, are considerably less stable than polyacrylamides or cell constituents. The ester bond is hydrolyzed either chemically, through the use of buffers at either pH extreme, or by esterases found in the sample. Polyesters, therefore, have not found widespread application as support matrices, but can, under appropriate conditions, nevertheless, be reasonably adaptable to protein and cellular immobilization.

In addition to polyesters, nylon is a synthetic polymer with many applications in cell purification. Nylon fibers were among the earliest materials employed in cell purification (126). Both polyesters and nylon have the advantage of producing materials of great mechanical strength, albeit frequently of limited surface area and low capacity. This, however, may be more than compensated in particular applications by their mechanical durability, as well as the ability to be fabricated in bead, fiber, and membrane form and an unlimited number of shapes of solid materials adapted for particular uses such as in clinical analyzers.

Polystyrene is an extremely important matrix with limited, but significant, applications in clinical chemistry. Polystyrene readily absorbs antibody molecules and has been used widely as the basis for developing enzyme-linked immunoassay systems (127). Unfortunately, polystyrene also has the property of absorbing any protein with a hydrophobic portion on its surface. This nonspecific binding is a serious detriment to the application of polystyrene beyond immobilized antibodies where it has certainly a deserved and well-earned niche.

## 3 ACTIVATION METHODS

As a guideline in choosing activation methods, the simplest and most effective advice is that no single activation method, just as no single matrix, will give a complete solution to all matrix problems. Rather, a variety of matrices, with a variety of activating systems, needs to be considered. The choice of which method to use is a matter of trial and error, although some activating methods are better than others for specific applications. For example, the nucleophilic displacement reactions are superior when the leakage of the small amount of the immobilized cell or biomolecule would have severe consequences on the outcome of the particular experiment or treatment. On the other hand, cyanogen bromide activation, which has the drawback of potential leakage, is simple and very well studied. Where trace leakage is not a problem, activation with cyanogen bromide should certainly be tried. Similar arguments apply for choosing between each of the many different activation methods currently available.

### 3.1 Hydroxylic Matrices

Probably the best series of immobilization methods for matrices containing alcohol functions are based on nucleophilic displacement reactions that result in immobilization by means of a secondary amine bond (or, in the case of hydroxyl/sulfhydryl functions, an ether/thioether bond). There are a variety of these reactions, most of which work best with primary hydroxyl groups on the matrix. The number of freely available primary hydroxyl groups, relative to the total number of hydroxyl groups, on most polysaccharide matrices is often small. This results in modifying secondary hydroxyl functions with reagents that cannot subsequently be removed from the matrix, either during coupling proteins or other biomolecules or in subsequent "deactivation" with high concentration of strong nucleophiles. An example of this is the use of tosyl chloride for activation. Tosyl, tresyl, and similar sulfonyl chlorides (128) are excellent activating agents that work very well for the immobilization of protein. However, if the degree of activation is too high, a vast proportion of the reacted tosyl ester groups will not be displaced by the nucleophile, regardless of concentration or temperature used. In fact, elimination, if any reaction at all, will predominate with the secondary sulfonyl ester derivatives. For this reason, Gribneau (129), who was the first to record using sulfonyl esters as activation agents, gave up on the process quickly. It appeared to Gribneau that the tosyl and mesyl esters he was using could only be marginally displaced by nucleophiles from the matrix. The problem was that he had derivatized the matrices under extreme conditions, namely high temperature, high concentrations of base, and high concentrations of activating agent. Independently, Nilsson and Mosbach (128) used tosyl and tresyl chlorides as activating agents under mild conditions for short periods of time and were very successful in displacement and the subsequent immobilization of the desired biomolecules. Scouten et al. (130) followed this by using colored sulfonyl esters whose displacement could be readily followed, permitting numerous, easily performed, experiments. The results of their work show that conditions even milder than those employed by Nilsson and Mosbach (128) are essential if the displacement of all the activating function is to be possible. Nevertheless, the sulfonyl ester method is one of the best to be employed, and this author recommends that any person who is immobilizing cells or biomolecules become well acquainted with this method.

One interesting variation of the sulfonyl chloride activation method is the use of ligand-containing sulfonyl chlorides as methods of activating hydroxyl matrices such as agarose. Initially, the use of colored sulfonyl esters (130) only had the advantage of permitting a rapid series of "quick and dirty" experiments to monitor the degree of displacement of the activating agent visually. Many of these dyes, however, have an affinity for proteins at their substrate-binding sites. Other materials can be synthesized which are not colored but which also contain a ligand connected to a reactive sulfonyl chloride. This permits affinity-directed immobilization of proteins (131), such that the protein is immobilized at or near a specific ligand-binding site.

An extension of this concept is oriented immobilization (132) of proteins, which is a very active area of research at present. Increased capacity may result when the protein is immobilized on the opposite end of the molecule from the active site. Similar immobilization procedures have been applied to the study of cell surfaces, and it would be expected that selected immobilization of certain portions of a mixed population of cells might be achievable using this general concept.

Another group of reagents that converts hydroxyls into derivatives readily displaced by nucleophilic attack are the fluoromethylpyridinium salts and their derivatives. That Ngo (133) at BioProbe International developed the original fluoromethyl-pyridinium (FMP)-activated system using the simple, readily available fluoromethylpyridinium tosylate as the activating agent (see Fig. 3). One potential advantage of FMP appeared to be the methylpyridinium group would be a substantially better leaving group than most sulfonyl esters. The converse has proved to be true in that the methylpyridinium group, like nonfluorinated sulfonyl esters, is not as good a leaving group as was expected, although no satisfactory explanation for this has been proposed. Nonetheless, FMP has proved to be a very popular and useful reagent that immobilizes proteins by the same, stable, secondary amine (or ether/thioether) bond obtained using sulfonyl esters. Interestingly, the positive charge on the methylpyridinium may, in fact, orient the immobilization of protein molecules in a well-defined fashion. This subject is under active investigation in the author's laboratory, as well as elsewhere, in the hope of finding good, inexpensive reagents for oriented protein immobilization.

Secondary amine linkages are also frequently obtained by two other routes. The simplest of these is to oxidize a polysaccharide matrix using

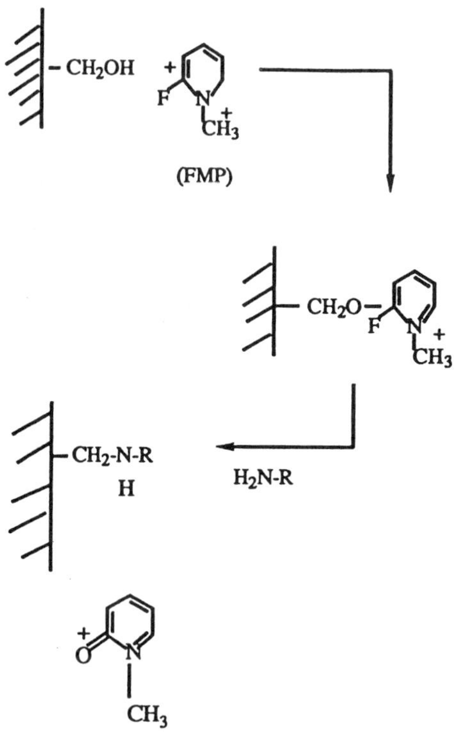


**Figure 3** Activation and coupling using fluoromethyl pyridinium salts.

periodate cleavage (134), which cleaves between vicinal hydroxyls and yields two aldehyde groups. These aldehyde groups, in turn, will bind amine groups to form imines, which can be reduced selectively and mildly with sodium cyanoborohydride to obtain high yields of tightly immobilized protein molecules.

Alternatively, epoxides are often used for protein immobilization (135). Treatment of agarose or cellulose with butane diglycidyl ether in the presence of strong base results in excellent activation levels with relatively uniformly reactive epoxide functions. Unfortunately, epoxides are very sluggishly reactive compared to tresyl esters and, therefore, residual epoxide residues remain after protein immobilization has been completed. These need to be carefully deactivated by prolonged treatment with amines such as Tris or glycine buffers or with thiols or similar nucleophiles.

An interesting side effect of this activation is the further crosslinking of the gel. Linear polymers of several units of epoxide protruding from the gel are also probably created. Thorough studies of this have never been made, nor has the exact chemical structure of epoxide immobilized biomolecules been elucidated.

Although relatively few publications exist, the epoxide or oxirane method has shown to be excellent for the immobilization of IgG. It should also be noted that the epoxide ring can be opened by other groups, making it possible to use this method to create other reactive functionalities. For example, the opening of the epoxide ring with *p*-aminothiolphenol results in an amine function that is readily diazotized and coupled to nucleic acid residues (136). This is the basis of the so-called Western blots.

Frequently, the activation of hydroxylic matrices is not performed by the conversion of the hydroxyl directly to a reactive function, but in many instances, the hydroxyl group itself serves as a point to *add* a reactive function. One of the most useful of these procedures is chloroformate derivativization (137). A variety of chloroformates, which are derivatives of phosgene, have been developed and are commercially available. These have a variety of uses in organic synthesis, but their application to the immobilization of proteins and similar biomolecules is rather similar in each case. The chloro group is the most reactive of the two functionalities on the chlorocarbonate. The chlorocarbonate reacts with the hydroxyl group of the matrix to form a derivative that has a second good leaving group, which now can be displaced by the attacking protein nucleophile, usually a lysyl amine. The resulting bond is reasonably stable, although not as stable as that created through nucleophilic displacement of tosyl esters and similar materials. The variety of commercially available chlorocarbonates offers some choice in immobilization. The author has been extensively involved in the synthesis of other chloroformates that can attract protein molecules to the surface in an oriented fashion for a subsequent immobilization. This approach is extremely interesting and promises to yield chemically directed immobilization of desired proteins.

Another similar activation procedure uses cyanuric chloride and its derivatives, termed triazines, to activate the gel (138). Cyanuric chloride contains three equally reactive chloride functions, the first of which reacts with amazing rapidity at very low temperatures with virtually any nucleophile. The addition of cyanuric chloride, thus, to a hydroxylic matrix under basic conditions results in rapid activation. The other two chlorides now have been made less reactive by the electron-donating effect of the oxygen that was formerly part of the hydroxyl. Therefore, the second chloride reacts slower but still relatively rapidly at modest temperatures with a half-time of 30 min at 25°C. Once this group has reacted, the third chloride reacts but at a substantially lower rate, requiring, perhaps, several hours at 60°C. For all purposes of protein immobilization, the third chloride group reacts too slowly to be of substantial value, since the conditions required could denature the protein.

Triazines are often first reacted with an amine and then subsequently with the matrix. The amine substitution is usually in the form of a colored dye, and colored triazines are the basis of a substantial portion of the commercial dye industry. For further information concerning triazine activation the reader is referred to Ref. 139.

### 3.1.1 Spacer Arms

Frequently, the initial step in the activation of a matrix is to attach a reactive group at some distance from the matrix itself. This group is held to the matrix by a spacer arm, which has been demonstrated to be necessary for optimal function of many proteins. Binding proteins directly against a matrix creates steric hinderance for normal folding and/or conformational change during activity (140). Further, ligands that are immobilized directly to a matrix may be too close to the matrix for the ligand to reach into the binding site of a protein that one may be purifying by affinity chromatography.

The importance of spacer arms cannot be overly emphasized. Too short a spacer arm and the reactive function may not reach the protein molecule and the immobilized protein molecules may have the steric impediments mentioned above; too long a spacer arm, if the spacer arm is hydrophobic, will result on its folding back on itself and having an *effective* spacer arm that is, once again, too short. Elongated spacer arms, if necessary, should be composed of hydrocarbon groups interrupted occasionally with hydrophilic residues.

Activation of spacer arms often involves different techniques than used for direct activation of matrices. The vast majority of spacer arms contain one of three functions. Some spacer arms (as mentioned previously) are formed from bisepoxides and these require no further activation. The other two frequently employed spacer arm functionalities terminate in either amines or carboxylic acids. Carboxylic acids are easily activated with a water-soluble carbodiimide. The water-soluble, carbodiimide-activated material may then be directly reacted with the nucleophilic ligand being immobilized. Alternatively, it may be reacted with a secondary leaving group such as *o*-nitrophenol to form an activated carboxyl that has a longer half-life in storage than carbodiimide-activated carboxyls.

When spacer arms terminating in an amine are used, a ligand containing a carboxyl function is activated with a water-soluble carbodiimide and subsequently reacted with the amine-containing matrix. The result will be a stable amide bond between the ligand and the spacer arm. Alternatively, amine spacer arms may be activated with triazines in virtually the same fashion as described for activation of hydroxylic matrices.

## 3.2 Metallic Oxides

Metallic oxides, including silicon oxides, are easily activated for covalent immobilization of proteins and similar biomolecules. The simplest method for activation of silica is by reaction with organo silanes. The most common activating agents are aminopropyltrimethoxysilane and propylglycidyltrimethoxysilane. These produce matrices containing amines or epoxides, respectively. The epoxides can be further ring-opened with aqueous dilute base or acid to produce a hydroxylic coating, which can be further activated as described above for hydroxylic materials. Alternatively, the epoxide ring may serve as an activated function for the immobilization of proteins and other biomolecules. This is frequently utilized for immobilization of antibodies in high-performance liquid immunoaffinity chromatography. Activation of aminopropyl silane-treated silica and/or porous glass can be done as described previously utilizing cyanuric chloride or by carbodiimide coupling to proteins.

Activation with silanes as described above can also be used on other types of metal oxides such as, for example, nickel oxide (141) or iron oxide. More commonly, however, these oxides are activated using titanium chloride (142) as the activating function. This is an excellent method, since it produces a bond that is very stable over the pH range in which immobilized proteins are normally employed.

## 3.3 Polystyrenes

Polystyrenes can be activated by only a few methods. More commonly, proteins are adsorbed to the surface of polystyrene. This is probably the most widely used method of protein immobilization in the world, since it forms the basic backbone of enzyme-linked immunoassay systems where polystyrene microtiter plates are used to adsorb antibodies. However, for the immobilization of most other biomolecules, the nonspecific adsorption of proteins by polystyrene is the most severe drawback to its application. This binding is simply too severe to permit utilizing polystyrene without any modification.

Several modifications of polystyrenes exist, which have potential for converting this otherwise undesirable material into one that can be reasonably employed in solid-phase biochemical systems. There are two types of modifications, those where the monomers used in the polymerization process are modified *before* the polystyrene is prepared, and those that involve postpolymerization modification, usually at the surface. The latter is the most widely used and will be the one we will discuss here.

In one such process, the polystyrene surface is nitrated, and the nitro group is reduced to an amine, which is subsequently used for immobilization, usually by diazo coupling of proteins (143).

An alternative postpolymerization modification process involves plasma etching and oxidation, generally converting the surface of the polystyrene into a highly carboxylated, partly degraded form. The carboxyl groups at the surface can subsequently be activated as described above. Alternatively, some postpolymerization plasma processes yield a hydroxylic surface in which case ligands can be immobilized as described previously for hydroxylic matrices.

## 3.4 Polyacrylamides

Polyacrylamides are formed by the free radical polymerization of various acrylic acid derivatives. A number of different types of monomers, other than polyacrylamide itself, have been utilized in the preparation of the polyacrylamide material. Particularly useful is Trisacryl, made from tris hydroxymethyl aminomethane derivatives of acrylic acid. Also very useful are the hydroxyethyl acrylamides. The latter are particularly useful for the preparation of immobilized enzymes, antibodies, and similar materials, because they are easily activated for the attachment of proteins using the same methods as described with cyanogen bromide. Trisacryl can be similarly activated and is an equally useful matrix.

## 3.5 Other Polymers

Various other polymers have been utilized from time to time, including polyurethane foams, maleic anhydride derivatives, and other organic polymer materials containing easily functionalized and activated groups. A wide variety of these materials are commercially available. Table 2 lists many of these polymers and the commercial sources from which they can be obtained. This particular table will be useful for the preparation of covalently immobilized materials for cell adhesion studies from a wide variety of matrices.

## REFERENCES


1. Kennedy, J.F., and Cabral, J.M.S. (1981). in *Solid Phase Biochemistry* (W.H. Scouten, ed.), Wiley Interscience, New York, p. 253.
2. Scouten, W.H. (1987). *Meth. Enzymol.*, *135*: 30–65.
3. Mattiasson, B. (1983). *Immobilized Cells and Organelles*, Vols. 1 and 2, CCR Press, Boca Raton, FL.
4. Mosbach, K. (1987). Immobilized enzymes and cells, *Meth. Enzymol.*, *135*.
5. Sharma, S.K., and Mahendroo, P.P. (1980). *J. Chromatogr.*, *184*: 471.
6. Soderman, D.D., Germershansen, J., and Katzen, H.M. (1973). *Proc. Natl. Acad. Sci. USA*, *70*: 792.
7. Rutishauser, U., D'Eustachio, P., and Edelman, G.M. (1973). *Proc. Natl. Acad. Sci. USA*, *70*: 3894.
8. Rutishauser, U., Millette, C.F., and Edelman, G.M. (1972). *Proc. Natl. Acad. Sci. USA*, *69*: 1596.
9. Abdon, N.I., and Richter, M. (1969). *J. Exp. Med.*, *130*: 141.
10. Trump, G.N. (1973). *Biochem. Biophys. Res. Commun.*, *54*: 544.
11. Robbins, J.B., and Schneerson, R. (1974). *Meth. Enzymol.*, *34*: 703.
12. Woff, H., Gatringer, C., and Wick, G. (1978). in *Affinity Chromatography*, Vol. 25 (O. Hoffmann-Ostenhof, M. Breitenbach, F. Koller, D. Kraft, and O. Scheiner, eds.), Pergamon Press, Elmsford, NY, pp. 313–316.
13. Woff, H., Gattringer, C., and Wick, G. (1979). *J. Immunol. Meth.*, *25*: 77.
14. Tangen, O., Berman, H.J., and Marfey, P. (1971). *Thromb. Diath. Haemorrh.*, *25*: 268.
15. Wigzel, H., Sundqvist, K.G., and Yoshida, T.O. (1972). *Scand. J. Immunol.*, *1*: 75.
16. Edelman, G.M., and Rutishauser, U. (1974). *Meth. Enzymol.*, *34*: 195.
17. Truffa-Bachi, P., and Wofsy, L. (1970). *Proc. Natl. Acad. Sci. USA*, *66*: 1685.
18. Flanagan, S.D., and Barondes, S.H. (1975). *J. Biol. Chem.*, *250*: 1484.
19. Flanagan, S.D., Taylor, P., and Barondes, S.H. (1975). *Nature* (*Lond.*), *254*: 441.
20. Ventor, B.R., Ventor, J.C., and Kaplan, N.O. (1976). *Proc. Natl. Acad. Sci. USA*, *73*: 2013.
21. Wallenbrock, M.A., and Killion, J.J. (1978). *J. Cell. Physiol.*, *95*: 151.
22. Mage, M., Evans, W.H., and Peterson, E.A. (1969). *J. Immunol.*, *102*: 908.
23. Evans, W.H., Mage, M.G., and Peterson, E.A. (1969). *J. Immunol.*, *102*: 899.
24. Edelman, G.M., Rutishauser, U., and Millette, C.F. (1971). *Proc. Natl. Acad. Sci. USA*, *68*: 2153.
25. Fey, H., and Wetzstein, H.P. (1975). *Med. Microbiol. Immunol.*, *161*: 73.
26. Boldt, D.H., and Lyons, R.D. (1979). *Cell. Immunol.*, *43*: 82.

27. Hellstrom, U., Dillener, M.L., Hammerstrom, S., and Perlmann, P. (1976). *J. Exp. Med.*, *144*: 1386.
28. Light, N.D., and Tanner, M.J. (1978). *Anal. Biochem.*, *87*: 263.
29. Johnson, A.H., Scott, D.W., McKeown, P.T., Pool, P.A., and Amos, D.B. (1977). *Transplant Proc.*, *9*: 145.
30. Wigzell, H., and Anderson, B. (1969). *J. Exp. Med.*, *129*: 1969.
31. Killion, J.J., and Kollmorgan, G.M. (1976). *Nature (Lond.)*, *259*: 674.
32. Campbell, P.A., and Grey, H.M. (1972). *Cell. Immunol.*, *5*: 171.
33. Anderson, B. (1972). *J. Exp. Med.*, *135*: 312.
34. Henry, C., Kimura, J., and Wofsy, L. (1972). *Proc. Natl. Acad. Sci. USA*, *69*: 34.
35. Walsh, F.S., Barber, B.H., and Crumpton, M.J. (1976). *Biochemistry*, *15*: 3557.
36. Jacobson, B.S., and Branton, D. (1977). *Science*, *195*: 302.
37. Jacobsen, B.S. (1977). *Biochim. Biophys. Acta*, *471*: 331.
38. Lotscher, H.R., Schwerzmann, K., and Carafoli, E. (1977). *FEBS Lett.*, *99*: 194.
39. Wofsy, L., Kimura, J., and Truffa-Bachi, P. (1971). *J. Immunol.*, *107*: 725.
40. Greaves, M.F., and Bauminger, S. (1972). *Nature (Lond.), New Biol.*, *235*: 67.
41. Binz, H., and Weigzell, W. (1975). *J. Exp. Med.*, *142*: 1231.
42. Rutishauser, U., and Edelman, G.M. (1972). *Proc. Natl. Acad. Sci. USA*, *69*: 3774.
43. Barker, C.R., Worman, C.P., and Smith, J.L. (1975). *Immunology*, *29*: 765.
44. Crum, E.D., and McGregor, D.M.C. (1976). *Cell Immunol.*, *23*: 211.
45. Wallach, D.F.H. (1974). *Meth. Enzymol.*, *34*: 171.
46. Wallach, D.F.H., and Lin, P.S. (1973). *Biochim. Biophys. Acta*, *300*: 211.
47. Brunner, G., Ferber, E., and Resch, K. (1976). *Differentiation*, *5*: 161.
48. Resch, K., Laracher, A., Mahler, B., Stoeck, M., and Rode, H.N. (1978). *Biochim. Biophys. Acta*, *511*: 176.
49. Edelman, G.M., Rutishauser, U., and Millette, C.F. (1971). *Proc. Natl. Acad. Sci. USA*, *68*: 2153.
50. Wigzell, H., Huber, C., and Schirrmarcher, V. (1972). *Haematologia*, *6*: 369.
51. Matthews, N., Rolland, J.M., and Nairn, R.C. (1976). *J. Immunol. Meth.*, *9*: 323.
52. Godfrey, H.P., and Gell, P.C. (1976). *Immunology*, *30*: 695.
53. Nicola, N.A., Burgess, A.W., Metcalf, D., and Battye, F.L. (1978). *Aust. J. Exp. Biol. Med. Sci.*, *56*: 663.
54. Kimura, A., Wigzell, H., Holmquist, G., Ersson, B., and Carlsson, P. (1979). *J. Exp. Med.*, *149*: 473.
55. Ghetie, V., Mota, G., and Sjoquist, J. (1978). *J. Immunol. Meth.*, *21*: 131.
56. Alice, M.J., and Varon, S. (1979). *Exp. Cell Res.*, *120*: 269.
57. Palacios, R., Sullivan, D., Summers, N.M., Kiley, M.L., and Schimke, R.T. (1973). *J. Biol. Chem.*, *248*: 540.
58. Hammarstrom, S., Hellstrom, U., Dillner, M., Perlman, P., Perlman, H., Axrlsson, B., and Robertsson, E.S. (1978). in *Affinity Chromatography*,

Vol. 25 (O. Hoffmann-Ostenhof, M. Breitenbach, F. Koller, D. Kraft, and O. Scheiner, eds.), Pergamon Press, Elmsford, NY, pp. 273–286.
59. Jacobsen, B.S., Cronin, J., and Branton, D. (1978). *Biochim. Biophys. Acta*, *506*: 81.
60. Kailsh, D.I., Cohen, C.M., Jacobsen, B.S., and Branton, D. (1978). *Biochim. Biophys. Acta*, *506*: 97.
61. Cohen, C.M., Douglas, I.K., Jacobsen, B.S., and Branton, D. (1977). *J. Cell. Biol.*, *75*: 119.
62. Wallach, D.F.H., Kranz, B., Ferber, E., and Fischer, H. (1972). *FEBS Lett.*, *21*: 29.
63. Zachowski, A., and Baraf, A. (1974). *Biochem. Biophys. Res. Commun.*, *57*: 787.
64. Brunner, G., Ferber, E., Hansen, K., and Resch, K. *Proc. Arbeitstagung uber Leukozytenkulturn, Erlangen*, March 1974, p. 5.
65. Brunner, G., Ferber, E., and Resch, K. (1977). *Anal. Biochem.*, *80*: 420.
66. Brunner, G., Ferber, E., Resch, K., and Golecki, J. (1978). in *Affinity Chromatography*, Vol. 25 (O. Hoffmann-Ostenhof, M. Breitenbach, F. Koller, D. Kraft, and O. Scheiner, eds.), Pergamon Press, Elmsford, NY, pp. 303–306.
67. Sidorovo, E.V., Abakumova, O. Yu., Skvorston, T.V., and Lerman, M.T. (1973). *J. Mol. Biol.*, *7*: 533.
68. Sidorovo, E.V., Trudolyubova, M.G., and Lerman, M.I. (1974). *Mol. Biol. Rep.*, *1*: 401.
69. Taylor, J.M., and Tse, T.P.H. (1976). *J. Biol. Chem.*, *251*: 7461.
70. Melcher, U. (1975). *Anal. Biochem.*, *64*: 461.
71. Faust, C.H., Vassalli, P. Jr., and Mach, B. (1972). *Experientia*, *28*: 745.
72. Thompson, E.B. (1974). *Meth. Enzymol.*, *34*: 294.
73. De Jonge, H.R. (1975). *FEBS Lett.*, *53*: 237.
74. Helm, R.M., Conrad, D.H., and Froese, A. (1979). *Int. Arch. Allergy Appl. Immunol.*, *58*: 90.
75. Lee, T., and Heintz, R.L. (1975). *Arch. Biochem. Biophys.*, *168*: 35.
76. Le Coffic, F., Baca, B., and Moreau, N. (1974). *FEBS Lett.*, *41*: 69.
77. Miller, J.V., Cuatrecasas, P., and Thompson, E.B. (1971). *Proc. Natl. Acad. Sci. USA*, *68*: 1014.
78. Haller, O., Gidlund, M., Hellstrom, U., Hammerstrom, S., and Wigzell, H. (1978). *Eur. J. Immunol.*, *8*: 765.
79. Ljungsteadt-Pahlman, I., Seiving, B., and Sjoholm, I. (1977). *Exp. Cell Res.*, *110*: 191.
80. Dovorak, J.J., Gipps, E., and Kidson, C. (1978). *Nature (Lond.)*, *271*: 564.
81. Campbell, C.W.B. (1977). *Biochem. Soc. Trans.*, *5*: 389.
82. Garcia-Seguar, L.M., Martinez-Rodriguez, R., and Martinez-Murillo, R. (1978). *Experientia*, *34*: 1598.
83. Sharma, S.K., unpublished results.
84. Zisapel, N., and Zurgil, O.N. (1979). *Brain Res.*, *178*: 297.
85. Kinzel, V., Richards, J., and Kubler, D. (1977). *Exp. Cell Res.*, *105*: 389.
86. Eshhar, Z., Waks, T., and Bustin, M. (1974). *Meth. Enzymol.*, *34*: 750.
87. Irle, J., Piquet, P.F., and Vassali, P. (1978). *J. Exp. Med.*, *148*: 32.

88. Belitsina, N.V., Elizarov, S.M., Glukhova, M.A., Spirin, A.S., Butorin, A.S., and Vasilenko, S.K. (1975). *FEBS Lett.*, *57*: 262.
89. Romanowska, E., Lugonski, C., and Mulczyk, M. (1976). *FEBS Lett.*, *66*: 82.
90. Sundberg, L., and Hoglund, S. (1973). *FEBS Lett.*, *37*: 70.
91. Kenyon, A.J., Gander, J.E., Lopez, C., and Good, R.A. (1973). *Science*, *179*: 187.
92. Yoon, J.W., Kenyon, A.J., and Good, R.A. (1973). *Nature* (*Lond.*), *New Biol.*, *245*: 205.
93. Matheka, H.D., and Mussgay, M. (1969). *Arch. Ges. Virusforsch.*, *27*: 13.
94. Sweet, C., Stephen, J., and Smith, H. (1974). *Immunochemistry*, *11*: 295.
95. Oroszlan, S., Long, C.W., and Gilden, R.V. (1976). *Virology*, *72*: 523.
96. Ladipo, S.L., and de Zoeten, G.A. (1971). *Virology*, *46*: 567.
97. Wood, K.R., Stephen, J., and Smith, H. (1968). *J. Gen. Virol.*, *2*.
98. Galvez, G.E. (1966). *Virology*, *28*: 171.
99. Parikh, I., March, S., and Cuatrecasas, P. (1974). *Meth. Enzymol.*, *44*: 77.
100. Arnott, S. (1974). *J. Mol. Biol.*, *90*: 269.
101. Porath, J., and Axen, R. (1976). *Meth. Enzymol.*, *44*: 19.
102. Porath, J. (1973). *Biochemie*, *55*: 943.
103. Bloch, R., and Burger, M.M. (1974). *Biochem. Res. Commun.*, *58*: 13.
104. Vorlop, K-D., and Klein, J. (1981). *Biotechnol. Lett.*, *3*: 9.
105. Cooper, T. (1977). *Tools of Biochemistry*, Wiley Interscience, New York, p. 169.
106. Knight, C.S. (1967). *Adv. Chromatogr.*, *4*: 61.
107. Stamberg, J. (1988). *Separ. Pur. Meth.*, *17*: 155.
108. Ohlson, S., Larsson, P.-O., and Mosbach, K. (1979). *J. Appl. Microbiol. Biotechnol.*, *7*: 103.
109. Gisby, P.E., and Hall, D.O. (1980). *Nature*, *287*: 251.
110. Smitsrod, O., and Skjak-Braek, G. (1990). *Trends Biotechnol.*, *8*: 71.
111. Chibata, I., Tosa, T., Sato, T., and Takata, I. (1987). *Meth. Enzymol.*, *135*: 189.
112. Parascandola, P., and Scardi, V. (1982). *Biotechnol. Lett.*, *4*: 753.
113. Mosbach, R., Koch-Schmidt, A-C., and Mosbach, K. (1972). *Meth. Enzymology*, *44*: 53.
114. Konecny, J., and Sieber, M. (1980). *Biotechnol. Bioeng.*, *22*: 2013.
115. Husain, S.S., Lipinski, B., and Gurewich, V. (1981). *Proc. Natl. Acad. Sci. USA*, *78*: 4265.
116. Haller, W. (1983). *Solid Phase Biochemistry* (W.H. Scouten, ed.), Wiley Interscience, New York, p. 535.
117. Epton, R., McLaren, J.V., and Thomas, T.H. (1972). *Carbohydr. Res.*, *22*: 301.
118. Scouten, W.H., and Firestone, G.C. (1976). *Biochim. Biophys. Acta.*, *453*: 277.
119. Voivodov, K.I., Galunski, B.R., and Dyankoy, S.S. (1979). *J. Appl. Biochem.*, *1*: 442.
120. Haller, W. (1983). in *Solid Phase Biochemistry* (W.H. Scouten, ed.), Wiley Interscience, New York, p. 535.

121. Hermentin, P., Doenges, R., Franssen, U., Bieva, C., Vander Bruggher, F.J., Stryckmans, P., Friesen, H.J., Optacey, B., Schneiden, S., Ax, W., Enssle, K.H., Kurrle, R., and Seiler, F.R. (1990). *Bioconjug. Chem.*, *1*: 411.
122. Ogumtimein, G.B., and Reilly, P.J. (1980). *Biotechnol. Bioeng.*, *22*: 1127.
123. Taylor, R.F. (1985). *Anal. Chim. Acta.*, *172*: 241.
124. Miron, T., and Wilchek, M. (1985). *Appl. Biochem. Biotechnol.*, *11*: 445-456.
125. Insolmer, Accurate Chemical and Scientific Co., Westbury, NY.
126. Edelman, G.M., and Rintshauer, U. (1974). *Meth. Enzymol.*, *34*: 195.
127. Kricka, L.J. (1985). *Ligand-Binder Assays*, Marcel Dekker, New York.
128. Nilsson, K., Norrlow, O., and Mosbach, K. (1981). *Acta Chem. Scand.*, *35*: 19.
129. Gribnau, T.C.J. (1977). Ph.D. thesis, University of Nijmegen, Nijmegen, The Netherlands.
130. Scouten, W.H., and van der Tweel, W. (1984). *Ann. NY Acad. Sci.*, *434*: 249.
131. Scouten, W.H., Rines, R., Mulder, A.H.L., and Loy, R. (1988). *Ann. NY Acad. Sci.*, *542*: 204.
132. Turkova, J., Petkov, L., Sajdok, J., Kas, J., and Benes, M. (1990). *J. Chromatogr.*, *500*: 585.
133. Ngo, T. (1986). *Biotechnology*, *4*: 134.
134. Stults, N.L., Lin, P., Hardy, M., Lee, Y.C., Uchida, Y., Tsukada, Y., and Sugimori, T. (1983). *Anal. Biochem.*, *135*: 392.
135. Porath, J., and Sundberg, L. (1972). *Nature New Biol.*, *238*: 261.
136. Seed, B. (1982). *Nucleic Acids Res.*, *10*: 1299.
137. Miron, T., and Wichek, M. (1987). *Meth. Enzymol.*, *135*: 84.
138. Lily, M.D. (1982). *Meth. Enzymol.*, *44*: 46.
139. Clonis, Y.D., Atkinson, A., Bruton, C.J., and Lowe, C.R. (1987). *Reactive Dyes in Protein and Enzyme Technology*, Stockton Press, New York.
140. Jirka, V., and Turkova, J. (1987). *Meth. Enzymol.*, *135*: 341.
141. Munro, P.A., Dunnill, P., and Lilly, M.D. (1972). *Biotechnol. Bioeng.*, *19*: 101.
142. Kennedy, J.F., and Cabral, J.M.S. (1987). *Meth. Enzymol.*, *135*: 117.
143. Grubhoter, N., and Schleith, L. (1953). *Naturwissenschaften*, *40*: 508.

# Index